Canada's Cold Environments

CANADIAN ASSOCIATION OF GEOGRAPHERS SERIES
IN CANADIAN GEOGRAPHY

Canada's Cold Environments
Hugh M. French and Olav Slaymaker, editors

Canada's Cold Environments

EDITED BY
HUGH M. FRENCH AND
OLAV SLAYMAKER

McGill-Queen's University Press
Montreal & Kingston • London • Buffalo

© McGill Queen's University Press 1993
ISBN 0-7735-0925-9

Legal deposit first quarter 1993
Bibliothèque nationale du Québec

Printed in Canada on acid-free paper

McGill-Queen's University Press acknowledges the finan-
cial support of the Government of Canada through the
Canadian Studies and Special Projects Directorate of the
Department of the Secretary of State of Canada.

Canadian Cataloguing in Publication Data

Main entry under title:
  Canada's cold environments
  (Canadian Association of Geographers series in Canadian
  geography)
  Includes index.
  ISBN 0-7735-0925-9
  1. Cold regions. 2. Canada – Geography. I. French, Hugh
  M. II. Slaymaker, Olav, 1939–   . III. Series.
  GB132.N6C36 1992   574.5'2621   C92-090379-7

This book was typeset by Typo Litho composition inc.
in 10/12 Times.

Contents

Tables

Figures

Contributors

ROGER G. BARRY, Professor, Cooperative Institute for Research in Environmental Sciences and Department of Geography, University of Colorado, Campus Box 449, Boulder, Colorado 80309-0449, U.S.A.

JOHN J. CLAGUE, Research Scientist, Geological Survey of Canada, 100 West Pender Street, Vancouver, British Columbia V6B 1R8

DEREK C. FORD, Professor, Department of Geography, McMaster University, Hamilton, Ontario L8S 4K1

HUGH M. FRENCH, Dean of Science and Professor, Departments of Geography and Geology, and Ottawa-Carleton Geoscience Centre, University of Ottawa, Ottawa, Ontario K1N 6N5

JAMES S. GARDNER, Vice-President (Academic) and Provost, University of Manitoba, Winnipeg, Manitoba R3T 2N2

ELLSWORTH F. LEDREW, Professor, Department of Geography, and Director, Earth-Observations Laboratory, Institute for Space and Terrestrial Science, University of Waterloo, Ontario N2L 3G1

JAMES C. RITCHIE, Professor of Botany, Scarborough College, University of Toronto, 1265 Military Trail, Scarborough, Ontario M1C 1A4

WAYNE R. ROUSE, Professor, Department of Geography, McMaster University, Hamilton, Ontario L8S 4K1

OLAV SLAYMAKER, Professor, Department of Geography, University of British Columbia, Vancouver, British Columbia V6T 1W5

MICHAEL W. SMITH, Associate Professor, Department of Geography, and Ottawa-Carleton Geoscience Centre, Carleton University, Ottawa, Ontario K1S 5B6

MING-KO WOO, Professor, Department of Geography, McMaster University, Hamilton, Ontario L8S 4K1

Preface

Coldness is a pervasive Canadian characteristic, part of the nation's culture and history. Cold takes many forms of which low air temperatures are merely one; others include exposure and wind, frozen ground, snow, lake and river ice, sea ice, glaciers, and darkness.

Canada's cold environments include approximately three-quarters of the area of a nation over 10 million km^2 in extent. The vastness of these environments, and their geological and physiographical complexity, make their description a daunting, yet essential task if we are truly to understand Canada's geography. Canada has a long prehistory in which people coped with coldness with few technological buffers. The European beginnings in Canada were also conditioned by seasonality and cold; the names David Thompson and Alexander Mackenzie, Jean de Brébeuf and Pierre de La Vérendrye and his sons, and Sir John Richardson and Sir John Franklin are all associated with the forging of Canada and the struggle for survival in its cold environments. Eventually, traditions of art and writing have emerged in which the dangers and landscapes of cold places figure prominently.

In this volume, we, and our contributors, have attempted to provide an authoritative, yet readable scientific statement about the physical nature of Canada's coldness. However, we have not attempted a comprehensive geographic coverage. Neither is this volume an encyclopaedia of the various parameters of Canada's coldness. Instead, we have focused on the distinctive attributes of Canada's cold environments, their temporal and spatial variability, and the constraints that coldness places on human activity. Environment plays a determining role in these hostile climes. Consider, for example, mid-December, with air temperatures of $-50°C$ and barely adequate illumination, or heavy snowfall in July at 2,000 m a.s.l., or continuous twenty-four-hour intense rainfalls in November, with air temperatures of $+2°C$ and onshore winds of 30–50 km/hr; these conditions are real barriers to creative cultural activity. Nevertheless, they are all part of a normal year's experience in Canada's cold regions.

Our objectives in undertaking this task have been threefold. The first has been to provide an insight into the ways in which cold affects biophysical processes of change, at a range of scales. The second has been to provide a biophysical context for an understanding of the human geography of Canada. The third objective relates to predicted global changes which, if they occur, will have a profound and special effect on the cold regions of Earth.

We wish to thank the authors of individual chapters, all of whom provided material in a timely and efficient manner and achieved, we believe, the desired mix of authoritative information and accessible style. We also thank R. Cole Harris, senior editor of this series, for his initiative and encouragement during a long gestation period. Our typists, Terry Goldberger, Anne-Marie Landry, and Sandy Lapsky, have provided brilliant support, as have our cartographers.

Any lack of coherence, and errors of fact or interpretation, are our responsibility, and we ask for your indulgence.

Hugh M. French, Ottawa
Olav Slaymaker, Vancouver
February 1992

PART ONE

Cold Lands, Cold Seas

CHAPTER ONE

Canada's Cold Land Mass

HUGH M. FRENCH AND
OLAV SLAYMAKER

DIMENSIONS OF COLDNESS

Canada is dominated by cold. Snow and sub-freezing temperatures are rare only along the maritime lowland fringes of the Pacific coast. However, the areas along the southern borders where most of the population resides are largely temperate in climate. There, seasonal agriculture is possible, plant and animal productivity is relatively high, and the constraints of cold are temporarily forgotten during the summer months. By contrast, in the more remote areas of the country the constraints imposed by coldness persist throughout the year and dominate both the landscape and socio-economic activities, especially in more northerly latitudes and at higher elevations.

The dimensions of coldness with which we are concerned include not only low absolute temperatures but also exposure to wind-chill, snow, ice, and permafrost, as well as the magnitude, duration, and frequency of their occurrence. It is not easy, therefore, to provide an unambiguous, quantitative definition of the nature and extent of Canada's cold environments.

Physical Attributes

Canada is unusually cold because of its latitude and the influence of the Cordillera on the movement of air masses (Smith 1989). This mountain barrier hampers the flow of the westerlies and confines maritime influences to the Pacific coast. The Cordillera also creates a permanent wave in the upper westerlies which move northward as they approach the Pacific coast and then southward over central and eastern Canada. This meridional (i.e. north–south) flow is especially well developed in winter, and as a result cold arctic air frequently flows southward into central Canada.

Eastern Canada has a continental climate less extreme than central Canada's, the result of cyclonic conditions in the westerlies that draw mild air northward from the Atlantic. These systems reduce the annual temperature range and spread precipitation more evenly throughout the year. The Great Lakes also exert a moderating influence and promote snow accumulation in their lee.

Northward, away from inhabited Canada, these moderating influences disappear and the climate becomes progressively colder. In the high arctic islands, cold air masses persist throughout the year and westerlies penetrate only as far as the subarctic, or boreal zone, in summer. Because of the meridional trend of the westerlies and the associated importance of arctic air masses, the boreal zone extends progressively southward as one moves east across Canada.

Precipitation also correlates in a broad sense with Canada's physiography. Much falls as snow. The highest amounts are in the Cordillera, and, in general, heavy snowfalls characterize most of the highlands of Canada. An exception is the High Arctic, where intense cold and the relative absence of cyclonic disturbances give low, often desert-like, amounts. The low precipitation notwithstanding, low evaporation rates and the presence of permafrost, which impedes drainage, produce humid surface conditions.

Seasonality, another fundamental characteristic of Canada's climate, has two principal consequences for the country's cold landmass. The first is the importance of freeze-thaw, or frost action, and its influence on rock weathering. Seasonal frost action results in significant damage to crops, highways, and buildings every year. The second is the seasonality of climate, resulting in storage of moisture during winter and its dramatic release in spring as snowmelt and ice melt. During spring flood, the rivers of Canada are enormously active agents of erosion and sediment transport. Numerous studies indicate that between 40 and 70 per cent of annual runoff, and between 60 and 90 per cent of total sediment transport, occur in the spring.

The severity of Canada's climate is best illustrated by the comparison of freezing degree–days and growing degree–days for some of Canada's largest cities with similar data for a few typical northern locations (Table 1.1). While all the locations in Table 1.1 would be considered cold by most non-Canadian readers, the three "northern" localities stand out, with over 3,000 freezing degree–days and fewer than 1,000 growing degree–days.

There are many other ways to illustrate the coldness of Canada's climate. For example, the value 0°C is fundamental, since it is the threshold temperature at which thawing occurs. The ice nucleation temperature (i.e. freezing point) is also closely related to 0°C. The important changes of phase between water, snow, and ice are critical to cold environments. Figure 1.1 shows the mean number of days with minimum temperatures of 0°C or below and mean annual total snowfall. Taken together, such data leave no doubt that coldness is an overwhelming characteristic of most of Canada's landmass.

Nordicity

'Nordicity,' a term coined by the Quebec geographer Louis-Edmond Hamelin in his book *Nordicité canadienne* (1978), is closely linked to the concept of coldness.

Table 1.1
Climatic statistics for selected Canadian stations

Station	Latitude (N)	Longitude (E)	Mean annual temperature (°C)	GDD*	Frost-free period (days)	Snow cover (days)	FDD†	Annual total precip. (mm)
A "Non-northern"								
Vancouver	49°11'	123°10'	9.8	2,019	212	7	45	1,068
Edmonton	53°34'	113°31'	2.8	1,516	127	117	1,501	447
Saskatoon	52°10'	106°41'	1.6	1,618	110	128	1,977	353
Winnipeg	40°54'	97°14'	2.3	1,791	118	122	1,903	535
Toronto	43°40'	79°24'	8.9	2,434	192	59	438	790
Ottawa	45°19'	75°40'	5.8	2,069	137	117	1,040	851
Montreal	45°30'	73°35'	7.2	2,301	183	117	814	999
Quebec City	46°48'	71°23'	4.4	1,729	132	140	1,153	1,089
Halifax	44°38'	63°30'	6.8	1,708	183	63	422	1,381
B "Northern"								
Churchill	58°45'	94°04'	− 7.3	688	81	213	3,791	397
Baker Lake	64°18'	96°00'	− 12.3	326	61	240	5,206	213
Resolute	74°43'	94°59'	− 16.4	36	9	283	6,238	136

Source: Hare and Thomas (1979).
* GDD = growing degree–days (base temperature +0°C).
† FDD = freezing degree–days (base temperature 0°C).

Hamelin argued that a number of factors contribute to the 'northern' character of a region, including not only latitude but also other natural and/or locational factors such as summer heat, total precipitation, and vegetation. Several human factors are also involved, such as accessibility, provision of transportation, and characteristics of the population. When each factor is assigned a value, they can be summed to indicate the degree of 'nordicity.' Following Hamelin, Table 1.2 gives representative nordicity (i.e. polar) values for various cities in Canada. According to Hamelin, the southern limit of the north corresponds to a polar value of 200. This limit ranges from 60°N at the Yukon-BC border to 53°N in Labrador.

Arctic and Alpine

The terms *arctic* and *subarctic* are often used to refer to northern or cold environments (e.g. Bird 1967; Armstrong, Rogers, and Rowley 1978; Young 1989), while for mountain areas the terms *alpine* and *subalpine* are used (e.g. Love 1970; Ives and Barry 1974). Such terms are useful but imprecise because their 'warm' limits are not defined.

Many physical scientists define the Arctic and the subarctic ecologically (e.g. Ritchie and Hare 1971; Ritchie 1984). For example, *boreal forest* is the term

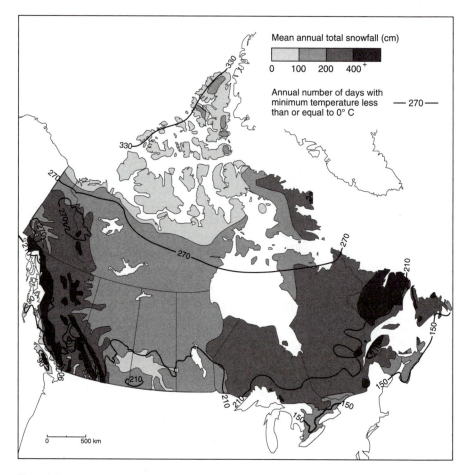

Figure 1.1
Winter climate of Canada, as illustrated by mean annual number of days with minimum temperatures of 0°C or below and mean annual total snowfall. *Source*: Atmospheric Environment Service (1987).

usually applied to the vast forests of the subarctic, and *treeline* refers to the transition zone between the northern limits of the boreal forest and the treeless *tundra*, or Arctic, further north. In the alpine context, the term *timberline* is used to refer to the upper altitudinal limit of closed crown cover of trees and separates *alpine* from *subalpine* (e.g. Arno and Hammerly 1984). In high latitudes and at high altitudes, trees trap snow and insulate the ground from the extremes of cold in the winter. They also shield people and animals from the wind; north of treeline and above timberline the wind is largely unimpeded, and exposure can become critical to plants, animals, and people.

Table 1.2
Zonal nordicity values for various Canadian locations

Zone	Latitude (N)	Location	Polar value
Extreme north	78°	Isachsen, NWT	925
	70°	Barnes Ice Cap, NWT	804
	62°	St Elias Mts, Yukon	856
Far north	74°	Resolute, NWT	775
	72°	Sachs Harbour, NWT	764
	67°	Old Crow, Yukon	624
	64°	Frobisher Bay, NWT	584
	68°	Aklavik, NWT	511
Middle north	58°	Churchill, Man.	450
	56°	Winisk, Ont.	435
	62°	Yellowknife, NWT	390
	61°	Whitehorse, Yukon	283
	58°	Fort Nelson, BC	282
	56°	Thompson, Man.	258
	53°	Goose Bay, Lab.	218
'Non-northern'	55°	Grande Prairie, Alta	198
	54°	The Pas, Man.	185
	57°	Fort St John, BC	183
	48°	Cochrane, Ont.	137
	49°	Edmonton, Alta	125
	47°	St John's, Nfld	115
	43–46°	Montreal-Ottawa-Toronto	40–45
	49°	Vancouver, BC	35

Source: Hamelin (1978).

People and Cold

Human activities in Canada, be they agriculture, building construction, work schedules, leisure activities, or transportation, to name but a few, are all influenced by climatic conditions (Hare and Thomas 1979). As just one illustration, Figure 1.2 summarizes these physical and logistical constraints from the viewpoint of oil and gas exploration activity in the High Arctic. One can devise similar schemes for various activities in the Arctic and subarctic and for the mountains, be they road/rail transportation, mining, skiing, or hiking. In all cases, the pattern of physical and logistical constraints associated with coldness is clear.

People are especially vulnerable in these cold environments as the combination of strong winds and low temperatures produces severe winter conditions. It is not uncommon, for example, in parts of Keewatin and the arctic islands, for air temperatures to fall below −40°C with winds in excess of 32 km/hr (Maxwell 1980). An extreme situation was recorded at Chesterfield Inlet in January 1935, when the

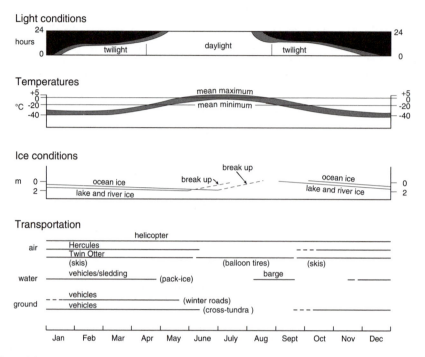

Figure 1.2
Typical planning schedule used for oil and gas exploration in high arctic islands of Canada. *Source*: French 1979.

mean daily temperature was −43°C and the mean wind speed was 16 km/hr for the entire month. To deal with these conditions, a 'wind-chill' factor, which is based on the cooling effect of the wind on naked flesh, is widely used in Canada and elsewhere (Figure 1.3). This graph illustrates that it is convenient to think in terms of an equivalence – the temperature of still air that would give the same sensation of cold as the combination of actual air temperature and wind.

The concept of wind-chill applies not only to northern, polar, and alpine regions but also to environments not normally associated with extreme cold. For example, although air temperatures on the Queen Charlotte and Vancouver Island ranges are commonly mild, high winds ensure that wind-chill index ii conditions (see Figure 1.3) are regularly experienced. On the exposed outer coast at Cape St James, where timberline is as low as 900 m a.s.l., mean and maximum hourly wind speeds are consistently higher than on both the east coast of the Queen Charlotte Islands at Sandspit and in the Coast Mountains at Alta Lake (Table 1.3).

Blizzards are another climatic hazard in cold environments. Federal agencies responsible for transportation in the north, such as the Atmospheric Environment Service (AES) and Transport Canada, consider that there are blizzard conditions when, simultaneously, snow or blowing snow is associated with wind speeds above

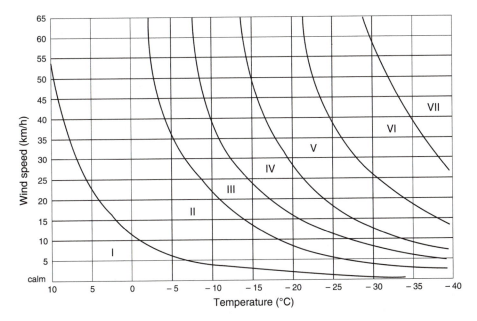

Figure 1.3
Temperature/wind-chill index: nomogram used by US Air Force at Goose Bay Air Base, Labrador,
Sept. 1963.

Read the temperature horizontally and the wind velocity vertically. The point of intersection is the wind-
chill factor. For example, if the wind were 30 km/h and the temperature −25°C, the wind-chill index
would be V.

    I Comfortable with normal precaution.
   II Work and travel become more uncomfortable unless properly clothed.
  III Work and travel become more hazardous unless properly clothed. Heavy outer clothing necessary.
  IV Unprotected skin will freeze with direct exposure over prolonged period. Heavy outer clothing
     becomes mandatory.
   V Unprotected skin can freeze in one minute with direct exposure. Multiple layers of clothing mandatory.
     Adequate face protection becomes important. Work and travel alone not advisable.
  VI Adequate face protection becomes mandatory. Work and travel alone prohibited. Supervisors must
     control exposure times by careful work scheduling.
 VII Personnel become easily fatigued. Buddy system and observation mandatory.

NOTE: Proper clothing simply means protecting all skin areas from direct wind with sufficient thickness
to prevent undue coldness.

40 km/hr, visibility is less than 0.8 km, and temperature is below −12°C. Then,
most outdoor activities stop. Blowing snow itself is a major problem, occurring, for
example, almost one-quarter of the time in midwinter in Keewatin. In the high arctic
islands, where the presence of open water in summer results in relatively small
differences between air and dewpoint temperatures, low cloud and fog are common.

Table 1.3

Mean and maximum wind speeds (km/hr) recorded at Queen Charlotte Islands and Coast Mountains, BC, 1965–75

	J	F	M	A	M	J	J	A	S	O	N	D
Mean hourly												
Cape St James	25.7	27.3	18.4	18.7	18.5	19.6	16.1	15.9	15.5	20.9	25.4	27.1
Sandspit	13.5	13.9	12.0	12.2	13.7	12.0	11.5	11.4	10.8	17.8	14.7	14.3
Alta Lake	1.4	1.9	1.7	1.9	2.8	2.9	2.4	1.9	2.2	0.9	1.9	1.1
Maximum hourly												
Cape St James	88	75	75	62	46	54	43	41	38	81	72	68
Sandspit	50	46	39	32	36	29	31	31	24	60	52	47
Alta Lake	15	20	14	17	14	15	14	15	21	23	16	21

Sources: Monthly Record; Meteorological Observations in Canada, Atmospheric Environment Service, Environment Canada.

Elsewhere, in midwinter, in some of the larger settlements such as Inuvik in the Mackenzie delta, ice fog is a common characteristic brought about by the freezing of warm, moist air emanating from buildings and vehicle exhaust fumes.

COLD LANDS

For our purposes, we can divide Canada's cold landmass into northern and polar lands and mountain environments.

Northern and Polar Lands

It is common to equate Canada's northern and polar lands with Yukon and Northwest Territories. However, large parts of Ontario, of Quebec and Labrador, and of the western provinces should also be regarded as 'northern.' Yukon and Northwest Territories make up approximately one-quarter of the total area of Canada. When we add the northern parts of the western provinces, Ontario, Quebec, and Labrador, we can conclude conservatively that at least half of Canada's immense landmass is *both* northern *and* cold. Not surprising, a diverse range of biological, physical, and climatic conditions characterizes this vast area (e.g. see Prest 1983; Heginbottom 1989). Yet in winter there are common implacable realities: intense cold, little or no daylight, and less precipitation than most desert regions of the world.

Northern lands are commonly classified as tundra or boreal in nature. Table 1.4 outlines some climatic values regarded as typical of various northern vegetation zones and boundaries (see chapter 4). The arctic treeline is the conspicuous boundary between the two. In Canada, treeline extends from northern Yukon diagonally south and eastward toward the west coast of Hudson Bay, in northern Manitoba, follows

Table 1.4
Some climatic values typical of northern vegetation zones and boundaries

	Climatic values			
	M	*J*	*A*	*D*
Tundra zone				
Tundra, Queen Elizabeth Islands	5	480	5–10	75
Tundra, southern areas	10	445	10–15	90–120
Arctic treeline	10–13	450	15–20	120–145
Boreal zone				
Forest tundra and woodland	13–15	400–500	20–25	150–170
Northern forest line	15	450–500	30	168–175
Boreal forest	15–18	450–525	30–35	180–210
Southern boreal limit	16–18	475–525	34–40	200–225

Source: Hare and Thomas (1979).

Note:

M = mean daily temperature (degrees C), July.

J   = July net radiation (langleys per day), mean value.

A = annual net radiation (kilolangleys per year), mean value.

D = duration of thaw season (days) (temperature exceeds 0°C).

the southern coast of Hudson Bay to Great Whale River, Quebec, and runs thence eastward to Ungava Bay and the Labrador coast. North of treeline, the ground is perennially frozen and vegetation is characterized by tundra species and low-growing shrubs. South of treeline, northern (boreal) forests extend through the lowlands of the Mackenzie valley and into northern Ontario and Quebec and other parts of eastern Canada. In the higher elevations of the Mackenzie Mountains in Northwest Territories and in Yukon, boreal forest merges with either open woodland or forest tundra, as the case may be. A similar transition occurs in northern Quebec and in Labrador. Wherever elevations are sufficient, alpine tundra conditions occur. Figure 1.4 plots the treeline against other northern characteristics such as permafrost, vegetation types, daylight conditions, and the extent of sea-ice cover in adjacent Canadian arctic waters.

The scenery of Canada's northern lands varies from undulating, monotonous, and vast lowlands in Keewatin and Mackenzie districts of Northwest Territories, northern Ontario, northern Quebec, and northern Manitoba, through spectacular ice-covered mountains with deep fjords on Baffin and Ellesmere islands, to arid, desert-like landscapes on the extreme northern and western high arctic islands.

The arctic islands constitute a northern archipelago that reaches to 83°N. Remote, virtually uninhabited, largely ice-free in summer yet separated by straits and channels that are ice-covered for large portions of the year, and experiencing some of the harshest climatic conditions in Canada, the high arctic islands are a unique component of the nation's northern lands.

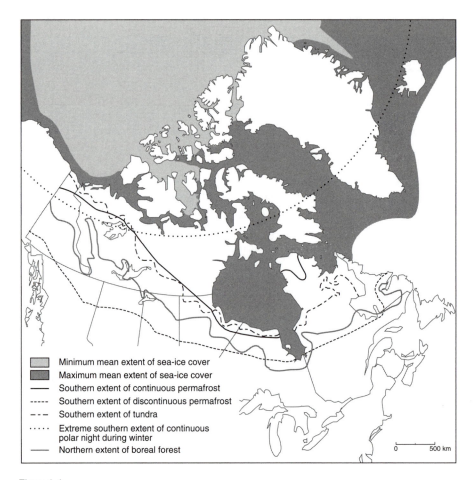

Figure 1.4
Northern Canada: 'northern' limiting parameters and extent of sea-ice cover.

Northern Physiographic Regions

It is relatively easy to divide northern Canada into four major physiographical regions, each of which can be subdivided (e.g. Bostock 1971; Bird 1980: 100–4). They are listed in Table 1.5 and illustrated in figure 1.5. The most extensive is the Canadian Shield (region ɪ), which occupies most of the arctic mainland of Northwest Territories, much of northern Ontario and northern Quebec, much of the central and eastern arctic islands south of Parry Channel, and sectors of Devon and Ellesmere islands further north. Igneous and metamorphic rocks are widespread. The terrain is rocky, undulating, and lake-strewn. Structurally, the Shield consists of a series of broad arches and basins, earlier of varied relief, since eroded to a more nearly level surface.

Table 1.5
Physiographic regions of northern Canada

I   CANADIAN SHIELD

a   Lowlands, hills, and plateaus: central Keewatin and Mackenzie District, plus Shaler Mountains, Victoria Island
b   Sandstone plains: central Keewatin
c   Folded sedimentary hills: northern Quebec and Labrador
d   Highland rim: eastern Baffin, Devon, and Ellesmere Islands
e   Hudson Bay Basin
f   Foxe Basin

II   NORTHERN PLATFORM LOWLANDS

a   Arctic islands: Victoria and Prince of Wales islands, Brodeur Peninsula
b   Mackenzie Lowlands

III   INNUITIAN PROVINCE

a   Uplands of Bathurst and Melville islands
b   Sverdrup Basin
c   Mountains of Axel Heiberg Island and northern parts of Ellesmere Island

IV   ARCTIC COASTAL PLAIN

a   Mackenzie delta and northern Yukon coastal plain
b   Western arctic islands

Source: Bird (1980).

Note: Figure 1.5 locates these regions and subregions.

In a few places, sedimentary rocks are preserved within broad shallow basins, such as Hudson Bay and Foxe Basin. For the most part, however, the Canadian Shield is an exhumed and eroded surface of great geological age and tectonic stability.

Flanking the Canadian Shield to the north and west are extensive areas of near-horizontal sedimentary rocks, mostly limestones, which form the Northern Platform Lowlands (region II). These undulating plateau-like areas make up much of Victoria and Prince of Wales islands and extend southward to form the Mackenzie Lowlands east of Great Slave and Great Bear lakes.

The Queen Elizabeth Islands form the Innuitian Province (region III). The rocks are gently dipping or folded. Faulting has led to formation of the various island groups. A broad through-channel, historically sought as the northwest passage and known in various parts as M'Clure Strait, Viscount Melville Sound, Barrow Strait, and Lancaster Sound, lies along a major structural line.

The islands of the Sverdrup Basin are underlain by soft rocks younger than those of either the Canadian Shield or the Innuitian Province. These rocks give rise to low-lying terrain, usually less than 600 m a.s.l. Hydrocarbons are known to exist in the basin, especially offshore of Lougheed Island. Elsewhere in the high arctic islands,

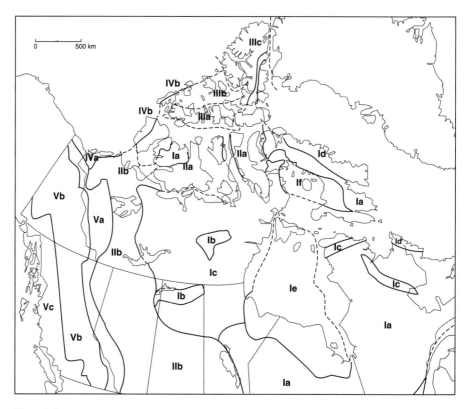

Figure 1.5
Physiographic regions of northern Canada, excluding Western Cordillera (key in Table 1.5). *Source*: Bird (1980).

older and harder rocks form barren upland terrain, rising in places, as on western Melville Island, to over 1,000 m a.s.l. Glacier-capped mountains, sometimes over 2,000 m a.s.l. on Axel Heiberg and Ellesmere islands, dominate the eastern rim of Canada's High Arctic.

The Arctic Coastal Plain (region IV), a low-lying zone adjacent to the Arctic Ocean, extends from Meighen Island through Prince Patrick and Banks islands to the Mackenzie delta and northern Yukon coast. It is underlain by perennially frozen sand, gravel, and silt, often possessing high ice contents, as in the Mackenzie delta and parts of western Banks Island. In the Mackenzie delta, and offshore beneath the Beaufort Sea, hydrocarbons are known to exist.

Mountains

One-quarter of the total area of Canada is mountain land, and of this approximately 40 per cent is also northern. Distinctive features of these mountain regions are their rugged topography and the "island" nature of their distribution (see chapter 8).

Canada's mountains form highland rims on the east, north, and west sides of a plain-like surface that includes the core of the inhabited parts of the country. In the west, the Cordillera is an almost unbroken mountain chain of great diversity; in the east, the Appalachian–Acadian Mountain system forms the Atlantic provinces, and the upturned southern edge of the Laurentide Highlands a lower, more undulating barrier. In the northeast, the Torngat Mountains of Labrador and the mountains of Baffin, Axel Heiberg, and Ellesmere islands are a more broken, but scenically spectacular, barrier. This geographical separation or "island" quality of mountain lands has influenced the evolution of mountain landscapes (e.g. separate and independent ice caps) as well as the revegetation and colonization of post–Ice Age mountain regions (e.g. the "refugia" of the Queen Charlotte Islands).

The scenery of Canada's mountain lands varies from the well-known castellated cliffs of near horizontally bedded sedimentary strata of the Banff-Jasper highway region and the snow- and ice-draped slopes of the massive Mt Logan (5,951 m a.s.l.) in the St Elias Mountains to the less familiar stark, fjord landscapes of Baffin Island and the lush rainforest-covered slopes of the Queen Charlotte Ranges.

The Queen Charlotte Ranges, on the west coast, and the Shickshock Mountains, in the Gaspé Peninsula of Quebec, represent islands of coldness extending further west or south than Canada's other mountain lands. Notable features are the low timberline in the Queen Charlotte Islands (900 m a.s.l.) and the low limit of permafrost (1200 m a.s.l.) in the Shickshock Mountains.

Physiographical Regions

Canada's mountains have four main physiographical regions: the Canadian Shield, the Appalachian-Acadian, the Innuitian, and the Cordilleran. The Canadian Shield (Figures 1.6A and 1.6B: units 1–5) includes the Davis Highlands of Baffin, Devon, and southern Ellesmere islands (unit no. 1); the Labrador Highlands (unit 2); the Mealy Mountains (unit 3); the Laurentian Highlands (unit 4); and the Shaler Mountains (unit 5). The bulk of the Shield consists of an exhumed erosion surface: the first four of these mountain ranges are located at the uptilted southeast and northeast margins, and the fifth (the Shaler Mountains) is an up-arched inlier of Proteozoic rock in the arctic lowlands.

The Appalachian-Acadian Region (Figure 1.6B: units 6–10) includes five mountain and highland regions arranged in a series of northeast- and southwest-trending ranges. These mountains retain the imprint of the Caledonian orogeny and, because of their long erosional history are less rugged than the other mountain regions considered here. The region includes Newfoundland (unit 6), the Gaspé Peninsula of Quebec (unit 7), and the highlands of New Brunswick (unit 9) and Nova Scotia (unit 10).

The Innuitian Region (Figure 1.6A: units 11–13) contains three substantial mountain regions – the Axel Heiberg (unit 11), Grantland (unit 12), and Victoria and Albert (unit 13) Mountains. One-third of Ellesmere and Axel Heiberg islands is covered with ice, and there are twelve large ice caps, each with an area over

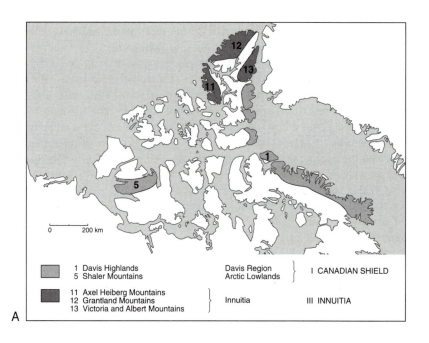

1 Davis Highlands
5 Shaler Mountains
    Davis Region
    Arctic Lowlands } I CANADIAN SHIELD

11 Axel Heiberg Mountains
12 Grantland Mountains
13 Victoria and Albert Mountains } Innuitia   III INNUITIA

A

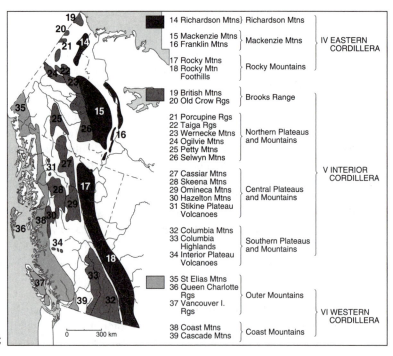

14 Richardson Mtns} Richardson Mtns
15 Mackenzie Mtns
16 Franklin Mtns } Mackenzie Mtns } IV EASTERN CORDILLERA
17 Rocky Mtns
18 Rocky Mtn Foothills } Rocky Mountains

19 British Mtns
20 Old Crow Rgs } Brooks Range

21 Porcupine Rgs
22 Taiga Rgs
23 Wernecke Mtns
24 Ogilvie Mtns
25 Petty Mtns
26 Selwyn Mtns } Northern Plateaus and Mountains

27 Cassiar Mtns
28 Skeena Mtns
29 Omineca Mtns
30 Hazelton Mtns
31 Stikine Plateau Volcanoes } Central Plateaus and Mountains } V INTERIOR CORDILLERA

32 Columbia Mtns
33 Columbia Highlands
34 Interior Plateau Volcanoes } Southern Plateaus and Mountains

35 St Elias Mtns
36 Queen Charlotte Rgs
37 Vancouver I. Rgs } Outer Mountains } VI WESTERN CORDILLERA

38 Coast Mtns
39 Cascade Mtns } Coast Mountains

C

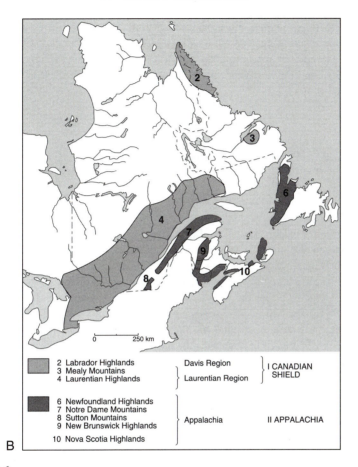

B

	2 Labrador Highlands	Davis Region		I CANADIAN
	3 Mealy Mountains			SHIELD
	4 Laurentian Highlands	Laurentian Region		
	6 Newfoundland Highlands			
	7 Notre Dame Mountains			
	8 Sutton Mountains	Appalachia		II APPALACHIA
	9 New Brunswick Highlands			
	10 Nova Scotia Highlands			

Figure 1.6
Canada's mountain regions: (A) northern Canada, (B) eastern Canada, (C) Cordillera.

2,500 km². Local relief of up to 1,200 m, with the highest summits around 2,500 m a.s.l., provides the setting for some of the harshest environments on Earth.

The Cordilleran Region (Figure 1.6C; units 14–39) is the most complex mountain region. The Eastern Cordillera (units 14–18) includes the Richardson (unit 14), Mackenzie (15 and 16), and Rocky Mountains (17 and 18) and extends over approximately 300,000 km². It is composed almost entirely of folded sedimentary strata and reaches its highest elevation in Mt Robson (3,953 m a.s.l.) in the Continental Ranges of the Rockies. The Interior Cordillera (units 19–34) includes three general groups of plateaus and mountains – Northern (21–26), Central (27–31), and Southern (32–34). The Columbia Mountains in the south (unit 32) contain the highest summits, such as Sir Wilfrid Laurier (3,580 m a.s.l.), but volcanic cones are visually the most impressive. The Western Cordillera (units 35–39) includes the Outer Mountains – the great mountain ranges of southwestern Yukon (St Elias Mountains: unit 35) and the island ranges (36 and 37) – and the Coast Mountains (37 and 38). These ranges extend over approximately 220,000 km^2 and contain the largest icefields in the Cordillera and the highest peaks in Canada – Mt Logan (5,959 m a.s.l.) and St Elias (5,489 m a.s.l.).

COLD WATERS

Fresh-water bodies and marine waters exert a strong influence on Canada's landmass, modifying its climate and extending its land surface in winter (see chapter 2).

Cold Seas

The Arctic Ocean, Hudson Bay, Foxe Basin, Baffin Bay, and the various straits and channels between the arctic islands are large, often deep water bodies. They are frozen for much or all of the year, forming pack ice several metres thick. The pack ice becomes, to all intents and purposes, an extension of the land in winter.

The Arctic Ocean is by far the largest of these bodies of water. Most of its surface is permanently frozen, the pack ice moving slowly in a clockwise direction. Open water develops only in late summer off the west coast of Banks Island and in the Beaufort Sea. M'Clure Strait (the western exit of the northwest passage) and the channels separating the more northerly of the arctic islands remain clogged with large, semi-permanent ice floes. As Canada does not have large ice-breaking vessels (as exist in Russia), marine navigation in the arctic islands is confined largely to waters south and east of Barrow Strait and to a period of six to eight weeks in August and September. There is a slightly more extended shipping season in the Beaufort Sea and Amundsen Gulf of the western Arctic. Further south, Hudson Bay freezes completely by the end of December and does not begin to clear until the following July. The port of Churchill, in northern Manitoba, has a shipping season of only five to six months.

Ice in the arctic waters of northern Canada can be divided into two categories: ice of freshwater origin, which includes icebergs, and sea ice.

Icebergs, which are most frequent in Baffin Bay and off the east coast of Canada as far south as Newfoundland, result from the breakup and disintegration of glaciers when they reach the sea. The glaciers originate largely in the Greenland and Ellesmere Island ice sheets. If icebergs drift into the North Atlantic, they pose a threat to shipping.

Ice islands are exceptionally large icebergs consisting of individual pieces of tabular freshwater ice, often 10 to 60 m thick, which circulate in the Arctic Ocean. They originate in the ice shelves off the northern coast of Ellesmere Island. In 1962–63, for example, the Ward Hunt Ice Shelf lost 675 km^2 by the breaking off of five large ice islands. One of these was used for a number of years by the government of Canada as a floating scientific research station in the Arctic Ocean.

Distribution and thickness of sea ice are extremely variable in both the short and the long term. The maximum and minimum mean extents of sea-ice cover in northern Canada are indicated in Figure 1.4. Sea ice occurs in varying stages of development: currently forming new or young ice is generally less than 30 cm thick, first-year ice is generally between 30 and 200 cm thick, and multi-year ice, representing ice floes that have persisted for more than one year, may be more than 200 cm thick. Because sea ice moves in response to currents, tides, and wind, tensions and pressures develop within the ice. Pressure ridges, consisting of broken ice several tens of metres high, can form when large ice floes converge. In addition to hindering travel, pressure ridging can threaten artificial islands and offshore drilling structures, as in the Beaufort Sea. In areas where ice moves apart, leads of open water develop, seriously disrupting travel across the ice.

Cold Lakes and Rivers

Numerous lakes and an intricate network of surface drainage characterize northern Canada. For the most part, the rivers are dominated by nival (i.e. snow) effects which are discussed more fully below in chapters 5 (northern rivers) and 8 (mountain rivers). Hydrologically, spring is particularly active. Snowmelt produces a high runoff, which is usually associated with the highest water level during breakup. The prediction of the time of breakup and the magnitude of the flood are important annual concerns for settlements along Canada's northern rivers.

The largest river in Canada is the Mackenzie (Figure 1.7A). It is appropriate to examine this drainage system in some detail in this introductory chapter not only because of its size but also because it integrates the effects of both mountain and northern hydrology. The simplicity of the flow hydrograph masks the complexity of the tributary sources.

Figure 1.7B shows the 1986 daily hydrographs of the Mackenzie River above its junction with Arctic Red River and of several of the other major tributaries. The Hay River (drainage area 47,900 km^2) drains the northern lowlands and has the "normal" attributes of winter low flow, brief spring-snowmelt high flow, and summer recession interrupted by rainstorms. The Liard River (drainage area 33,000 km^2), typical of Cordilleran rivers, also has low winter flow, but high runoff is sustained for a longer period by the snowmelt and ice melt from a range of elevations. Large runoff per unit area is another feature of this mountainous basin. By contrast, the Camsell River (drainage area 30,900 km^2) drains the Canadian Shield, where low precipitation is impounded by a myriad of lakes; lake storage evens out the flow over the year so that snowmelt runoff is withheld and a moderate level of winter flow is maintained. Although these sub-basins have distinct hydrological regimes of either nival or mountain character, the Mackenzie's annual flow is the sum total of these tributary contributions combined with regulation of flow provided by dammed and natural lakes.

The Mackenzie is navigable through its length from Hay River on the south shore of Great Slave Lake to Tuktoyaktuk on the coast of the Beaufort Sea. The river plays

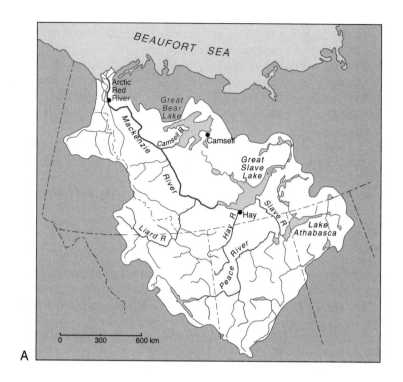

A

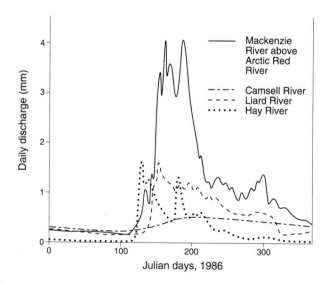

B

Figure 1.7

Mackenzie River: (A) extent of drainage, with major tributaries; (B) daily discharge, 1986, above Arctic
Red River and Camsell, Liard, and Hay rivers.

Table 1.6
Mean freeze-up and breakup dates, Mackenzie River, 1946–55

Location	Distance from Great Slave Lake (km)	Freeze-up	Breakup
Fort Providence	80	24 Nov.	18 May
Fort Simpson			
Mackenzie above			
Fort Simpson	335	27 Nov.	15 May
Mackenzie below			
Fort Simpson	351		11 May
Fort Norman	825	15 Nov.	14 May
Norman Wells	909	10 Nov.	15 May
Fort Good Hope			
Ramparts	1,094	5 Nov.*	22 May
Settlement	1,101	12 Nov.	15 May
Arctic Red River Settlement			
Arctic Red River	1,445	8 Oct.	25 May
Mackenzie River	1,445	1 Nov.	24 May
Lang Trading Post	1,561	9 Oct.†	26 May
Aklavik	1,607	9 Oct.†	28 May
Reindeer Station	1,615	18 Oct.*	27 May

* Seven years of record.

† Nine years of record.

a vital role in the annual resupply of settlements in the western Arctic. During the navigation season (Table 1.6), barges frequent the waterway. However, the season is short: breakup begins in May just north of the Liard River junction and progressively moves downriver, while freeze-up begins during early October in the Mackenzie delta.

In contrast to Russia, Canada has few large, navigable rivers, and this fact has substantially affected northern development. Many areas outside the Mackenzie valley and western arctic coast can be supplied only by air or by overland haul across snow roads in winter.

Breakup on large deep rivers that retain water under the ice all winter usually begins with flows of local meltwater over the ice surface. Then the ice rises as flow increases underneath and shore leads develop. Breakup is usually rapid. Shallow rivers that freeze to the bottom begin to flow over the ice in spring and erode channels through the bottom-fast ice. Large volumes of ice rarely come free rapidly to cause major ice-jams.

Among mountain rivers (commonly small), the spring snowmelt flood is likely to be prolonged because heavy snowfall in the mountains usually permits high rates of snowmelt to be maintained over large areas for considerable periods of time.

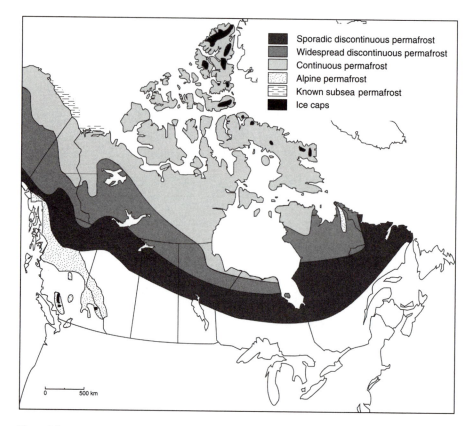

Figure 1.8
Permafrost distribution in Canada. *Source*: Associate Committee on Geotechnical Research (1988).

PERMAFROST

A consequence of prolonged winter cold and the short summers is the formation of perennially frozen ground, or permafrost (chapters 6 and 12). Over half of Canada's land surface is underlain by permafrost of one sort or another (Figure 1.8). At a broad scale, there are two permafrost zones: continuous and discontinuous. Alpine and sub-sea variants are also recognized. In the continuous zone, permafrost exists everywhere beneath the land surface and varies in thickness from approximately 100 m at the southern limit to over 1,000 m in the extreme north. Treeline is an approximate boundary between zones of continuous and discontinuous permafrost. In the discontinuous permafrost zone, some areas are free of permafrost and unfrozen zones, or taliks, are common. In general, the southern limit of continuous permafrost coincides with a mean annual air temperature of −8°C and the southern limit of discontinuous permafrost with −1°C.

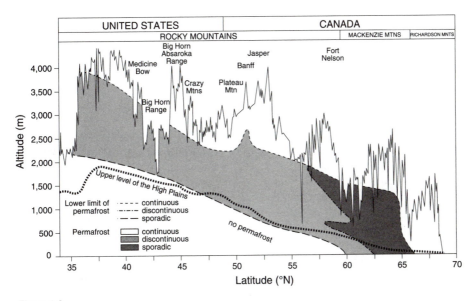

Figure 1.9
Permafrost zone variations, Rocky Mountains (north–south section). *Source*: Harris (1988).

In the mountains of western Canada, permafrost conditions vary according to latitude, elevation, and aspect (Figure 1.9). Usually, permafrost comprises a continuous zone at high elevations below which permafrost is discontinuous. In the Banff and Jasper regions of southern Alberta, the lower limit of continuous permafrost is approximately 2,500 m a.s.l. This elevation decreases further north as the climate becomes colder and rises to the west as increasing depth of snowcover and proximity to Pacific air masses raise ground temperatures. In the southern Mackenzie Mountains, the lower limit of permafrost is approximately 1,800 m a.s.l., but in the southern Richardson Mountains, at latitude 65°N, or when the timberline and northern limit of trees coincide, continuous permafrost is the rule whatever the elevation or aspect.

In practical terms, the importance of permafrost in Canada's cold regions cannot be overemphasised. Permafrost influences virtually all aspects of life in the regions that it underlies (e.g. see Brown 1970). In theory, because permafrost is a thermal condition, its occurrence has little relation to the presence of moisture in the ground. However, ice is an important component of permafrost and gives rise to distinct geotechnical and engineering problems associated with ground heaving or melting of ice-rich permafrost (e.g. see French 1976; Williams and Smith 1989) (see chapter 6).

THE QUATERNARY LEGACY

The landscapes of Canada's cold landmass frequently bear the imprint of events and conditions that have prevailed during the last two million years, a period of time known as the Quaternary. These events include the advance and retreat of ice sheets and glaciers and isostatic changes in land and sea level (e.g. Andrews 1989a; 1989b; Clague 1989a; 1989b).

Contemporary Glaciers

Nearly all of northern Canada has been glaciated at some time during the last two million years. Today, however, only 3 per cent of Canada's landmass is covered by glaciers. With a few exceptions, present-day ice caps and large glaciers are located in either the highland northeastern rim of the Canadian Shield, especially on Baffin, Devon, and Ellesmere islands, or in the St Elias Mountains of the Western Cordillera. Two ice caps, the Barnes and the Penny, exist on Baffin Island, and numerous calving glaciers extend to the sea on Ellesmere Island.

Glaciers in the Canadian Rockies are largely of highland or alpine type. Icefields such as the Columbia and Wapta are extensive névés from which outlet valley glaciers flow in two or more directions. Alpine glaciers such as Castleguard, Peyto, and Saskatchewan have been widely researched. Since the opening of the Canadian Pacific Railway in the 1880s, mountains and glaciers accessible to visitors have been much publicized, but the total area of Rocky Mountain glaciers is estimated at less than 2,000 km^2. In the interior of British Columbia, small glaciers occur in the Cassiar, Columbia, Hazelton, Omineca, and Skeena Mountains, and in the Mackenzie and Selwyn Mountains north of 60°N. But the total area of ice in all these regions is only several hundred square kilometres.

The Boundary Ranges of the BC Coast Mountains and Yukon's St Elias Mountains contain approximately 29,000 km^2 of glacier ice, the largest such concentration in Canada outside the arctic islands. This figure compares with 37,000 km^2 on Baffin Island and 107,000 km^2 in the Queen Elizabeth Islands.

There is widespread evidence of more extensive glaciation in the past (Prest 1983) – for example, in the deeply eroded valleys and fjords of Baffin Island and the mountain and fjord topography of the Western Cordillera. By contrast, but equally revealing, much of the Canadian Shield and interior lowlands is mantled with a veneer of glacial till, coarse outwash, or ice-contact deposits left by retreating ice sheets. Even today, parts of Ellesmere Island, Foxe Basin, and eastern Hudson Bay are experiencing rapid isostatic rebound (e.g. Andrews 1989a; Egginton and Andrews 1989), following earlier depression of the Earth's crust beneath the tremendous weight of these ice sheets. Raised beaches and marine sediments are widespread throughout much of Keewatin, the arctic islands, and northern Quebec.

Evidence from tidal records and geodetic relevelling indicates a consistent pattern of uplift in the outer coastal areas of the Western Cordillera of up to 2 mm per year (Clague 1989a). Elsewhere, the St Elias Mountains and the Boundary Ranges of the Coast Mountains are rising at rates of several millimetres to a few centimetres per year. It is likely that the latter rates are an isostatic readjustment to loss of ice cover, whereas the lower rates are probably tectonic in origin.

Patterns of De-Glaciation

Only the most northern and western arctic islands, parts of the Mackenzie delta, and northern interior of Yukon escaped glaciation during the Quaternary. During the last ice advance, which reached its maximum extent approximately 18,000 years ago, most of Canada lay beneath either Cordilleran or Laurentide ice, sometimes several kilometres thick (e.g. Dyke, Dredge, and Vincent 1983; Andrews 1989a). By 7500 yr BP, the Laurentide ice had retreated to two centres of ice dispersal – in northern Quebec-Labrador and in central Keewatin.

In the mountains of western Canada, the pattern of Cordilleran glaciation was more complex (see chapter 10). The British Mountains, Mackenzie Mountains, Okanagan Range (Cascade Mountains), Old Crow Range, Richardson Mountains, and Taiga Range either escaped glaciation altogether or experienced only alpine and/or valley glaciation. At the close of each non-glacial episode during the Quaternary, glaciers were restricted to the high mountains of the Cordillera. Classically, a four-phase model of glacier growth has been generally accepted (alpine, intensive alpine, mountain ice sheet, and continental ice sheet). A recent model of the growth and decay of Cordilleran ice sheets is illustrated in Figure 9.2 (below). The actual history is more complex than indicated, however, because periods of growth were interrupted by intervals during which glaciers stabilized or receded. It is thought that decay of the ice sheets involved a combination of complex frontal retreat in the coastal region and downwasting accompanied by widespread stagnation throughout much of the interior of the Cordillera.

CONCLUSION

Low temperatures, wind-chill, snow, ice, and permafrost were the primary dimensions of Canada's cold environments during the past two million years and remain so to the present day. The frigidity of the greater part of the country has deeply influenced, directly or indirectly, the past evolution of Canada's scenery, as well as the processes of settling its rural landscapes and the present interaction of industrial society with its biophysical environment.

Global warming, if it occurs, will ameliorate this condition of coldness with respect to mean temperature conditions, but its impact on temperature extremes, wind-chill, distribution and amount of snow, advance and retreat of glaciers, and permafrost degradation are the subject of intensive research, the highlights of which are reviewed in the following chapters.

REFERENCES

Andrews, J.T. 1989a. "Postglacial emergence and submergence." In Fulton (1989) 546–62.
– 1989b. "Nature of the last glaciation in Canada." In Fulton (1989) 544–6.
Armstrong, T., Rogers, G., and Rowley, G. 1978. *The Circumpolar North*. London: Methuen.

Arno, S.F., and Hammerly, R.P. 1984. *Timberline: Mountain and Arctic Forest Frontiers*. Seattle: The Mountaineers.

Associate Committee on Geotechnical Research. 1988. *Glossary of Permafrost and Related Ground-Ice Terms*. National Research Council Canada, Ottawa, Technical Memorandum 142.

Atmospheric Environment Service (AES). 1987. *Climatic Atlas Climatique – Canada*. Canada Department of the Environment, Atmospheric Environment Service, Canadian Climate Program, Map Series 1 – Temperature and Degree Days, and Map Series 2 – Precipitation.

Bird, J.B. 1967. *The Physiography of Arctic Canada*. Baltimore: Johns Hopkins Press.

– 1980. *The Natural Landscapes of Canada*. Second edition. Toronto: John Wiley.

Bostock, H.S., 1971. "Physiographic Subdivisions of Canada." In R.J.W. Douglas, ed., *Geology and Economic Minerals of Canada*, Ottawa: Queen's Printer, 10–30.

Brown, R.J.E. 1970. *Permafrost in Canada and Its Influence upon Northern Development*. Toronto: University of Toronto Press.

Clague, J.J. 1989a. "Quaternary Sea Levels (Canadian Cordillera)." In Fulton (1989) 43–7.

– 1989b. "Cordilleran Ice Sheet." In Fulton (1989) 40–2.

Dyke, A.S., Dredge, L.A., and Vincent, J-S. 1983. "Canada's Last Great Ice Sheet." *Geos* 12 no. 4: 6–10.

Egginton, P.A., and Andrews, J.T. 1989. "Sea Levels Are Changing." *Geos* 18 no. 2: 15–22.

French, H.M. 1976. *The Periglacial Environment*. London and New York: Longman.

– 1979. "Oil and Gas Exploration in the High Arctic Islands: Problems and Prospects." *Marburger Geographische Schriften* 79: 13–26.

Fulton, R.J., ed. 1989. *Quaternary Geology of Canada and Greenland*. Geological Survey of Canada, Geology of Canada, No. 1.

Hamelin, L-E. 1978. *Canadian Nordicity: It's Your North, Too*. Translation of *Nordicité canadien*. Montreal: Harvest House.

Hare, F.K., and Thomas, M.K. 1979. *Climate Canada*. Toronto: J. Wiley and Sons.

Harris, S.A. 1988. "The Alpine Periglacial Zone." In M.J. Clark, ed., *Advances in Periglacial Geomorphology*. UK: John Wiley and Sons, 369–413.

Heginbottom, J.A., co-ordinator. 1989. "A Survey of Geomorphic Processes in Canada." In Fulton (1989) 575–643.

Horner, R.B., Lamontagne, M., and Wetmiller, R.J. 1987. "Rock and Roll in the N.W.T.: The 1985 Nahanni Earthquakes." *Geos* 16 no. 2: 1–4.

Ives, J.D., and Barry, R.G., eds. 1974. *Arctic and Alpine Environments*. Methuen, Harper and Row Publishers.

Love, D. 1970. "Subarctic and Subalpine: Where and What?" *Arctic and Alpine Research* 2: 63–73.

Maxwell, J.B. 1980. *The Climate of the Canadian Arctic Islands and Adjacent Waters*. Atmospheric Environment Service, Climatological Studies 30. Ottawa: Environment Canada.

Prest, V.K. 1983. *Canada's Heritage of Glacial Features*. Geological Survey of Canada, Miscellaneous Report 28.

Ritchie, J.C. 1984. *Past and Present Vegetation of the Far Northwest of Canada*. Toronto: University of Toronto Press.

Ritchie, J.C., and Hare, F.K. 1971. "Late-Quaternary Vegetation and Climate near the Arctic Tree Line of Northwestern North America." *Quaternary Research* 1: 331–41.

Smith, M.W. 1989. "Climate of Canada." In Fulton (1989) 577–81.

Williams, P.J., and Smith, M.W. 1989. *The Frozen Earth: Fundamentals of Geocryology*. Cambridge: Cambridge University Press.

Young, S.B. 1989. *To the Arctic: An Introduction to the Far Northern World*. New York: John Wiley and Sons.

CHAPTER TWO

Canada's Cold Seas

ROGER G. BARRY

Canada's arctic and subarctic seas form a vital element of the northern environment. The ice-dominated waters fostered the distinctive Inuit culture, with a subsistence basis adapted to the marine environment. These same waters acted as a barrier to exploration for a Northwest Passage and in the last two decades have necessitated adoption of specialized engineering technologies for development of seabed resources. Marine transportation requires ice-strengthened vessels and icebreaker assistance for much of the year; off the east coast, icebergs present a hazard to shipping and drilling platforms. The common element linking the bodies of water with which we are concerned is their seasonal cover of sea ice.

WATER MASSES

The primary physical characteristics of sea water are its salinity, temperature, and density. Polar oceans are unusual in that variations in salinity play the major role in determining density, unlike elsewhere in the open ocean, where temperature is decisive. For salinities exceeding 24.7‰ (parts per thousand), the density of sea water increases as temperatures decrease toward freezing point. Hence, the resulting convection in the surface layer may delay formation of sea ice until this layer is cooled throughout to near freezing.

The annual cycle of ice growth and decay significantly affects salinity and temperature characteristics of the upper layers of the polar oceans and, therefore, their stratification. For example, surface salinities over the Beaufort Sea continental shelf increase by 2 to 3‰ during autumn freeze-up (Melling and Lewis 1982). However, this is not a simple process. The surface waters in the Arctic Ocean are modified by thermohaline effects, advection, and ocean-atmosphere energy transfers. The most significant of these transfers and effects (Melling et al. 1984) are (1) inflow and mixing of Pacific waters, (2) export of arctic water and ice via the East Greenland Current, (3) runoff from the land, (4) absorption of solar radiation in summer, (5) turbulent heat losses through leads and polynyas in winter, and (6) density increase through salt being excluded from the ice during freezing. Convection in the surface layers involves vertical currents set up by the large heat losses in winter from open leads. This process transports excess salt downward in the uppermost layer.

The Arctic Ocean

The Arctic Ocean is divided into two deep basins – the Eurasian and the Canada – by the transpolar Lomonosov Ridge. The Canada sub-basin lies south of the Alpha Cordillera and reaches depths in excess of 3,500 m. The continental shelf is narrow off most of arctic North America, extending only 50 to 100 km from the coast, except in the southeastern Beaufort Sea, where it is some 150 km offshore.

In the Canada Basin, a shallow well-mixed water layer overlies a strong halocline at 25 to 100 m depth (Figure 2.1). Salinities in this layer average 31–32‰ in the Beaufort Sea in winter, decreasing to 28–30 in late summer as a result of surface runoff and seasonal melt (Melling and Lewis 1982). Inflow from the Mackenzie River is the main cause of low summer salinities on the Shelf. These processes also maintain surface temperatures near the freezing point (-1.5 to $-1.8°$C). The halocline prevents deep convective overturning and heat flux from the warmer saline waters below about 200 m. The halocline appears to be maintained primarily by lateral injection of brine from the continental shelves in winter (Aagaard, Coachman, and Carmack, 1981).

In the Canada Basin, temperatures increase slightly at 50–100 m, with salinities of 32–33‰ (Carmack 1986), because of influx from the Bering Sea. It overlies a layer of lower salinity that is thought to relate to water from the Bering Sea and from adjacent shelf regions. At depths between 200 and 1,000 m, there is Atlantic Water with temperatures of 0° to 0.5°C and salinities approaching 34.9‰. The absence of any water with properties intermediate between Arctic and Atlantic water masses indicates that there is no direct vertical mixing in the Arctic Basin. Aagaard, Foldvik, and Hillmann (1987) indicate that Atlantic Water is transformed into Arctic Intermdiate Water within 600 km of Fram Strait by turbulent heat fluxes to the atmosphere of up to 200 W m^2 in winter and by some freshening in the Arctic Ocean.

Baffin Bay/Labrador Sea

Northern Baffin Bay is connected to the Arctic Ocean by Nares Strait (sill depth of 250 m) and Barrow Strait (sill depth 130 m), both of which are important for water exchange; the passage through Jones Sound seems to be less significant (Coachman and Aagaard, 1974). The surface waters of western Baffin Bay consist of cold, low-salinity Arctic Water (Muench 1971). The general characteristics are temperatures of 0 to $-1.8°$C and salinities 31–34‰. However, the upper 50 m or so is greatly modified in summer by solar heating, ice melting, and wind mixing, giving temperatures between $-1°$C and 3°C and salinities of 28–32.5‰ (Fissel, Lemon, and Binch 1982). Beneath Arctic Water, at depths between 200 and 1,300 m, lies warmer

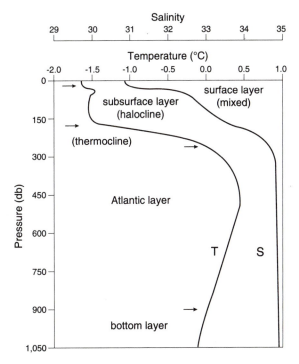

Figure 2.1
Water mass characteristics, Canada Basin, from typical hydrographic profiles. T = temperature; S = salinity. *Source*: Melling and Lewis (1982).

and more saline Atlantic Water (0° to ≥2°C, 34.2–34.5‰), while below that is Baffin Bay Deep Water (near 0°C, 34.5‰).

Canadian Arctic Archipelago

Numerous channels connect the Arctic Ocean and the Atlantic Ocean through the Canadian Archipelago. The western and eastern parts of the archipelago are separated by channels with sill depths of only 85–140 m, but sill depths between the Beaufort Sea and western archipelago are 350–450 m.

The water mass forming the upper 250–300 m in the archipelago is Arctic Water with temperatures near freezing for most of the year in the surface mixed layer (de Lange Boom, Melling, and Lake 1987). The water structure is similar to that in Figure 2.1. Below Arctic Water is Arctic Intermediate Water of Atlantic origin to 900 m depth, with temperatures above 0°C and salinities of 34.65–34.85‰ in the northwestern channels. Surface salinities increase from around 31.0‰ in the Arctic Ocean to 33.0‰ over the central sills.

Summaries of scattered data (Collin and Dunbar 1964; LeBlond 1980) indicate southeastward drift (typically 0.1–0.2 m/s) of Arctic Water. As water moves southeastward into the western archipelago, it is modified by the diffusion of heat from underlying Atlantic Water, producing slight warming of the halocline (0.25C° warmer in Parry Channel). This compares with water over the Arctic Ocean continental shelf, where ice growth produces downward lateral mixing of cold saline water from the surface (Melling et al. 1984). At the same time, upward removal of heat slightly cools underlying Atlantic Water in the archipelago. The major transformation of Arctic Water is produced by mixing caused by strong tidal flows and by the loss of sensible heat in winter through thin ice and open water.

CIRCULATION

The Beaufort Sea and Canada Basin

The surface circulation of the Arctic Ocean is dominated by the clockwise Beaufort Gyre centred in the Canada Basin. This ocean gyre, and corresponding motion of the pack ice, are primarily a response to winds associated with a mean anticyclonic pressure field during the winter. The current velocity averages only 0.01–0.04 m/s, so ice may take between three and twelve years to circulate around the gyre (Colony and Thorndike 1984). Ice velocities are lowest in April, when the ice is most compact. Correspondingly, maximum drifts are observed in August and September, when ice concentrations are lower. The Transpolar Drift Stream also affects the circulation of the surface water and pack ice. Polar water and ice in the Transpolar Drift Stream exit from the Arctic Basin as the East Greenland Current through Fram Strait.

Sub-surface circulation in the Arctic occurs mainly in narrow boundary currents along the marginal boundaries of the Eurasian and Canada basins. The circulations are cyclonic, counter to the upper ocean drift. In the Canada Basin, kinetic energy is concentrated mainly in mesoscale eddies, 20 km or so in diameter, which appear to have a time scale of a year or more.

The North Atlantic sector is the main area of both inflow to and outflow from the Arctic Basin. Table 2.1 illustrates the great magnitudes of the West Spitsbergen Current and the East Greenland Current. Fram Strait, with a sill depth of 2,600 m, accounts for about three-quarters of the mass exchange and 90 per cent of the heat exchange between the world ocean and the Arctic (Aagaard and Greisman 1975). The numbers in Table 2.1 must be considered tentative because of large interannual variations in the West Spitsbergen Current and in the flow through Bering Strait (Coachman and Aagaard 1981; 1988; Aagaard 1982), and there are major uncertainties in the deep-water contributions.

If one assumes a surface area of 10^7 km^2, the freshwater budget of the Arctic Ocean can be estimated (Table 2.2). The inflows are runoff (R) from rivers, the net atmospheric contribution of precipitation (P) minus evaporation (E), and the freshwater component of Pacific low-salinity water which enters the Chukchi Sea through

Table 2.1
Mass budget of the Arctic Basin

Inflow	Units: $10^6 \ m^3/s$	Outflow	Units: $10^6 \ m^3/s$
Bering Strait	0.8	Arctic Archipelago	− 1.7
West Spitsbergen Current	3.6	East Greenland: Polar Water	− 1.8
Spitsbergen-Franz Joseph Land	0.7	Atlantic Water	− 3.0
Runoff	0.1	Ice	− 0.1
	5.2		− 6.6

Sources: Aagaard and Greisman (1975); Coachman and Aagaard (1988); Fissel et al. (1988); and Foldvik, Aagaard, and Torresen (1988).

Table 2.2
Freshwater budget of the Arctic Ocean

	Units: $10^3 \ m^3/s$
Inflow	
Precipitation − evaporation*	44
River runoff:† Eurasia	91
North America	20
Inflow from Bering Sea‡	60
Subtotal	+ 215
Outflow	
Ice export§	− 103
	to − 158
Freshwater export‖	− 72
Subtotal	− 175
	to − 230
Net result (inflow minus outflow)	+ 40
	to − 15

* Burova (1981).

† Treshnikov (1985).

‡ Assumes Pacific inflow of $25 \times 10^3 \ km^3/yr$ with a salinity of 32.4‰ mixing with Arctic Water with a mean salinity of 34.9‰, producing a 19-cm layer (equivalent) of freshwater (Aagaard, personal communication, 1988).

§ Wadhams (1983); Vinje and Finnekasa (1986).

‖ Aagard and Carmack (1989).

Bering Strait (flux: F). The runoff is equivalent to a 36-cm thick layer of fresh water, of which the Mackenzie River contributes about 9 per cent. The runoff estimates are unlikely to be too high, whereas the $(P - E)$ term and Bering Strait inflow may be overestimates in view of the imbalance shown in Table 2.2. These inflows are assumed to be balanced by export of ice and freshwater in the East

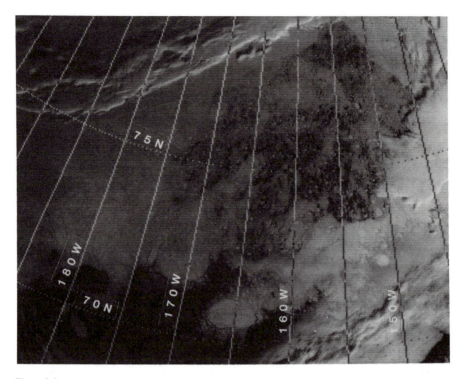

Figure 2.2
Cloud-free visible-band image of Beaufort Sea for 2 September 1988, showing areas of reduced ice concentration caused by cyclonic activity. Defense Meteorological Satellite Program image (0.6-km resolution), courtesy National Snow and Ice Data Center, University of Colorado, Boulder.

Greenland Current. The ice export is probably near the upper end of the proposed range (158×10^3 m^2/s), which implies a small excess of freshwater outflow.

Ice motion in the Canada Basin is measured routinely by a network of drifting buoys that transmit their position to a satellite data link. These measurements show that in late summer the clockwise ice motion in the Beaufort Gyre commonly reverses for about a month, affecting ice concentration. The reversal is caused when slow-moving cyclone systems become centred over the Canada Basin (Serreze and Barry 1988; Serreze, Barry, and McLaren 1989). Beneath a low pressure system, ice in free drift (which is the approximate situation in summer) moves at nearly 20 degrees to the right of the geostrophic wind, as a result of the Coriolis effect. The resulting divergence of about 0.5 per cent per day reduces ice concentration. Areas of low-concentration ice are detected in this sector in late summer (Barry and Maslanik 1989) and can be inferred by cloud-free visible-band images (Figure 2.2).

Circulation on the Beaufort Shelf in summer interacts with flow in the Canada Basin and adjacent Chukchi Sea. Offshore circulation is determined by the Beaufort

Gyre, while Pacific Water moves northward into the Chukchi Sea, via the Bering Strait, and then continues eastward along the shelf break as a warm undercurrent. Aagaard (1984) considers this to be a boundary current of the large-scale circulation in the Canada Basin. Over the inner shelf, summer circulation is mainly wind-driven. Buoy trajectories in the southern Beaufort Sea indicate that surface currents are predominantly alongshore, associated with long-term geostrophic winds (St Martin 1987). Changes of wind direction can cause major differences in local surface salinities in the eastern Beaufort Sea by deflecting the Mackenzie River discharge. For example, O'Rourke (1974) cites values of only 5‰ off Richards Island during weak westerlies, contrasted with 15–25‰ during moderate easterlies.

Canadian Arctic Archipelago

Water circulation in the Arctic Archipelago has only recently been investigated systematically. Fissel et al. (1988) assess residual (non-tidal) currents based on under-ice current meter measurements during 1982–85. These confirm earlier suggestions of a southeastward flow but show significant regional and seasonal variability. In the deep channels of the Queen Elizabeth Islands and the western archipelago (Viscount Melville Sound), residual currents are weak and variable. There is southward flow (0.05–0.13 m/s) toward Parry Channel, via Byam Martin Channel and Penny Strait, and stronger eastward flows on the south side of Barrow Strait (0.15–0.25 m/s). In the broad southern channels of the archipelago (M'Clintock Channel, Prince Regent Inlet) and in the passages connecting with the Arctic Ocean, the few available measurements show southward residual flows (Figure 2.3).

The currents have a summertime maximum in the eastern half of the archipelago, but there is an autumn-winter maximum further west (Fissel et al. 1988). This difference may be related partly to dates of ice clearance but also involves regional differences in the seasonal regimes of sea-level gradient. There is a north–south gradient of sea level between Alert and Resolute which is strongest in autumn-winter and weakest in summer; a west–east sea-level gradient between the Beaufort Sea and Barrow Strait shows the opposite tendency (Prinsenberg and Bennett 1987). The latter accounts for greater water transport through Barrow Strait in summer.

Several of the wider channels exhibit counter-currents. For example, along the north side of Lancaster Sound, westward motion from an intrusion of the Baffin Current extends as far as eastern Parry Channel; counter-currents occur also in Prince Regent Inlet and Hudson Strait. Such channels are wide enough to permit two geostrophically balanced flows in the upper layer. LeBlond (1980) shows that if the radius of curvature of the coast is sufficiently large, a geostrophic flow can follow the coastline without separation.

Because of shallow sills, only surface waters penetrate from the Arctic Ocean into Baffin Bay, although relative contributions to the annual volume transport through the major sounds – Nares Strait–Smith Sound and the interconnected Viscount Melville Sound–Barrow Strait–Lancaster Sound (forming the Northwest Passage) – are

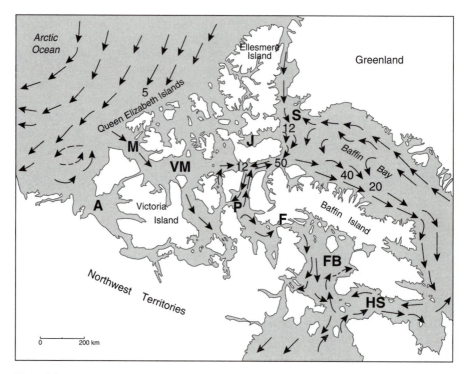

Figure 2.3
Surface circulation, eastern Canadian Arctic. Surface currents are in cm/s. Letters denote straits and basins: F = Fury and Hecla Strait; FB = Foxe Basin; HS = Hudson Strait; J = Jones Sound; L = Lancaster Sound; M = M'Clure Strait; P = Prince Regent Inlet; S = Smith Sound; VM = Viscount Melville Sound.

still uncertain. Fissel et al. (1988) give annual volume estimates of 67×10^4 m^3/s southward via Nares Strait, $29 \pm 5 \times 10^4$ m^3/s southward via Parry Strait and Wellington Channel into Lancaster Sound, and $45 \pm 25 \times 10^4$ m^3/s eastward through Barrow Strait. These values, and an estimated contribution via Jones Sound, form the basis for the outflow through the archipelago shown in Table 2.1.

Baffin Bay, Hudson Bay, and Eastern Arctic

The East Greenland Current, carrying pack ice southward along the east coast of Greenland, flows around Cape Farewell and, together with an Atlantic component (the Irminger Current), continues northward as the West Greenland Current (Figure 2.3), with temperatures of 2°–5°C and salinities of 33 to 34‰. The Atlantic component of the West Greenland Current is best developed in July through September (Collin and Dunbar 1964). This water branches off into Baffin Bay, with a well-developed southward return flow along the east coast of Baffin Island.

Baffin Bay, which has a maximum depth of more than 2,300 m, is linked to the Labrador Sea by Davis Strait, which is only about 600 m deep. By contrast, the Labrador Sea adjoins the 5,000-m–deep western North Atlantic. Southward transport in the Baffin Current far exceeds that northward in the West Greenland Current. Volume continuity is achieved by the inflow of Arctic Water via the channels in the Canadian Archipelago.

The Baffin Current, extending along the length of western Baffin Bay, is first detectable off Devon Island. In summer and autumn, it penetrates 50 km or more westward along the north side of Lancaster Sound before moving south and then east to continue down the east coast of Baffin Island (Sanderson and LeBlond 1984). It is generally within 100 km of the coast. In winter and spring, the near-surface flow, measured off Devon Island and northern Baffin Island, diminishes by about 50 per cent and synoptic current variations are damped, apparently because of the presence of sea ice (Lemon and Fissel 1982). The Baffin Current divides at Hudson Strait (63°N), with one branch entering Hudson Strait and the other joining the Labrador Current, together with cold-fresh outflow from the south side of Hudson Strait. According to LeBlond et al. (1981), this flow-splitting is related to seabed rise northeast of Resolution Island.

The Labrador Current is made up of 80 per cent water from the Irminger/West Greenland Current, which is deflected by the sill in Davis Strait, and 20 per cent from the Baffin Current. The Atlantic Water joins the latter well offshore. Lazier (1982) shows that temperatures are very uniform (near $-1.0°C$ at 50 m depth) over the shelf, with a sharp increase over the continental slope of about $5C°$ to the offshore waters. Measurement of seasonal variations in the current averaged over the shelf (between the coast and the 1,000-m isobath) shows a temperature range in the mixed layer of between $-1.5°C$ (February–March) and 6.7°C (August) and salinity variations of between 31‰ (July–August) and 33.2‰ (February–March). Judging by long-term ship observations (1946–80), the annual sea-surface temperature range increases from about 6.6°C off central Labrador to 12–13°C over the Grand Banks (Thompson, Loucks, and Trites 1988).

The Baffin Bay–Labrador Current greatly depresses temperatures along the eastern coast of northern Canada. As a consequence, pack ice and icebergs are transported far southward in late winter and spring. The maximum southward extent of sea ice occurs in February and March, when it reaches 47°N, on average, and icebergs commonly drift as far south as 44°N over the Grand Banks in April and May.

The Labrador Sea is an important region of meridional water and heat transport and, in many winters, of deep water formation. Ikeda (1987) estimates annual southward freshwater transport of $10^{12}(3.15 \times 10^3 \text{ m}^3/\text{s})$, relative to a base salinity of 34.8‰, which is comparable to the flow of Arctic Water into Baffin Bay or to summer runoff into Hudson Bay. There is corresponding northward heat flux of 2×10^{13} W caused by southward transport of ice and cold water.

Hudson Bay and its southern extension into James Bay form an extensive, but shallow, inland sea covering $1.2 \times 10^6 \text{ km}^2$, with depths generally less than 150 m.

Its remoteness from the Atlantic Ocean and its subarctic location give rise to a severe climate with long, cold winters and cool summers. Hudson Bay is almost completely ice-covered from mid-December until early July, but late-summer water temperatures are between 5° and 9°C, as a result of net monthly heat inputs of $\geqslant 150$ W/m^2 during June–August (Danielson 1969). Inshore areas have lower temperatures because mixing brings up colder, deeper water. Offshore, mixing is prevented by strong vertical stratification, with deep-water temperatures near -1.5°C. As a result of this stratification, the residence time of the deep water is of the order of 5 to 15 years (Roff and Légendre 1986).

In summer, the surface layer of Hudson Bay has a weak counter-clockwise (cyclonic) circulation, driven by wind and runoff (Prinsenberg 1982). The mean outflow is a seasonal response to the input of freshwater. The water is brackish, with surface salinities of 29‰ in the centre and 24–28‰ near the coasts. Mean annual runoff contributes 21×10^3 m^3/s to Hudson Bay, equivalent to a 78-m layer of freshwater (Prinsenberg 1988). Runoff is especially significant in James Bay, where depths are mostly less than 50 m. Because of the many rivers that enter it, James Bay receives 60 per cent of the total runoff, and salinities may be as low as 10‰ off major river mouths in summer.

The melting of the ice cover is equal to, or greater than, runoff as a component of the summer freshwater budget, as shown in Figure 2.4 (Prinsenberg 1984). From freshwater budget calculations, it seems that more ice grows and decays annually than observed ice thicknesses suggest. The winter-to-summer change in the freshwater layer (ΔW) is balanced by runoff (R), precipitation minus evaporation ($P-E$), water and ice flux ($F_w + F_i$), and (residual) icemelt (M):

$$\Delta W + (R + P-E) + (F_w + F_i) + M = 0.$$

In freshwater equivalent units (m), estimated values for this expression for Hudson Bay are as follows:

$$-2.9 + (1.1) + (-0.24) + 2.0 = 0.$$

Observations indicate that level ice reaches a maximum thickness of about 1.6 m during April and May (Markham 1986). The 0.4-m discrepancy is attributed to ridging, as discussed below.

There is also a net annual loss of moisture by evaporation from Hudson Bay, relative to precipitation, whereas over James Bay precipitation and evaporation are in balance (Prinsenberg 1986a). When net ($P-E$) figures are combined with runoff, there is an annual freshwater input equivalent to a 0.6-m layer over the whole area.

The hydrography and oceanography of Foxe Basin are not well known because of the long ice season, which lasts from mid-November through mid-August in the centre of the basin and even longer in the north. Water depths are about 50 m in the

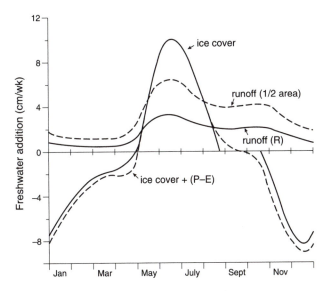

Figure 2.4
Freshwater addition (cm/wk) to Hudson Bay caused by ice cover, runoff
(*R*), precipitation (*P*), and evaporation (*E*), assuming 1.6-m maximum
ice thickness. Curve "runoff (½ area)" shows that effect of melting ice
is equivalent to runoff contribution even if runoff covers only half of
surface area. *Source*: Prinsenberg (1988).

north and east but exceed 200 m off Southampton Island in Foxe Channel. Summer
ice motion and current observations indicate a southward surface flow of about
0.5 m/s along Melville Peninsula and northern Southampton Island, representing a
continuation of the drift of arctic waters through the channels of the Arctic Archipelago
(Prinsenberg 1986b). This drift is facilitated by runoff and ice melting forcing an
outflow of surface water from the basin. Little is known of the winter circulation.
The annual volume transport (eastward) through Fury and Hecla Strait is only
4×10^4 m^3/s (Fissel et al. 1988).

Water temperatures in Foxe Basin are between $-1.7°C$ and (in Foxe Channel)
3°C in summer. Salinities vary widely because of ice melt and runoff, becoming
more uniform between 29 and 32‰ in winter, with corresponding summer values
of 0.5–0.7°C and 31–32‰ (Prinsenberg 1982). This lower-salinity water tends to
flow over the more dense Foxe Basin water, except where mixed by tidal action.
The tidal range in the shallow eastern part of the basin reaches 4 to 8 m.

Hudson Strait is a channel 300–400 m deep connecting Hudson Bay and Foxe
Basin to the Labrador Sea. It is some 750 km long and averages 150 km in width,
although it is extended about 200 km southward by Ungava Bay. The surface circulation consists of a 0.1 m/s inflow from Davis Strait, as a branch of the Baffin
Current that continues along the south coast of Baffin Island, and a 0.3 m/s south-

eastward outflow from Hudson Bay and Foxe Basin along the Quebec coast (Drinkwater 1988).

Tides in the North Atlantic force large semidiurnal tides in Hudson Strait and Ungava Bay. Mean tidal ranges at the eastern entrance average 4–8 m, but large tides at the head of Ungava Bay and off the south coast of Baffin island, near Big Island, are between 6 and 14 m (Fisheries and Marine Services 1977). As a result, there are strong tidal currents of up to 2 m/s (Drinkwater 1988).

The summer patterns of both surface temperature and salinity show that isolines trending from northwest to southeast produce across-strait gradients. These are of the order of 1–2°C and directed from south to north for surface temperature and 2–3‰ from north to south for salinity. River runoff produces salinities below 30‰ in southern Ungava Bay. Annual runoff is approximately one-quarter of that received by the much larger volume of Hudson Bay. Drinkwater (1988) notes that ice melt, assuming 135 cm of ice, gives a freshwater contribution of almost 4×10^4 m^3/s between mid-May and the end of July. This exceeds summer runoff into Ungava Bay.

Seasonal runoff, ice melt, and outflow of low-salinity water from Hudson Bay and Foxe Basin strongly stratify density in southern Hudson Strait and Ungava Bay. However, at the eastern end of the strait in September, there is a vertically well-mixed water mass of low temperature (about 0.5°C) and relatively high (32.5‰) salinity. Convergence and mixing in this area produce a Labrador Shelf water mass (Dunbar 1951) which can be identified all along the Labrador coast (Lazier 1982).

SEA-ICE CONDITIONS

Physical Processes

The freezing point of water decreases from 0°C for freshwater to −2°C for water of 35‰ salinity. Moreover, the temperature of maximum density decreases from 4°C for freshwater to −1.3°C (freezing point) at 24.7‰ salinity (Figure 2.5).

When sea water freezes, a skim of pure ice crystals (grease ice) forms on the surface. As freezing continues and more pure ice forms, surface salinity increases. Platelets of frazil ice become extensive and aggregate into slush ice and then nilas. Rapid crystal growth traps brine in pockets, giving young ice a salinity of 12–15‰. This value decreases as brine drains under gravity through the ice via drainage tubes. Finally, during melt, flushing of brine occurs, so that multi-year ice has surface salinity values near zero, increasing with depth to 2–4‰ in the floe. Consequently, this ice has traditionally been used for drinking water by Inuit and others travelling the pack ice.

The rate of thickening of an ice cover depends basically on the surface temperature of the ice (or snow) and the thickness of the ice (and snow). The basal ice temperature is generally at about −1.8°C. Assuming a linear temperature gradient in the ice, an analytical relationship for ice growth can be written:

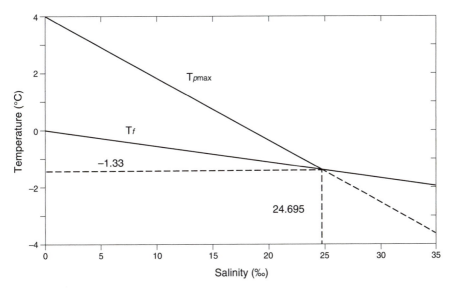

Figure 2.5
Effects of salinity: relationships with freezing point of water (T_s) and with temperature of maximum water density (T_{max}).

$$-\rho_i \times L \times \frac{dH}{dt} = F_c + F_w,$$

where ρ_i is ice density (900 kg/m^3),
     L is latent heat of fusion (0.335 $\times$ 10^6 J/kg),
     H is ice thickness,
     F_c is conductive heat flux through the ice, and
     F_w is oceanic heat flux to the ice at the ice/water interface.

Expansion of equation 1 to calculate F_c and heat flux from the ice surface into the atmosphere allows derivation of an expression for ice growth in relation to cumulative freezing degree–days, if oceanic heat flux is neglected. Figure 2.6 compares this relationship for bare ice ($h_s = 0$) with empirical formulae based on freezing degree–days. It shows that the derived expression overestimates the rate of sea-ice growth. Although the derived equation can be modified by incorporating a term for conduction through a snow layer (Maykut 1986), a more serious error probably involves the neglect of F_w. This may be of the order of 15–40 W/m^2 in the first month of ice growth, decreasing subsequently to about 10 W/m^2. More complete treatment of sea-ice growth requires a sophisticated thermodynamic ice model.

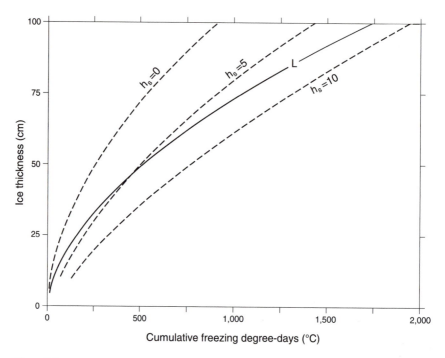

Figure 2.6
Rates of ice growth predicted by analytical relation (see text) and by empirical formula based on freezing degree–days. *Source*: Maykut (1986).

Climate Characteristics and Annual Ice Cycles

The general seasonal regimes of sea ice in the Canadian Arctic are illustrated in Figure 2.7 for typical mid-winter conditions (A) and for mid-August (B) of a "favourable" year. Severity of ice conditions in the western part of the Arctic Archipelago and also Foxe Basin is related to the persistence of temperature inversions and low air temperatures during winter and spring because of a predominance of anticyclonic conditions (Maxwell 1981). Conversely, in summer frequent low cloud and fog inhibit melt. In contrast, the Baffin Bay–Davis Strait region has year-round cyclonic activity, and the climate is more maritime, with cool, cloudy summers. Figure 2.8 illustrates the variety of climatic conditions in the major regions of the Arctic Archipelago identified by Maxwell (1981) and in the other ice-covered seas discussed below.

A broad overview of the annual ice cycles in three marine regions of northern Canada is provided in Figure 2.9, based on four years of passive microwave data, 1973–76 (Parkinson et al. 1987). Each figure shows the area covered by ice (of greater than 15-per-cent concentration) and the "pseudo-ice area" (excluding any open water). In the Canadian Archipelago, half of the total ocean area of

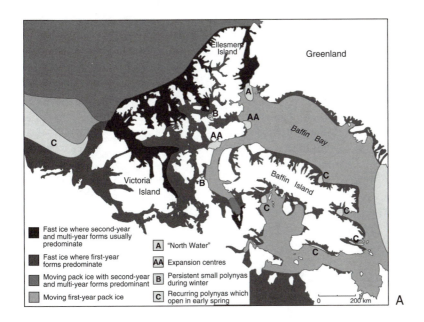

Fast ice where second-year and multi-year forms usually predominate

Fast ice where first-year forms predominate

Moving pack ice with second-year and multi-year forms predominant

Moving first-year pack ice

A "North Water"

AA Expansion centres

B Persistent small polynyas during winter

C Recurring polynyas which open in early spring

Ellesmere Island

Greenland

Baffin Bay

Victoria Island

Baffin Island

0   200 km

A

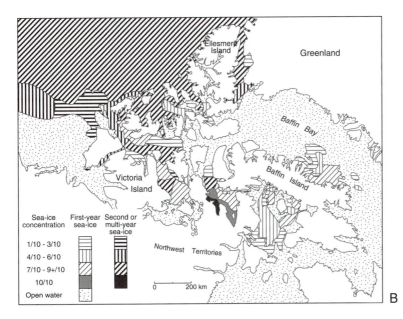

Ellesmere Island

Greenland

Baffin Bay

Victoria Island

Baffin Island

Northwest Territories

Sea-ice concentration | First-year sea-ice | Second or multi-year sea-ice

1/10 - 3/10

4/10 - 6/10

7/10 - 9+/10

10/10

Open water

0   200 km

B

Figure 2.7
Sea-ice conditions, Canadian Arctic: (A) midwinter (typical); (B) mid-August ("favourable" year).
Source: Maxwell (1981).

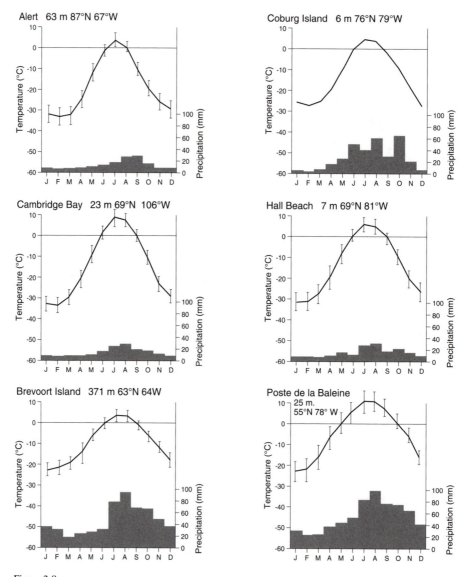

Figure 2.8

Mean annual temperature and precipitation regimes at six stations in arctic and subarctic Canada: Alert (1950–70), Coburg Island (1972–76, incomplete record), Cambridge Bay (1941–70), Hall Beach (1957–70), Brevoort Island (1959–70), and Poste de la Baleine (1931–60). *Sources*: Wilson (1968); Müller et al. (1977); and Maxwell (1981).

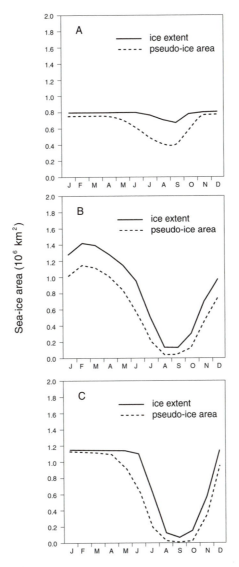

Figure 2.9
Ice extent (>15-per-cent concentration) and "pseudo-ice area" (excluding any open water), annual cycles (1973–76): (A) Canadian Archipelago, (B) Baffin Bay–Davis Strait–Labrador Sea, and (C) Hudson Bay. *Source*: Parkinson et al. (1987).

800,000 km^2 is open water in August and September. Likewise, the vast area of Hudson Bay (1,161,000 km^2) is frozen over from December through June. However, areas of open water increase rapidly in early summer, and the Bay is almost ice-free from August through October. The Baffin Bay–Davis Strait–Labrador Sea region has a seasonal maximum of sea ice in February, but, with 15-per-cent open-water areas within the pack-ice boundaries, largely ice-free conditions exist in August and September.

Regional variability of sea-ice cover is examined in more detail below.

The Beaufort Sea

Sea ice is present throughout the year in the Beaufort Sea, except near the mainland coast in summer. Away from the coast, the ice cover is close to 100 per cent for nine or ten months of the year, moving slowly westward in the Beaufort Gyre system. Drift speeds average 20 km/month in winter and four or five times faster in summer. The pack ice in the Canada Basin rotates clockwise approximately about an axis in the northern part of the Basin.

The predominantly multi-year sea ice of the northeastern Beaufort Sea comprises large floes, with a mean diameter of 700 m (Hudson 1987). Heavily hummocked and ridged floes, according to Hudson, account for about 1 per cent of the multi-year ice. In the eastern Canada Basin (72°40′N; 138°15′E), sea ice attains a mean thickness of 3.9 m in April (Wadhams and Horne 1980). Wadhams (1981) reports 44 per cent ice more than 5 m thick and a mean thickness of 6.1 m off northern Ellesmere Island and Greenland in October. Off the western coasts of the archipelago, the gyre motion produces thick, heavily ridged multi-year ice. Thicknesses average 6–8 m in this area (Bourke and Garrett 1987).

Over the Beaufort Sea continental shelf, there is a complex cross-shelf zonation of ice types (Reimnitz, Toimil, and Barnes 1978). The occurrence of different ice zones in the southeastern part of the Beaufort is shown schematically in figure 2.10. Close to shore there is bottom-fast ice, while beyond the 2-m isobath an extensive zone of floating fast ice covers most of the shelf; the 20-m isobath is generally between 25 and 60 km offshore. The seaward limit of landfast ice is commonly demarcated by grounded pressure ridges, known as stamukhi, formed in waters 15–40 m deep. Ridging is caused by wind and wave action early in the winter and occasional storm surges. Onshore winds and storm surges may also cause ice pile-up and ride-up along the Beaufort Sea coast (Kovacs and Sodhi 1980). When strong onshore winds coincide with high tides, sea ice can be pushed more than 100 m inland on beaches.

In the zone between the 15- and 45-m isobaths, numerous linear furrows in the seabed display gouging by ice. These furrows occur out to a depth of 50–60 m and are 0.5–6 deep (Lewis 1977); they are believed to be produced by fragments of ice islands and deep grounded ridges. Their frequent occurrence implies a high risk of ice damage to seabed structures such as wellheads or oil and gas pipelines. Off the

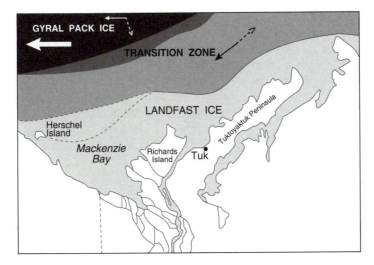

Figure 2.10

Ice zones of southeastern Beaufort Sea shown schematically; *Source*: Marko (1975). Heavy arrow denotes mean ice motion in Beaufort Gyre; dominant ice motion is also shown for winter-spring (—►) and summer-fall (⋯►). Variable margin of landfast ice in western Mackenzie Bay is indicated by dashed lines.

Alaskan coast, they are caused by ice moving from northwest to southeast; there is also an eastward current drift in summer (Reimnitz et al. 1988).

The seasonal regime of landfast ice along Alaska's Beaufort Sea coast is reasonably well known (Barry, Moritz, and Rogers 1979). New ice forms at the beginning of October, becoming continuous by mid-October. From November to January or February, landfast ice is modified by storm events. Subsequently, it becomes stable inside the 15-m isobath until April or May. Flooding of the fast ice by rivers occurs about 25 May, and the first melt pools appear about 10 June. Openings and ice movement take place at the end of June, and by the end of July the nearshore area is largely free of fast ice. A similar pattern undoubtedly occurs along Canada's Beaufort Sea coast.

The southeastern sector of the Beaufort Sea has a number of distinct characteristics. During March and April, there is commonly a well-developed north–south lead along the edge of the fast ice off Banks Island (Smith and Rigby 1981). During intervals of strong and persistent easterly winds, a large polynya may also form in this location in May or June. Frequently, a major lead extends westward from Amundsen Gulf toward Point Barrow, following roughly the 500-m isobath. Nevertheless, smaller leads may be either parallel to or nearly normal to the coastline, according to short-term atmospheric forcing. In summer, plumes of freshwater enter Mackenzie Bay from the Mackenzie River. Typically, flows move either northward along the Tuktoyaktuk Peninsula or westward to the north of Richards Island, but in years of heavy ice runoff may be contained largely by pack ice until late summer (Marko 1975).

Canadian Arctic Archipelago

There are significant contrasts between ice conditions in the northern and western parts of the Arctic Archipelago and those elsewhere (Lindsay 1969). For example, Sverdrup Basin is predominantly 10/10 multi-year ice throughout the year, with ice 2–3 m thick (de Lange Boom, Melling, and Lake 1987). Usually ice opens and moves in the eastern archipelago only during late summer and early autumn (Marko 1977). In contrast, ice breaks and clears early in Lancaster Sound, where the ice is 1.0–1.5 m thick; it does the same on the northern side of Viscount Melville Sound.

Freeze-up in the northern and central channels begins in October but is delayed 3–4 weeks in Amundsen Gulf and Coronation Gulf, which have predominantly first-year ice (Parkinson et al. 1987). Lancaster Sound – an area of ice divergence, with ice being continually advected eastward into Baffin Bay (Dey 1980) – also has mostly first-year and landfast ice. Multi-year ice accounts for up to 50 per cent of the western part of Parry Channel (M'Clure Strait and Viscount Melville Sound) and 60–90 per cent of the Sverdrup Basin. Conditions in M'Clure Strait are similar to those over the adjacent eastern Beaufort Shelf in winter and more severe in summer because of persisting multi-year and ridged ice (McLaren, Wadhams, and Weintrant 1984) (see Table 2.3). M'Clure Strait usually has more ice ridges and more extensive areas of ice <1 m thick in summer than in winter.

The Arctic Archipelago has numerous recurrent polynyas (areas of open water and new or young ice <0.3 m thick) and flaw leads. Their locations are summarized in Figure 2.11. Those in Penny Strait, Queens Channel, Cardigan Strait, and Martin Byam Channel (Smith and Rigby 1981), all located near shallow sills, are attributable to turbulent motion in narrow channels and to the upward mixing of sensible heat from waters of Atlantic origin, as discussed earlier (Melling et al. 1984). Daily turbulent heat loss from these open areas in winter may exceed 300 W/m^2, of which about 60 per cent is sensible heat (den Hartog et al. 1983). For the Smith Sound polynya, Steffen and Ohmura (1985) estimate a mean November–March flux of 178 W/m^2 as observed by aircraft remote sensing. Eighty per cent of this flux is caused by enthalpy in the sea water, and the remainder results from the release of latent heat of fusion on freezing. The two factors principally responsible for polynyas are *mobile ice*, caused especially by persistent offshore winds in the lee of islands and also by local tidal currents, and *convective overturning*, which brings warm water to the surface. Studies by Steffen (1986) on the North Water of northern Baffin Bay demonstrate that, contrary to earlier views, open water makes up only a small per-centage of the polynya surface. For the Smith Sound polynya, 84–100 per cent of the area was covered by new and young ice (<0.3 m thick) during winters 1978–79 and 1980–81.

Surface melt begins in the southern channels during April and by June has extended to all of the archipelago. Breakup similarly appears first in the south (from Amundsen Gulf to Queen Maud Gulf) and from Baffin Bay westward toward Barrow Strait.

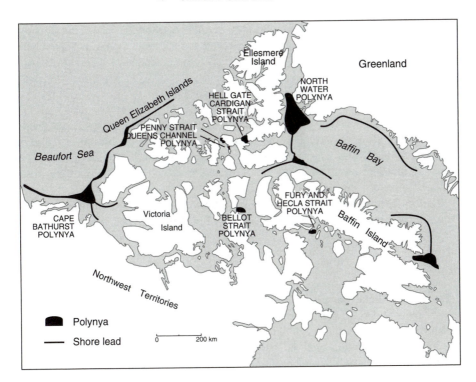

Figure 2.11
Locations of recurrent polynyas and shore leads. *Source*: Smith and Rigby (1981).

Table 2.3
Ice topography in M'Clure Strait and over the Beaufort Shelf

	Eastern Beaufort Sea		M'Clure Strait	
	February	*August*	*February*	*August*
Mean draft (m)	2.41	4.76	4.19	4.41
Ice (%): <0.5 m	9.3	0.4	4.2	0.40
0.5–2 m	41.3	12.8	19.0	32.00
2–5 m	41.0	48.6	49.0	36.00
>5 m	8.3	38.1	28.0	32.00
Polynyas* (%)	31.2	4.2	13.9	2.40
Level ice (%)	63.8	53.1	50.0	60.00

Source: McLaren, Wadhams, and Weintraub (1984).

* A working definition of no ice drafts exceeding 1 m was adopted.

The ice percentages total 100%; polynyas, level ice, and ridged ice (not given) separately total 100%.

Openings appear on the southern and eastern sides of Melville Island in late July and August (Marko 1977). Open water reaches its maximum extent in September, but there are large interannual differences (Parkinson et al. 1987; Markham 1984).

Hudson Bay and Eastern Arctic

Ice forms in October in Foxe Basin and by mid-November has spread southward over western Hudson Bay. By mid-December, almost all of the bay is covered (Markham 1986). However, open water along the east coast provides a source of heat and moisture until December, as illustrated by the climatic records from Poste de la Baleine (see Figure 2.8).

The observed maximum thickness of ice in spring is about 1.6 m in Hudson Bay–Hudson Strait and 2.0 m in Foxe Basin. The freshwater budget, as determined by Prinsenberg (1988), indicates that 40 to 90 per cent more ice is produced than is shown by data on ice thickness. Much of the discrepancy may be caused by ridge formation.

As in Hudson Bay, the ice cover of Foxe Basin consists almost entirely of first-year ice. The Foxe Basin's ice is distinctive (Prinsenberg 1986b): constant motion by currents and tides makes the surface unusually rough and hummocked for an inland water body, and high sediment content gives it a brownish colour. The sediment is derived from bottom material, stirred up by tides and storms, which becomes frozen into the ice and later concentrated at the upper surface by melting and sub-limation (Campbell and Collin 1958).

Freeze-up in the northern part of the basin takes place in the second half of October, spreading southward rapidly. There is regularly a large polynya off Hall Beach, and smaller openings occur on the lee (southeastern) sides of islands. Ice begins to break up and disperse in Foxe Channel during July, and the polynyas enlarge. Melting and transport of the ice into Hudson Strait continue until by late September the basin becomes essentially free of ice.

Ice covers Hudson Strait and Ungava Bay from late November through at least late July, although the pack is largely unconsolidated. Ice is advected into the western part of the strait in late October, with landfast ice forming along the coasts during November. In late November, ice from Davis Strait drifts westward along the south coast of Baffin Island, and the central part of the strait becomes ice-covered, from west to east (Drinkwater 1986). Shore leads form off Baffin Island throughout winter because of the prevailing northwesterly winds (Catchpole and Faurer 1983). Winds and currents move the ice continually. The large tides also contribute to this movement and lead to formation of a prominent ice foot along the shorelines, which persists after the seasonal decay of the floating fast ice.

Breakup begins in late May or early June. Leads develop along the coasts, and by late July the eastern half of Hudson Strait is largely ice-free. Ungava Bay and the western part of the strait are generally clear from early August, although in 'late years' ice may linger in Ungava Bay until the end of September (Crane 1978). The timing of ice retreat is related to the atmospheric circulation in early summer; warm southerly airflow associated with low pressure to the west or south accelerates melting, whereas cold northerlies, with lows over Davis Strait–Baffin Island, delay it.

In Baffin Bay, freeze-up starts in October in the northern part, and by December the ice edge extends from about 67°N on the Greenland coast southwestward to northern Labrador. This pattern reflects the ocean circulation described earlier. Nearly all of the ice is first-year ice, but some multi-year floes enter Baffin Bay from Smith Sound (Marko 1982), especially in years when the "ice dam" across the sound breaks.

The North Water polynya recurs during winter and spring in northern Baffin Bay and Smith Sound. Its existence was first reported by William Baffin in 1616, but it remained unexplored until the mid-nineteenth-century naval expeditions of Inglefield, Kane, Hall, and others. Its seasonal and interannual behaviour have been described by satellite and airborne remote sensing (Ito 1982). However, Steffen and Lewis (1988), using low-level airborne infrared thermometry, show that on nearly all flights 50 per cent of the transect was covered by white ice (>30 cm thick), not by the ice types characteristic of a polynya (nilas and grey ice or open water). The North Water "polynya" is, therefore, much smaller than shown by Smith and Rigby (1981), especially in late winter. Nevertheless, it has an ameliorating effect of 5–10°C on winter temperatures and greatly increases precipitation in October and November at Coburg Island (see Figure 2.8) (Müller et al. 1977).

The North Water polynya is attributable partly to mechanical removal of ice southward by wind and currents, as suggested by Muench (1971). Steffen (1986) estimates an ice drift of 17 km/day in Smith Sound, where northerly winds predominate, with two or three storm events in each winter month. Additionally, however, infrared thermometry has identified cells of warm water – typically <5 km across but occasionally 30 km or more long, with temperatures above the freezing point – that seem to be upwellings of Atlantic water. Upwelling could supply approximately 3×10^{12} W, sufficient to maintain the ice-free areas, where sensible heat flux from surface to atmosphere has a daily rate of about 140 W/m^2 in winter (Steffen 1985b). Nevertheless, since total heat flux over the entire North Water area is approximately 7×10^{12} W, coastal upwelling may not be the only mechanism maintaining the North Water.

Retreat of ice in Baffin Bay and Davis Strait proceeds both northward, from the vicinity of Hudson Strait, and southward, from Smith Sound (Dey 1980). Hence the area where ice lingers longest, occasionally until freeze-up, is off Home Bay, Baffin Island (68°N), which is consequently an occasional source area for second-year ice.

There is considerable interannual variability in the timing and rate of seasonal ice advance and retreat in Davis Strait. Crane (1978) has illustrated the role of atmospheric forcing in this variability for the period 1964–74. Early ice advance is associated with low pressure systems over southern Greenland and Davis Strait which bring cold northwesterly winds. These transport ice southward in Baffin Bay and from Foxe Basin–Hudson Strait into Davis Strait, and they also promote freezing. Conversely, low pressure systems over Quebec and Labrador give rise to warm easterly airflows that delay freeze-up. Similarly, ice melt and retreat are accelerated by warm southerly winds associated with low pressure systems over northwestern Quebec and the Foxe Basin. In such years, the ice tends to clear westward from the Labrador Sea and Davis Strait. In years with frequent northerly airflow in early summer, delayed retreat tends to progress from souuth to north. On a longer time scale, the date of ice clearance appears to be related to tropospheric circulation. For example, during 1964–73, the mean 700-mb-level trough moved eastward over Baffin Bay, giving rise to northerly airflow, cool summers, and severe ice conditions compared with the period 1951–60 (Barry 1981).

Ice in the southern Labrador Sea is derived primarily from advection from the north. However, based on mean drift rates, ice from northern Baffin Bay in October would not arrive off Newfoundland until June, when ice there has already melted (Bursey et al. 1977). Symonds (1986) estimates that bare ice undergoes bottom growth of about 1 cm/day during winter (December–March) at 55°N off Hopedale, whereas off St John's, Newfoundland, there is bottom ablation of 5 cm/day. Actual growth rates depend strongly on the depth of snow cover and the open-water fraction.

Ice appears off northern Labrador in early December, and by the end of the month it extends nearshore from Belle Isle Strait to about 53°N, then widening northward. At maximum, from mid-March through mid-April, a belt of ice extends 300 km or more off the entire Labrador coast (Sowden and Geddes 1980). The average dates of ice appearance and disappearance at 52°N during 1963–78 are 1 January and 4 June, respectively. Almost all the ice is first-year ice, with a mean thickness of 1.8 m. A small percentage of slower-melting multi-year ice of the order of 6–8 m thick, is present in summer (Zakrezewski 1988). In especially cold years with a long ice season such as 1972, ice may reach as far south as the Grand Banks.

Fluctuations in Ice Extent

Evidence of changes in sea-ice conditions is limited. Consistent published data are available only from 1953, based on summaries of ice concentration originally prepared by J.E. Walsh. An analysis of those data in terms of regional monthly ice anomalies for the Baffin Bay–Labrador Sea region (Figure 2.12) shows positive anomalies in the early 1970s and the 1980s, with smaller negative anomalies in the intervening period and in the 1960s. The ice anomalies (smoothed by a twenty-five-month running mean) are in phase with temperature in the Labrador Sea, whereas ice lags the temperature values by four months over Hudson Bay (Manak and Mysak 1987). This reflects the respective durations of ice cover in the two areas. The anomalies apparently propagate from the Greenland Sea into the Labrador Sea, following the surface currents, with similar velocity, over a 4–5-year interval. As yet, the cause of such fluctuations is undetermined, and no identifiable long-term trend has emerged in these data.

Recent investigations of fluctuations in ice extent and salinity anomalies in the northwestern North Atlantic suggest some significant interconnections. Measurements of salinity at Ocean Weather Station "Bravo" in the Labrador Sea show marked

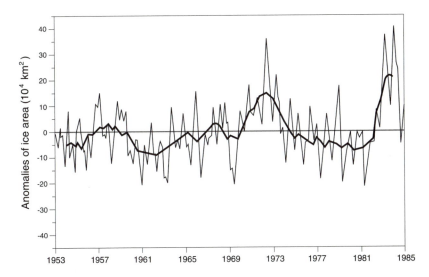

Figure 2.12
Monthly anomalies (10^4 km^2) of sea-ice extent in Baffin Bay and Labrador Sea shown as 3-month ($-$) and 25-month ($-$) running means. *Source*: Manak and Mysak (1987).

freshening in the upper layers, beginning in the late 1960s, and extending below a depth of 200 m (Lazier 1982). This represented part of the "great salinity anomaly" that affected the North Atlantic subpolar gyre, circulating with it, through the early 1980s. Mysak and colleagues (Mysak, Manak, and Marsden 1990; Mysak and Power 1991) propose that this anomaly originated with several years of excess runoff into the western Arctic Ocean and concurrent above-average Bering Strait inflow; Siberian runoff was low during the late 1960s, according to Cattle (1985). This freshwater anomaly enhances ice formation in the Beaufort Sea, which is exported into the Greenland Sea 4–8 years later, creating a freshwater excess there in years 5–9, with suppression of convective overturning in the Greenland Sea. Figure 2.13 illustrates a putative negative-feedback loop (Mysak, Manak, and Marsden 1990), which includes possible further interaction between winter leads and cyclogenesis in the Arctic. Mysak and Power (1991) suggest that such climatic oscillations affect the Arctic with a period of 15–20 years. For example, large positive ice anomalies affected the Greenland Sea in the 1960s and the late 1980s. These propagated into the Labrador Sea and the east coast of Newfoundland 3–5 years later. If substantiated further, such oscillations could have useful predictive value for east coast sea-ice conditions.

ICEBERGS

Eastern Canadian waters, from Baffin Bay to the Grand Banks, are almost unique in the Northern Hemisphere for the high frequency of large icebergs formed by the calving of tidewater glaciers. Robe (1980) identifies nine major glacier basins in

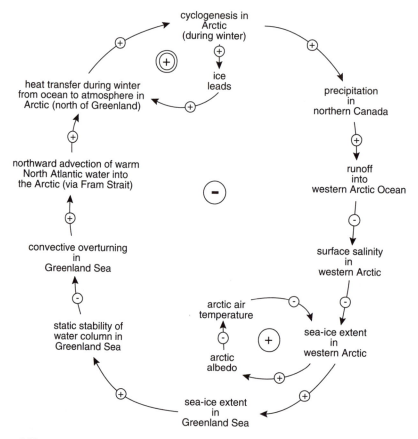

Figure 2.13
Possible negative-feedback loop linking arctic cyclonic activity, precipitation, runoff, salinity, sea-ice extent, oceanic stability, convective overturning, poleward oceanic heat transport, and heat flux into atmosphere. Lower small positive-feedback loop involving arctic air temperature is familiar ice-albedo feedback mechanism which could help cool troposphere if ice extent increases. Upper small feedback loop involving ice leads intensifies cyclonic activity if area of ice leads increases (Maslanik and Barry 1989). *Source*: Mysak, Manak, and Marsden (1990).

West Greenland and estimates that more than 30,000 medium-sized icebergs (of 5×10^6 tonnes) are produced annually. Statistics for 1968–70 show more than 100 icebergs per 1,000 km^2 off northwest Greenland and over 50 per 1,000 km^2 in eastern Baffin Bay between 69° and 70°N (Marko 1982).

As a direct result of the *Titanic* ship disaster in 1912, an International Ice Patrol Service was established to monitor shipping lanes off Newfoundland, providing an annual count of icebergs. The annual number observed south of 48°N off Newfoundland since 1900 averages 400 but ranges from zero to over 2,000 (Ketchen 1977); about 25 per year exceeding 50 m in height travel south of 52°N (Gustajtis 1979). Gustajtis and Buckley (n.d.) show more than 15 icebergs per square of 0.25 degree latitude extent off central Labrador in spring, with maximum frequencies between about 54° and 56°N.

Satellite tracking of icebergs in Baffin Bay shows that icebergs move in a well-defined zone over the continental slope and, following the Baffin Current, more or less along the 500-m isobath (Marko, Birch, and Wilson 1982). However, they follow the local bathymetry into coastal re-entrants, such as Clyde Inlet and Home Bay. As a result, they may become trapped in the zone of landfast ice for 9–10 months of the year. Most berg drafts are estimated to be between 30 m and 300 m, so that the zone of potential grounding is little more extensive than the limit of landfast ice, which may extend over waters at least 170 m in depth (Jacobs, Barry, and Weaver 1975). During August through November, when they are not grounded or trapped in fast ice, they typically drift about 10 km/day. Radar tracking off northern Labrador indicates average drift speed 2.5 times greater than the mean surface current (4 per cent of wind speed), with highly variable drift direction (Robe 1980). Consequently, prediction of their motion is a major problem.

The risk of iceberg collisions is not limited to transatlantic shipping but also includes offshore drilling platforms and drill ships. Iceberg scouring of the sea floor to 300 m depth is also a hazard for undersea oil and gas pipelines (Gustajtis 1979).

Frequency of iceberg occurrence is highly variable (Figure 2.14). Values were greatest in the period 1905–14 and the 1980s and lowest in the 1960s. Tabulation of rank by years from 1950 to 1988 shows 1984, 1972, 1974, 1983, and 1985 to be the most severe and 1966, 1958, and 1951 to be the mildest. Severity depends on frequency of calving in source region(s), drift rate, rate of melt, and frequency of grounding.

J. Newell (personal communication, 1988) finds that historical ships' archives indicate a downward trend in the size of the largest icebergs sighted over the last 150 years. Large tabular icebergs were much more common in the 1880s. For long-range forecasting, Walsh et al. (1986) show that the mode of the North Atlantic oscillation of sea-level pressure in January through March is useful in predicting overall iceberg severity. Severe iceberg conditions are indicated by northwesterly airflows, and mild conditions by southeasterly airflows. Approximately half of the annual count of icebergs is accounted for by the numbers in April and May, implying a lead time of several months in the pattern of sea-level pressure.

ACKNOWLEDGMENTS

I wish to thank Dr K. Aagaard for his helpful comments on a first draft and J. Newell and Dr E. Carmack. This work was supported by the Office of Naval Research under a University Research Initiative.

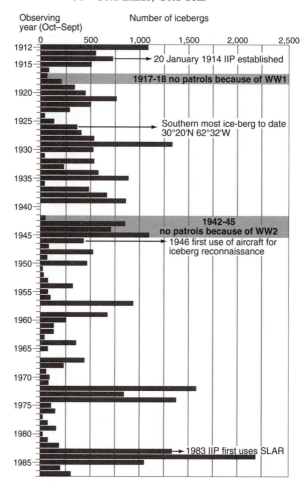

Figure 2.14
Number of icebergs estimated to cross 48°N off Newfoundland each year, 1912–87; courtesy M.A. Alfultis, United States Coast Guard. Observing years are Oct. 1899–Sept. 1900 (= 1900), and so on.

REFERENCES

Aagaard, K. 1982. "Inflow from the Atlantic Ocean to the Polar Basin." In L. Rey, ed., *The Arctic Ocean: The Hydrographic Environment and the Fate of Pollutants*, Monaco: Comité Arctique International, 69–82.

– 1984. "The Beaufort Undercurrent." In P.W. Barnes, D.M. Schell, and E. Reimnitz, eds., *The Alaskan Beaufort Sea: Ecosystems and Environments*, Orlando, Fla.: Academic Press, 47–71.

– 1989. "A Synthesis of the Arctic Ocean Circulation." *Rapport, Réunions et Procès-verbaux des Conseils Permanents Exploration Internationale de la Mer. Physical and Chemical Oceanography* 188: 11–22.

Aagaard, K., and Carmack, E.C. 1989. "The Role of Sea Ice and Other Fresh Water in the Arctic Circulation." *Journal of Geophysical Research* 94 no. 14: 485–98.

Aagaard, K., and Greisman, P. 1975. "Towards New Mass and Heat Budgets for the Arctic Ocean." *Journal of Geophysical Research*, 80: 3,821–7.

Aagaard, K., Coachman, L.K., and Carmack, E.C. 1981. "On the Halocline of the Arctic Ocean." *Deep-Sea Research* 28A: 529–45.

Aagaard, K., Foldvik, A., and Hillmann, S.R. 1987. "The West Spitsbergen Current: Disposition and Water Mass Transformation." *Journal of Geophysical Research* 92: 3,778–84.

Barry, R.G., 1981. "The Nature and Origin of Climatic Fluctuations in Northeastern North America." *Géographie physique et quaternaire* 35: 41–7.

Barry, R.G., Moritz, R.E., and Rogers, J.C. 1979. "The Fast Ice Regime of the Beaufort and Chukchi Sea Coasts, Alaska." *Cold Regions Science and Technology* 1: 129–52.

Barry, R.G., and Maslanik, J. 1989. "Arctic Sea Ice Characteristics and Associated Atmosphere–Ice Interactions in Summer Inferred from SMMR Data and Drifting Buoys: 1979–1984." *Geojournal* 18: 35–44.

Bourke, R.H., and Garrett, R.P. 1987. "Sea Ice Thickness Distribution in the Arctic Ocean." *Cold Regions Science and Technology* 13: 259–80.

Burova, L.P. 1981. "Vodnye resursy atmosfery nad Arkticheskim Basseinom." *Trudy Ark. Antarkt. Inst.* 370: 91–110.

Bursey, J.O., Sowden, J.W., Gates, A.D., and Blackwood, C.L. 1977. "The Climate of the Labrador Sea." *Proceedings of the Fourth International Conference on Ports and Ocean Engineering under Arctic Conditions* (POAC 77), Memorial University of Newfoundland, St John's, 938–51.

Campbell, N.J., and Collin, A.E. 1958. "The Discoloration of Foxe Basin Ice." *Journal of the Fisheries Research Bound of Canada* 15: 1,175–88.

Carmack, E.C. 1986. "Circulation and Mixing in Ice-covered Waters." In N. Untersteiner, ed., *The Geophysics of Sea Ice*, New York: Plenum Press, 641–712.

Catchpole, A.J.W., and Faurer, M.A. 1983. "Summer Sea Ice Severity in Hudson Strait, 1751–1870." *Climatic Change* 5: 115–39.

Cattle, H. 1985. "Diverting Soviet Rivers: Some Possible Repercussions for the Arctic Ocean." *Polar Record* 22: 485–98.

Coachman, L.K., and Aagaard, K. 1974. "Physical Oceanography of Arctic and Sub-Arctic Seas." In Y. Herman, ed., *Marine Geology and Oceanography of the Arctic Seas*. New York: Springer-Verlag, 1–72.

– 1981. "Reevaluation of Water Transports in the Vicinity of Bering Strait." In D.W. Hood and J.A. Calder, eds., *The Eastern Bering Sea Shelf: Oceanography and Resources*, vol. 1, NOAA, US Department of Commerce, University of Washington Press, Seattle, 95–110.

– 1988. "Transports through Bering Strait: Annual and Interannual Variability." *Journal of Geophysical Research* 93: 15,535–9.

Collin, A. 1963 "Waters of the Canadian Arctic Archipelago." In *Proceedings of the Arctic Basin Symposium*, Arctic Institute of North America, Washington, DC, 128–36.

Collin, A.E., and Dunbar, M.H. 1964. "Physical Oceanography in Arctic Canada." *Oceanographic and Marine Biology Annual Review* 2: 45–75.

Colony, R., and Thorndike, A.S. 1984. "An Estimate of the Mean Field of Arctic Ice Motion." *Journal of Geophysical Research* 89 C6: 10,623–9.

Crane, R.G. 1978. "Seasonal Variations of Sea Ice Extent in the Davis Strait–Labrador Sea Area and Relationships with Synoptic-Scale Atmospheric Circulation." *Arctic* 31: 434–47.

Danielson, E.W., Jr. 1969. *The Surface Heat Budget of Hudson Bay.* Marine Science Manuscript Report 9, McGill University, Montreal.

de Lange Boom, B.R., Melling, H., and Lake, R.A. 1987. *Late Winter Hydrography of the Northwest Passage: 1982, 1983 and 1984.* Canadian Technical Report Hydrography and Ocean Sciences, No. 79, Department of Fisheries and Oceans, Sidney, BC.

den Hartog, G., Smith, S.D., Anderson, R.J., Topham, D.R., and Perkin, R.G. 1983. "An Investigation of a Polynya in the Canadian Archipelago: 3. Surface Heat Flux." *Journal of Geophysical Research* 88: 2,911–16.

Dey, B. 1980. "Seasonal and Annual Variations in Ice Cover in Baffin Bay and Northern Davis Strait." *Canadian Geographer* 24: 368–84.

Drinkwater, K.F. 1986. "Physical Oceanography of Hudson Strait and Ungava Bay." In I.P. Martini, ed., *Canadian Inland Waters*, Amsterdam: Elsevier, 237–64.

– 1988. "On the Mean and Tidal Currents in Hudson Strait." *Atmosphere-Ocean* 26: 252–66.

Dunbar, M.J. 1951. *Eastern Arctic Waters. Fisheries Research Board of Canada*, Bulletin 88, Ottawa.

Fisheries and Marine Services. 1977. *Canadian Tide and Current Tables*, Vol. 4, *Arctic and Hudson Bay.* Ottawa.

Fissel, D.B., Lemon, D.D., and Birch, J.R. 1982. "Major Features of the Summer near Surface Circulation of Western Baffin Bay, 1978–1979." *Arctic* 35: 180–200.

Fissel, D.B., Birch, J.R., Melling, H., and Lake, R.A. 1988. *Non-tidal Flows in the Northwest Passage.* Canadian Technical Report, Hydrography and Ocean Sciences, No. 98, Department of Fisheries and Oceans, Sidney, BC.

Foldvik, A., Aagaard, K., and Torresen, T. 1988. "On the Velocity Field of the East Greenland Current." *Deep Sea Research* 35: 1,335–54.

Gustajtis, K.A., 1979. *Iceberg Scouring on the Labrador Shelf, Saglek Bank.* C-CORE Publication 79–13, Memorial University of Newfoundland, St John's.

Gustajtis, K.A., and Buckley, T.J. N.d. *Iceberg Density along the Labrador Coast.* From seasonal maps, 1:200,000. C-CORE, Memorial University of Newfoundland, St John's.

Hudson, R.D. 1987. "Multiyear Sea Ice Floe Distribution in the Canadian Arctic Ocean." *Journal of Geophysical Research* 92: 14,633–69.

Ikeda, M. 1987. "Salt and Heat Balances in the Labrador Sea Using a Box Model." *Atmosphere-Ocean* 25: 197–223.

Ito, H. 1982. *Sea Ice Atlas of Northern Baffin Bay*, Zürcher geographische Schriften No. 7, Department of Geography, Swiss Federal Institute of Technology (ETH), Zurich.

Jacobs, J.D., Barry, R.G., and Weaver, R.L. 1975. "Fast Ice Characteristics with Special Reference to the Eastern Canadian Arctic." *Polar Record* 17: 521–36.

Ketchen, H.G. 1977. "Iceberg Populations South of 48°N since 1900." In *Report of the International Ice Patrol Service in the North Atlantic Ocean (Season 1977)*, CG-188-32,

Bulletin No. 63, US Coast Guard, Department of Transportation, Washington, DC, C1–C6.

Kovacs, A., and Sodhi, D.S. 1980. "Shore Ice Pile-up and Ride-up: Field Observations, Models, Theoretical Analysis." *Cold Regions Science and Technology* 2: 209–88.

Lazier, J.R.N. 1982 "Seasonal Variability of Temperature and Salinity in the Labrador Current." *Journal of Marine Research* Supplement 40: 341–56.

Leblond, P.H. 1980. "On the Surface Circulation in Some Channels of the Canadian Arctic Archipelago." *Arctic* 33: 189–97.

Leblond, P.H., Osborn, T.R., Hodgins, D.O., Goodman, R., and Metge, M. 1981. "Surface Circulation in the Western Labrador Sea." *Deep-Sea Research* 28: 683–93.

Lemon, D.D., and Fissel, D.B. 1982. "Seasonal Variations in Currents and Water Properties in Northwestern Baffin Bay, 1978–1979." *Arctic* 35: 211–18.

Lewis, C.F.M. 1977. "The Frequency and Magnitude of Drift-Ice Groundings from Ice-Scour Tracks in the Canadian Beaufort Sea." *Proceedings of the Fourth International Conference on Port and Ocean Engineering under Arctic Conditions* (POAC 77), Memorial University of Newfoundland, St John's, vol. 2, 963–71.

Lindsay, D.G., 1969. "Ice Distribution in the Queen Elizabeth Islands." In *Ice Seminar*, Special Volume 10, Canadian Institute of Mining and Metallurgy, 45–60.

McLaren, A.S., Wadhams, P., and Weintraub, R. 1984. "The Sea Ice Topography of M'Clure Strait in Winter and Summer of 1960 from Submarine Profiles." *Arctic* 37: 110–20.

Manak, D.K., and Mysak, L.A. 1987. *Climatic Atlas of Arctic Sea Ice Extent and Anomalies, 1953–1984*. Climate Research Group Report 87–8, Department of Meteorology, McGill University, Montreal.

Markham, W.E. 1984. *Ice Atlas: Canadian Arctic Waterways*. Ottawa: Environment Canada.

– 1986. "The Ice Cover." In I.P. Martini, ed., *Canadian Inland Seas*, Amsterdam: Elsevier, 101–16.

Marko, J.R. 1975. *Satellite Observations of the Beaufort Sea Ice Cover*. Technical Report No. 34, Beaufort Sea Project, (BC) Department of the Environment, Victoria, BC.

– 1977. *A Satellite-Based Study of Sea Ice Dynamics in the Central Canadian Arctic Archipelago*. Contractor Report Series 77–4, Institute of Ocean Sciences, Sidney, BC.

– 1982. *The Ice Environment of Eastern Lancaster Sound and Northern Baffin Bay*. Environmental Studies No. 76, Indian and Northern Affairs Canada, Ottawa.

Marko, J.R., Birch, J.R., and Wilson, M.A. 1982. "A Study of Long-Term Satellite-Tracked Iceberg Drifts in Baffin Bay and Davis Strait." *Arctic* 35: 234–40.

Maslanik, J.A., and Barry, R.G. 1989. "Short-Term Interactions between Atmospheric Synoptic Conditions and Sea Ice Behavior in the Arctic." *Annals of Glaciology* 12: 113–17.

Maxwell, J.B. 1981. "Climatic Regions of the Canadian Arctic Islands." *Arctic* 34: 225–40.

Maykut, G.A. 1986. "The Surface Heat and Mass Balance." In N. Untersteiner, ed., *The Geophysics of Sea Ice*. New York: Plenum Press, 395–463.

Melling, H., Lake, R.A., Topham, D.R., and Fissel, D.B. 1984. "Oceanic Thermal Structure in the West Canadian Arctic." *Continental Shelf Research* 3: 233–58.

Melling, H., and Lewis, E.L. 1982. "Shelf Drainage Flows in the Beaufort Sea and Their Effect on the Arctic Ocean Pycnocline." *Deep Sea Research* 29: 967–85.

Muench, R.D. 1971. *The Physical Oceanography of the Northern Baffin Bay Region.* Baffin Bay–North Water Scientific Report No. 1, Arctic Institute of North America, Washington, DC.

Müller, F., and sixteen others. 1977. "Glaciological and Climatological Investigation of the North Water Polynya in Northern Baffin Bay." *Progress Report 1 October 1975 to 30 September 1976*, Swiss Federal Institute of Technology (ETH), Zurich, and McGill University, Montreal, 9–54.

Mysak, L.A., Manak, D.K., and Marsden, R.F. 1990, "Sea Ice Anomalies Observed in the Greenland and Labrador Seas during 1901–1984 and Their Relation to an Interdecadal Arctic Climate Cycle." *Climate Dynamics* 5: 111–33.

Mysak, L.A., and Power, S.B. 1991. "Greenland Sea Ice and Salinity Anomalies and Inter-decadal Climate Variability." *Climatological Bulletin* 25: 81–91.

O'Rourke, J.C. 1974. "Inventory of the Physical Oceanography of the Eastern Beaufort Sea." In J.C. Reed and J.E. Sater, eds., *The Coast and Shelf of the Beaufort Sea*, Arlington, Va., Arctic Institute of North America, 65–84.

Parkinson, C.L., Comiso, J.C., Zwally, H.J., Cavalieri, D.J., Gloersen, P., and Campbell, W.J. 1987. *Arctic Sea Ice, 1973–1976: Satellite Passive-Microwave Observations.* Washington, DC: NASA.

Prinsenberg, S.J. 1982. "Time Variability of Physical Oceanographic Parameters in Hudson Bay." *Le naturaliste canadien* 109: 685–700.

– 1984. "Freshwater Contents and Heat Budgets of James Bay and Hudson Bay." *Continental Shelf Research* 3: 191–200.

– 1986a. "Salinity and Temperature Distributions of Hudson Bay and James Bay." In I.P. Martini, ed., *Canadian Inland Seas*, Amsterdam: Elsevier, 163–86.

– 1986b. "On the Physical Oceanography of Foxe Basin." In I.P. Martini, ed., *Canadian Inland Seas*, Amsterdam: Elsevier, 217–36.

– 1988. "Ice-Cover and Ice-Ridge Contributions to the Freshwater Contents of Hudson Bay and Foxe Basin." *Arctic* 41: 6–11.

Prinsenberg, S.J., and Bennett, E.B. 1987. "Mixing and Transports in Barrow Strait, the Central Part of the Northwest Passage." *Continental Shelf Research* 7: 913–35.

Reimnitz, E., Barnes, P.W., Wolf, S.C., Graves, S.M., Kemperna, E.W., and Rearic, D.M. 1988. "The Deepwater Limit of Ice Gouging on the Beaufort Sea Shelf." In *Beaufort Sea: Information Update*, OCSEAP, US Department of Commerce, NOAA/NOS, Alaska Office, 21–6.

Reimnitz, E., Toimil, L., and Barnes, B. 1978. "Arctic Continental Shelf Morphology Related to Sea Ice Zonation, Beaufort Sea, Alaska." *Marine Geology* 28: 179–210.

Robe, R.Q., 1980. "Iceberg Drift and Deterioration." In S.C. Colbeck, ed., *Dynamics of Snow and Ice Masses*, New York: Academic Press, 211–59.

Roff, J.C., and Légendre, L. 1986. "Physico-Chemical and Biological Oceanography of Hudson Bay." In I.P. Martini, ed., *Canadian Inland Seas*, Amsterdam: Elsevier, 265–91.

St Martin, J.W. 1987. *Arctic Drifting Buoy Data, 1979–1985.* Report No. GD-D-10-87, US Coast Guard, Department of Transportation, Washington, DC.

Sanderson, B.G., and LeBlond, P.H. 1984. "The Cross-channel Plan at the Entrance of Lancaster Sound." *Atmosphere-Ocean* 22: 424–97.

Semtner, A.J., Jr. 1976. "A Model for the Thermodynamical Growth of Sea Ice in Numerical Investigations of Climate." *Journal of Physical Oceanography* 6: 379–89.

Serreze, M.C., and Barry, R.G. 1988. "Synoptic Activity in the Arctic Basin, 1979–1985." *Journal of Climate* 1: 1,276–95.

Serreze, M.C., Barry, R.G., and McLaren, A.S. 1989. "Seasonal Variations in Sea Ice Motion and Effects on Sea Ice Concentration in the Canada Basin." *Journal of Geophysical Research* 94: 10,955–70.

Smith, M., and Rigby, B. 1981. "Distribution of Polynyas in the Canadian Arctic." In I. Stirling and H. Cleator, eds., *Polynyas in the Canadian Arctic*, Occasional Paper No. 45, Canadian Wildlife Service, Ottawa, 7–28.

Sowden, W.J., and Geddes, F.E. 1980. *Weekly Median and Extreme Ice Edges for Eastern Canadian Seaboard and Hudson Bay*. Ottawa: Environment Canada.

Steffen, K. 1985a. *Surface Temperature and Sea Ice of an Arctic Polynya: North Water in Winter*. Geographical Institute, Swiss Federal Institute of Technology (ETH), Zurich.

– 1985b. "Warm Water Cells in the North Water, Northern Baffin Bay during Winter." *Journal of Geophysical Research* 90: 9,129–36.

– 1986. "Ice Conditions of an Arctic Polynya: North Water in Winter." *Journal of Glaciology* 32: 383–90.

Steffen, K., and Lewis, J.E. 1988. "Surface Temperatures and Sea Ice Typing for Northern Baffin Bay." *International Journal of Remote Sensing* 9: 409–22.

Steffen, K., and Ohmura, A. 1985. "Heat Exchange and Surface Conditions in the North Water, Northern Baffin Bay." *Annals of Glaciology* 6: 178–81.

Symonds, G. 1986. "Seasonal Ice Extent on the Northeast Newfoundland Shelf." *Journal of Geophysical Research* 91: 10,718–24.

Thompson, K.R., Loucks, R.H., and Trites, R.W. 1988. "Sea Surface Temperature Variability in the Shelf Slope Region of the Northwest Atlantic." *Atmosphere-Ocean* 26: 282–99.

Treshnikov, A.F., editor-in-chief. 1985. *Atlas Arktiki*. Arkt.-Antarkt. Nauchno-Issled. Inst., Glavnoe Upravlenie Geodizii i Kartografii Soveta Ministrov SSR, Moscow.

Vinje, T., and Finnekasa, O. 1986. *The Ice Transport through the Fram Strait*. Norsk Polar Institut, Skrifter 186, Oslo.

Wadhams, P. 1981. "Sea-Ice Topography of the Arctic Ocean in the Region 70 Degrees West to 25 Degrees East." *Philosophical Transactions of the Royal Society, London* A 302: 45–85.

– 1983. "The Sea Ice Thickness Distribution in the Fram Strait." *Nature* 305: 108–11.

Wadhams, P., and Horne, R.J. 1980. "An Analysis of Ice Profiles Obtained by Submarine Sonar in the Beaufort Sea." *Journal of Glaciology* 25 no. 93: 401–24.

Walsh, J.E., Wittman, W.I., Hester, L.H., and Dehn, W.S. 1986. "Seasonal Prediction of Iceberg Severity in the Labrador Sea." *Journal of Geophysical Research* 91: 9,683–92.

Wilson, C. 1968. *Notes on the Climate of Poste-de-la-Baleine, Québec*. Centre d'Études nordiques, Travaux divers 24, Université Laval, Quebec.

Zakrezewski, W.P. 1986. "Sea Ice Conditions in the Labrador Sea–Grand Banks Area." In G. Symonds and I.K. Peterson, compilers, *Proceedings, Canadian East Coast Workshop on Sea Ice, Jan 7–9, 1986*, Canadian Technical Report, Hydrographic and Ocean Sciences, No. 73, Fisheries and Oceans Canada, 60–87.

PART TWO

Northern and Polar Lands

CHAPTER THREE

Northern Climates

WAYNE R. ROUSE

Canada's northern and polar climates can be understood and modelled at a number of scales ranging from large to small. At the largest scale, the radiation balance of the Earth-atmospheric system is in perpetual deficit. This deficit fuels a vigorous general circulation of the atmosphere, also operating at the largest scale. The general circulation in turn interacts with the unique geographical environment which can be understood in terms of air and ground temperatures, precipitation, and wind. These elements can be combined with the radiation balance to define climatic zones which, on the largest scale, are categorized into high arctic, low and middle arctic, and subarctic. These zones do not parallel lines of latitude because they are greatly influenced by general atmospheric circulation, surface topography, and land-sea distribution. To a substantial degree, these climatic zones define, and are defined by, vegetation and permafrost characteristics.

Smaller-scale climatic processes and responses range from mesoscale through local scale to microscale. Mesoscale climate is especially associated with synoptic systems derived from the general circulation and with circulation patterns resulting from land-sea thermal differences during thaw season. At the local and microscales, land-lake and wet-dry thermal differences create small-scale advection which strongly influences heat and water balances.

The climate of the north is especially sensitive to change. This might be exerted through the "greenhouse effect," river diversion, and fire, in descending order of importance. The changes would be manifest at all scales, and climatic response would interact with the biosphere and lithosphere.

RADIATION BALANCE

The single most important factor controlling climate in high-latitude environments is the input of solar radiation and its absorption in the atmosphere and especially at the Earth's surface. Figure 3.1 shows the latitudinal distribution of potential solar insolation, I_o, for the Northern Hemisphere as it varies annually. This illustrates

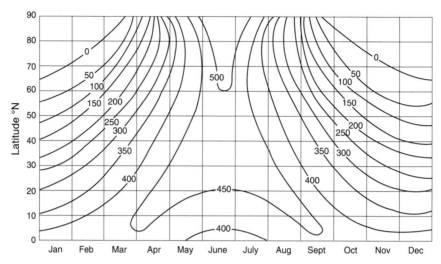

Figure 3.1
Latitudinal distribution of potential insolation, Northern Hemisphere. Units are in W/m². *Source*: List (1958).

several major features of the radiation environment. During the high-sun months of June and July, I_o is slightly greater in northern latitudes than in the tropics and the gradient of I_o between equator and pole is very small. In contrast, the steepest gradients occur in December and January. I_o changes much more rapidly during spring and fall in the north than in lower latitudes, especially north of the Arctic Circle. Thus at the latitude of Alert, NWT (82°30′N), the rate of change in I_o between the equinox and spring or fall solstice is 4.6 Wm²/day. This situation favours slow change in climate during midsummer and midwinter and rapid change over transition periods. We shall see later, though, that other processes at the surface complicate and sometimes obliterate any such regular sequence.

Patterns of potential insolation will rarely be duplicated by insolation received at the surface, $K\downarrow$, especially in Canada's north. Table 3.1 gives annual values of effective potential insolation, $K\downarrow/I_o$, for ten northern stations ranging from the low subarctic to the high Arctic and also for two stations in southern Canada for comparative purposes. Radiation received at the surface varies between 46 and 56 per cent of potential insolation, with values generally greater in higher than in lower latitudes. For stations in Table 3.1, the greater path lengths through which the solar beam must pass in higher latitudes are more than compensated for by fewer or thinner clouds, or both. Other components of the annual radiation balance at the surface are:

$$Q^* = K\downarrow (1-ab) + L\downarrow - L\uparrow \qquad \text{(1[a])}$$
$$K^* + L^* \qquad \text{(1[b])}$$

in which Q^* is net all-wave radiation, ab is surface albedo, $L\downarrow$ and $L\uparrow$ are incoming and outgoing long-wave terrestrial radiation respectively, and K^* and L^* are net solar

Table 3.1
Annual mean radiation data (W/m^2) for select arctic, subarctic, and temperate-latitude
stations

Station	Latitude (N)	Longitude (W)	I_o	$K\uparrow$	$K\downarrow/I_o$ (%)	K^*	L^*	Q^*
Arctic								
Alert, NWT	82°30′	62°20′	175	90	51.4	46	−43	3
Resolute, NWT	74°43′	94°59′	187	102	54.5	50	−45	5
Aklavik, NWT	68°14′	135°00′	202	104	51.5	69	−52	17
Aqaluit (Frobisher), NWT	64°45′	68°33′	215	119	55.3	62	−50	12
Baker Lake, NWT	64°18′	96°00′	217	118	54.4	64	−48	16
Subarctic								
Whitehorse, Yukon	60°43′	103°04′	232	118	50.9	82	−48	34
Churchill, Man.	59°45′	94°04′	242	132	54.5	90	−53	37
Schefferville, Que.	54°48′	66°49′	260	125	48.1	93	−57	36
Goose Bay, Nfld	53°19′	60°25′	268	128	47.8	93	−54	39
Moosonee, Ont.	51°16′	80°39′	278	130	46.8	96	−55	41
Temperate								
Winnipeg, Man.	49°54′	97°14′	283	156	55.1	118	−65	53
Toronto, Ont.	43°40′	79°24′	312	146	46.8	122	−66	56

Note: I_o is potential insolation; $K\downarrow$ and K^* are incoming solar and net solar radiation, respectively, at surface; L^* is net long-wave radiation; and Q^*, net all-wave radiation at surface.

and net long-wave radiation. These values are given in Table 3.1, and they are plotted against latitude in Figure 3.2, with trend lines joining the stations of Toronto and Alert.

A number of significant patterns emerge in Figure 3.2. Loss of net long-wave radiation shows small but regular decline with increasing latitude. As groups, the five subarctic stations and the five arctic stations have, as delineated in Figure 3.2, average L^* values that are 82 per cent and 74 per cent, respectively, of the average for the two temperate-latitude stations. The smaller L^* values in higher latitudes result primarily from surface radiative temperatures that are substantially colder relative to those of the overlying atmosphere than is the case for lower latitudes. Figure 3.2 also indicates that annual totals of I_o decrease regularly with increasing latitude, as do those for $K\downarrow$, but the latter show more scatter.

Latitudinal trends in K^* and Q^* indicate pronounced decreases with more northerly latitude. Table 3.2 indicates that while $K\downarrow/I_o$ shows little latitudinal trend, it responds to average cloud cover, which is greatest for the subarctic and least for the arctic stations. The percentage of net solar radiation (K^*) and the percentage of net all-wave radiation (Q^*) decline substantially with latitude, particularly in the arctic regions, where an average of only 7 per cent of potential insolation is realized as net radiation. A similar pattern is indicated by direct comparison of values for arctic, subarctic, and temperate locations. Arctic stations receive two-thirds of the potential insolation received by the temperate stations and little more than one-quarter of the

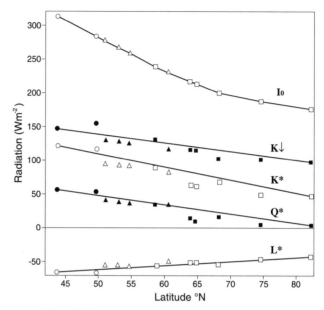

Figure 3.2
Latitudinal variation in radiation: potential insolation (I_o), solar radiation at surface ($K{\downarrow}$), absorbed solar radiation (K^*), net radiation (Q^*), and net long-wave radiation (L^*). Symbols: o temperate; △ subartic; □ arctic. Stations are listed in Table 3.1.

Table 3.2
Surface radiative characteristics for temperate, subarctic, and arctic stations

Average values	Temperate (two stations)	Subarctic (five stations)	Arctic (five stations)
For each zone			
$K{\downarrow}/I_o$	51	48	54
K^*/I_o	40	35	31
Q^*/I_o	18	15	7
Assuming temperate zone = 100%			
I_o	100	86	67
$K{\downarrow}$	100	83	73
K^*	100	75	53
Q^*	100	69	28

Note: I_o is potential insolation, $K{\downarrow}$ is solar radiation at surface, K^* is net solar radiation, and Q^* is net all wave radiation. Ratios are in %. Primary data on which data are based are found in Table 3.1.

Table 3.3
Annual radiation balance of Earth-atmosphere system in the Northern
Hemisphere

Latitude (°N)	W/m²
0–10	44
10–20	41
20–30	8
30–40	− 16
40–50	− 38
50–60	− 66
60–70	− 95
70–80	− 116
80–90	− 132

Source: Henderson-Sellers and Robinson (1986).

net radiation. This indicates clearly the potential for warming of the Arctic: it is not the paucity of solar radiation, but the high-albedo surfaces, which produce the very small net radiation. High-albedo surfaces result from the snow cover's extreme longevity and its high reflectivity. The latter is ameliorated in subarctic environments by open forest and other tall vegetation (Davies 1963; Rouse and Bello 1983; Rouse 1984a; Robinson and Kukla 1985). In the Arctic, however, cold temperatures maintain both the long snow-cover season and the scarcity of tall vegetation, so that any long-term warming will substantially increase net radiation at the Earth's surface.

The discussion to date has focused on lowland terrestrial stations, whereas much of the high Arctic involves the sea. The oceanic environment has been discussed in the previous chapter, and the radiation regime will be treated only briefly here. As developed by Vowinckel and Orvig (1970), where the pack ice is dominant throughout the year, as in the central polar basin, Q^* ranges around zero. This is a result of a year-round high-albedo surface and net long-wave radiation loss, L^*, which Vowinckel and Orvig estimated to be 20 per cent larger than for a corresponding land surface.

Where sea ice is seasonal, as in Baffin Bay, the Beaufort Sea, or in polynyas, annual Q^* tends toward small positive values because of the small surface albedo in summer. Even in the highest-latitude lowlands, the surface radiation balance is rarely negative (Maxwell 1980: 122). By contrast, the atmospheric radiation balance is strongly negative for all northern latitudes. For example, Vowinckel and Orvig (1964) calculated that over the central Polar Ocean annual surface radiation balance was less than 1 W/m², compared to an atmospheric radiation balance of − 103 W/m². Thus the Earth-atmospheric system has a strongly negative radiation balance, and this is true of all subarctic and arctic latitudes. This shows clearly in Table 3.3, which gives averages for ten degree latitudinal zones in the Northern Hemisphere.

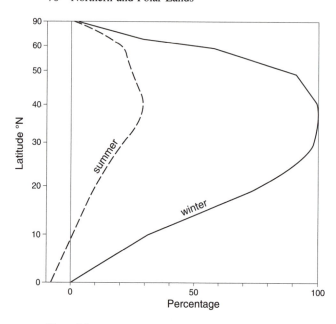

Figure 3.3
Poleward heat flux in summer and in winter expressed as percentage of maximum heat flux across 40° in winter. *Source*: Newell et al. (1969).

For all latitudes north of 40°, the surface-atmospheric system exhibits a radiation deficiency, and this grows steadily with increasing latitude. The radiation deficits shown in Table 3.3 represent the advectional energy that must be moved poleward in order to maintain an annual equilibrium heat budget in the Northern Hemisphere.

Figure 3.3 indicates the relative magnitude of this poleward heat flux during winter and summer. Winter energy flux is almost four times as great as its summer counterpart. Fluxes reach their maximum size in the middle latitudes. Whereas winter flux decreases rapidly in the higher latitudes, summer flux decreases only slowly. Two conditions result from this situation. First, poleward pressure and temperature gradients are much greater, and there is more storminess and higher wind speeds in winter than in summer. Second, the zone of maximum storm activity is located further poleward in summer than in winter (Petterson 1969).

GENERAL CIRCULATION OF
THE ATMOSPHERE

As noted by Hare (1968), the high latitudes do not display any unique atmospheric circulation patterns but are part of the global system of the circumpolar westerly vortex. This system is driven by the large latitudinal inequalities in radiation already noted and dominates all middle- and high-latitude zones. As a result, northern Canada, like the rest of Earth, has variable weather systems and is rarely dominated by any

SUMMER (JULY)

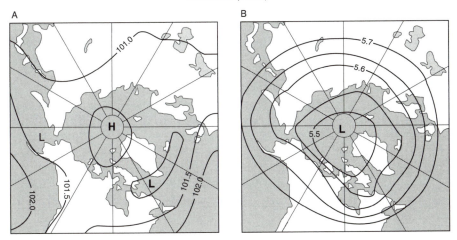

WINTER (JANUARY)

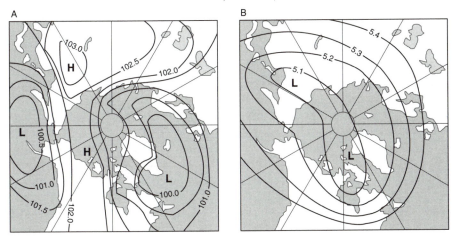

Figure 3.4
(A) Sea-level pressure (kPa); *Source*: Petterson (1969). (B) 50-kPa topography (km); *Source*: Hare (1968).

one system (LeDrew 1983). There are, however, distinctive patterns to the annual and seasonal patterns of atmospheric circulation. For example, within the circumpolar circulation there is a low-pressure, cold core. In the middle and upper troposphere, this air mass is relatively uniform in temperature and pressure characteristics and displays disorganized circulation patterns (Hare 1968).

As shown in Figure 3.4B, there is a single core centred on northern Ellesmere Island in summer, and the average circulation pattern is concentric around it. This arrangement favours a zonal (i.e. west-to-east) upper airflow across the Canadian north. In winter, however, a bipolar system comprises two cold cores, one centred on central Siberia and the other centred in the Canadian Arctic Archipelago. The wind circulation is eccentric and creates a well-developed trough over central North America. This results in the strong northwesterly airflow over the western and central Canadian Arctic and subarctic which dominates the climate of these regions.

Figures 3.4A shows that the corresponding surface pressure patterns are different and more complex. In summer, the average surface pattern indicates relative high pressure over the polar sea and low pressure over the North Atlantic and Pacific oceans. In winter, there is a surface pattern of intense lows over the North Atlantic and Pacific but high pressure over subarctic North America and Eurasia. These average surface patterns reflect the thermal influences from the cold polar sea in summer and the cold continents in winter. These patterns, however, reflect average conditions derived from a continuous succession of travelling cyclones and anti-cyclones.

Air masses in northern Canada, as elsewhere, move in response to the general circulation. The winter climate east of the Western Cordillera is dominated exclusively by cold, dry continental arctic air. This is divided from warmer and wetter maritime Pacific and Atlantic air masses and from southerly air by the arctic front (Figure 3.5). This front moves equatorward and poleward in concert with the seasons. Comparison of Figures 3.4 and 3.5 shows that the shape of the arctic front in January and July conforms to the airflow streamlines in those two months. The winter circulation trough clearly promotes the southward plunge of arctic air, whereas the zonal flow of July allows southern and maritime air to move into subarctic and arctic zones. In a thorough analysis of air mass frequencies in northern Canada derived from their trajectories and origins, Bryson (1966) produced results that are presented in Table 3.4. These data show that, in midsummer, only the High Arctic, as represented by Alert and Resolute, is totally dominated by continental arctic air. Within the Low Arctic, arctic air is still dominant in the central and western regions, as illustrated by Aklavik and Baker Lake, but maritime Pacific air becomes important; in the eastern regions, as at Iqaluit (Frobisher Bay), maritime Atlantic air is well represented.

In the subarctic, air-mass mixtures in general become more diverse. For example, maritime Pacific air affects all areas, but its influence naturally wanes as one moves eastward – from 85 per cent frequency at Whitehorse in Yukon to 9 per cent at Goose Bay, Labrador. Likewise, the influence of Atlantic air is limited to the eastern and central regions. Southern air does not influence the northwest subarctic; however, it becomes increasingly significant in the centre and south with, for example, a 40-per-cent representation at Moosonee, Ont., which lies at the southern boundary of the subarctic. Hudson Bay and James Bay act as a secondary, but not insubstantial, source region for air masses in the areas around and to the east of these inland seas. Finally, Churchill, Man., and Schefferville, Que., represent interior continental stations which, in midsummer, are substantially influenced by air masses from all source regions.

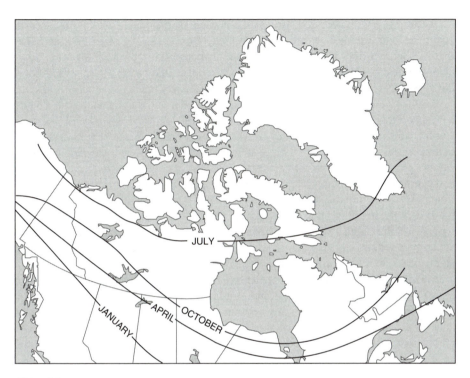

Figure 3.5
Continental arctic front: mean positions, various months. *Source*: Barry (1967).

Table 3.4
Frequency (percentage) of air masses, by source region, for stations listed in Table 3.1

| Station† | Air mass type* | | | | |
	Arctic	Pacific	Atlantic	Southern	Hudson Bay
Alert, NWT	100	–	–	–	–
Resolute, NWT	96	4	–	–	–
Aklavik, NWT	63	30	–	–	–
Aqaluit (Frobisher), NWT	45	9	35	5	6
Baker Lake, NWT	60	25	2	7	6
Whitehorse, Yukon	15	85	–	–	–
Churchill, Man.	40	33	5	15	7
Schefferville, Que.	22	14	35	22	7
Goose Bay, Nfld	16	9	48	22	5
Moosonee, Ont.	18	22	10	40	10
Winnipeg, Man.	10	44	–	41	5

Source: Bryson (1966).

* All frequencies ⩾30 per cent are underlined.

† Toronto is not included.

Bryson (1966) showed that the northern edge of the boreal forest (treeline) is congruent with the mean position of the arctic front in summer and indicated also that its southern limit is associated with the winter position of the front. This association probably represents causal links: precipitation, temperature, and evaporation are all substantially influenced by air mass frequencies.

Weather systems associated with general circulation in the Canadian arctic islands have been treated in some detail by Maxwell (1980). The frequency of cyclonic storm centres, in terms of time they overlie an area, ranges in winter from 5 per cent in the western islands to 15 per cent in the eastern ones and in summer from 7.5 per cent in the west to 20 per cent in the east. The greater frequency over the east results from the southwesterly flow of general circulation in the east as compared to northwesterly flow in the west. This circulation draws maritime Atlantic air into the eastern Arctic, promoting cyclonic activity along the arctic front. Activity reaches a peak over the Labrador Sea, Davis Strait, and Baffin Bay, where local cyclogenesis along the margins of open water areas, created by polynyas such as the North Water and by the Labrador Current, enhances regional storm development.

In winter in the low Arctic and subarctic, the greatest cyclonic frequencies occur in central Keewatin, the western and southern Hudson Bay area, and Labrador (Petterson, 1969). In summer, this pattern is maintained, but frequency generally doubles as the arctic front moves more frequently into this zone. Anticyclonic high-pressure conditions, as documented by Petterson (1969) and Maxwell (1980), are most frequent in winter but do not occupy any region more than 8 per cent of the time in any season. They are most common in the western and central Arctic and subarctic and least frequent in the eastern zone comprising Baffin Island, Labrador, the Ungava Peninsula, and their surrounding ocean areas.

In terms of overall average circumglobal patterns in the middle and high latitudes, western Hudson Bay and the eastern Arctic and subarctic rate high in storminess in all seasons, comparable to the North Atlantic Ocean. The western Arctic and western and central Keewatin rank high in terms of high-pressure clear conditions in winter.

Bradley and England (1979) carried out a detailed analysis of synoptic conditions experienced at Alert on northern Ellesmere Island and Isachsen on northern Ellef Ringnes Island, both located in the high Arctic. They identified twenty-two synoptic types based on average pressure fields: four occurred slightly more than 50 per cent of the time and none of the remainder more than 5 per cent of the time. The most frequent synoptic type comprises low pressure centred over Baffin Bay and high pressure northwest of Alert. This promotes a cold northerly air flow, is most common in winter, correlates negatively with maximum monthly temperature, and is associated with annual precipitation proportionate to its frequency. The second most frequent type displays low pressure over the North Pole and over Baffin Bay, with high pressure to the southwest centred over the Mackenzie valley, resulting in a strong northwesterly airflow. These two synoptic types bring more than 50 per cent of the annual precipitation to Alert but less than 25 per cent to Isachsen, where other synoptic types are important. Clearly the synoptic climate of the high Arctic is geographically and seasonally complex, and generalizations should not be readily made.

The complex results of changing synoptic conditions are brought out clearly by Alt (1978; 1979), who examined the summer synoptic conditions that control accumulation and ablation and hence the mass balance of the ice caps on both Meighen Island and Devon Island. Alt identified three synoptic systems for the Meighen Ice Cap. A type-I synoptic situation is an extension of the winter regime into summer, whereby high pressure to the west over the Polar Ocean and a cyclonic low to the east over Baffin Bay give a northerly flow off the ice pack. This suppresses melt over the ice cap and favours a steady-state mass balance. Type-II circulation features a cold low over the Polar Ocean north of Alaska. Cyclonic flow about this low directs moist air from the peripheral seas into the western Arctic, with accompanying frontal formations and frequent cyclonic storms. This synoptic type can lead to net ablation or accumulation over the ice cap, depending on whether storms consist primarily of rain or snow. Type-III circulation features a cold low north of Siberia which directs warm air from eastern Siberia into the western Arctic. Type-III flow promotes rapid melting of the ice cap.

Analysis for the Devon Island Ice Cap identified similar patterns and illustrated the sensitivity of ice caps to changing synoptic conditions. For example, one summer of ablation under type-III circulation can erase a positive glacier mass balance of five years.

The behaviour of ocean currents in the Arctic Archipelago and in Hudson Bay affects mesoscale climate, since the nature and movement of sea ice are intimately linked with these currents. The result is the flow of seasonal and multi-year ice packs both with surface currents and in response to upwelling of relatively warm subsurface waters, which maintain open-water polynyas. As indicated in Figure 2.4, large-scale surface flow through the Canadian Archipelago moves south and east (Leblond 1980). On the largest and most general scale, it is driven by circulation of circumpolar westerlies around the polar vortex. The net effect is to add pack ice from the Canada Basin on the northwest while subtracting ice on the southeast, through Davis Strait and Hudson Strait, to become incorporated in the Labrador Current. The flow of ice along the coast of Labrador represents an export of arctic conditions to temperate latitudes.

In an area as geographically complex as the arctic islands, many subcirculations make up this general flow. The best known, the Baffin Bay counter-clockwise circulation (see Figure 2.4), results in a warm current along the southwest coast of Greenland and a cold current along the northwest coast of Baffin Island. Hudson Bay also has its own distinctive counterclockwise circulation (Danielson 1971), which moves cold water and ice southward down the west coast, with a corresponding northward-flowing warm current along the east coast. In a fashion similar to the Labrador Current, the pack ice exports arctic conditions to temperate latitudes along the southwest coast of Hudson Bay.

Where bottom topography, current patterns, salinity gradients, or wind patterns favour it, upwelling of warmer bottom waters to the surface keeps the ocean ice-free for some or all of the winter. The resulting polynyas, along with other open-water leads, send heat and moisture in to the atmosphere and advective heat and moisture onto adjacent land. In winter, this process augments snowfall, which in turn affects the hydrologic cycle and the mass balance of glaciers. The North Water in northern Baffin Bay plays such a role by augmenting snowfall over the glaciers of nearby Ellesmere and Devon islands.

TEMPERATURE, PRECIPITATION, AND WIND

The vertical temperature profiles and surface temperature patterns are reasonably well understood for the terrestrial areas of northern Canada. They are, of course, dramatically different for winter and summer.

Figure 3.6 shows the vertical temperature structure of the troposphere in summer and winter averaged for latitudes 45°N, 65°N, and 85°N. These latitudes in Canada correspond roughly to temperate subarctic and arctic environments. Figure 3.6 illustrates two features relevant to the climates of northern Canada. First, the height of the troposphere decreases from summer to winter by about 4 km at latitude 45°N, but only by about 1 km at 65°N and 85°N. Second, the vertical temperature gradient in the troposphere averages about 25 per cent less in subarctic and arctic latitudes compared to middle latitudes. As a result, the difference between surface air temperatures, T_o, and average tropospheric air temperature, T_{TR}, at high latitudes is less than in upper-middle latitudes. For example, in summer $T_o - T_{TR}$ is 37C°, 24C°, and 16C°, respectively, at latitudes 45°N, 65°N, and 85°N; in winter, corresponding differences are 30C°, 10C°, and 10C°. Very large atmosphere-to-surface heat flux in the high latitudes is especially prominent in winter. This flux is perpetuated by persistent temperature inversion profiles in the lower troposphere during winter at latitudes 65°N and 85°N (see Figure 3.6).

In summary, the troposphere above northern Canada can be typified as compressed, with little seasonal difference in height; warm relative to land and sea surfaces beneath; and displaying a strong temperature inversion in winter. These conditions favour buildup of atmospheric aerosols, especially in winter, leading to "arctic haze" (Barrie 1986).

Seasonal variability in mean air temperatures is illustrated in Figure 3.7. In January, temperatures across the subarctic and Arctic range from −20°C to −35°C. Contrary to popular belief, winter isotherms do not parallel lines of latitude but display a strong southeasterly trend, from the western arctic coast to the west coast of Hudson Bay, and then a northeasterly trend to the Labrador-Ungava coast. For example, the −25°C isotherm in January crosses the Beaufort Sea coast at 71°N, the west coast of Hudson Bay at 55°N, and the Ungava coast at 63°N. This penetration of cold winter air into the low latitudes of central and eastern Canada accompanies the trough in the abovementioned westerly wind flow over Canada. It helps project arctic and subarctic environments into the Hudson Bay lowlands and is one cause of the northwest-to-southeast trend in the treeline through Keewatin (Bryson 1966).

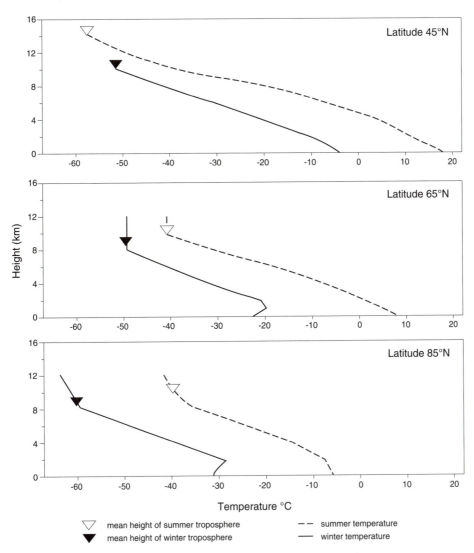

Figure 3.6
Vertical temperature profiles and height of troposphere in summer and winter at different northern latitudes.

In July, surface temperature differs relatively little over high latitudes, with a range from 15°C to 5°C. The southeast trend of isotherms from the Arctic coast to Hudson Bay is still well developed because of the long-lived sea-ice pack along the western and southern coasts of Hudson Bay, and it also help explain the southward plunge of subarctic and arctic environments in central Canada, the treeline's position, and the distribution of continuous and discontinuous permafrost (Rouse and Bello 1985; Rouse 1991). By October, most of the subarctic and Arctic have returned to subfreezing temperatures and winter conditions. In the southern Hudson Bay and James Bay

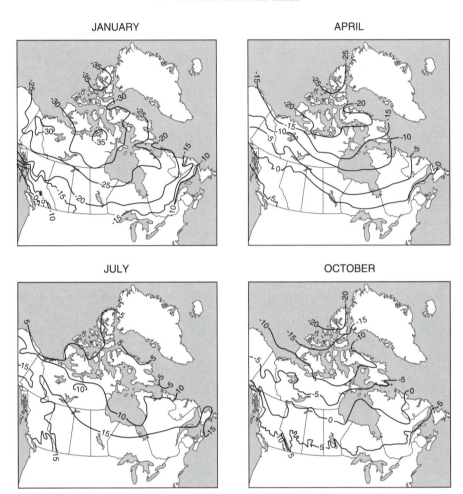

Figure 3.7
Surface air temperatures. *Source*: Hare and Hay (1974).

area, however, warm sea water releases heat to ameliorate atmospheric temperatures in the region, somewhat counteracting the cold effects of summer; but because it coincides with plant dormancy, short days, and low solar intensity, it has little influence on biological productivity and ground heating.

Canada's northern and polar lands are characterized by small precipitation, and some, especially in the High Arctic, can reasonably be regarded as cold deserts. This condition is a response, above all, to the small amounts of water vapour present in the atmosphere to facilitate precipitation during winter (Hay 1971). Figure 3.8A, for January, typifies winter. Nowhere is precipitable water vapour more than 4 mm, and for all of the Arctic Archipelago and central mainland Arctic it is less than 2 mm.

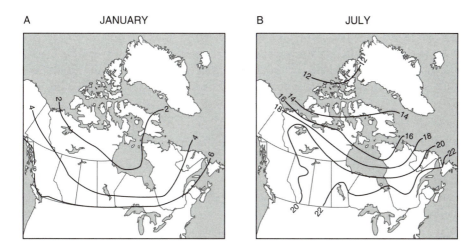

Figure 3.8
Precipitable water vapour (mm) in (A) January and (B) July. *Source*: Hay (1971).

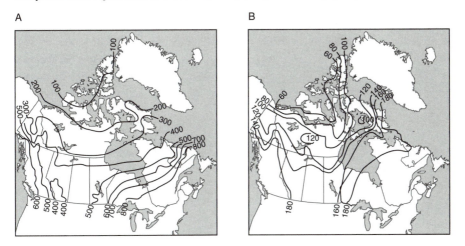

Figure 3.9
Annual mean (A) precipitation (mm) and (B) snowfall (cm). *Source*: Hare and Hay (1974).

In July (Figure 3.8B), in contrast, it ranges from 12 to 22 mm. In both seasons, water vapour capable of precipitation decreases inland from both coasts because of increasing distance from open oceanic bodies and intense winter cold.

Another factor creating desert-like conditions is precipitation efficiency – the percentage of precipitable water vapour that falls as precipitation on an average day. In Canada's northern lands it is about 20 per cent, compared to approximately 43 per cent and 62 per cent in the Great Lakes–St Lawrence Valley and the west coast, respectively (Tuller 1973). There is both less atmospheric moisture and less turnover than elsewhere. Mean annual precipitation steadily decreases poleward, from 700 mm in southern James Bay to 100 mm in the High Arctic (Figure 3.9A). The isohyets tend to parallel latitude except along the eastern coasts of Labrador, Baffin Island, and Ellesmere Island, where they indicate higher precipitation: local open water sources and/or adjacent mountainous areas favour winter orographic precipitation.

Northern snowfall is usually underestimated, often by a considerable amount (Findlay 1969; Hare and Hay 1971; Woo et al. 1983): instrumentation measures vertical snowfall, while under strong winds most snow is swept horizontally, and it is difficult to obtain representative measurements of snow accumulation. Nonetheless, the patterns shown in Figure 3.9B are noteworthy. The most snow accumulates in Labrador–Nouveau Québec, followed by southeastern Yukon and the eastern coastal areas of Baffin, Devon, and Ellesmere islands. In all cases, adjacent uplands and adjacent open water favour heavy snowfall. The least amounts of snow accumulate in central Keewatin, the western arctic coast, and the High Arctic islands: large distances from open water and flat terrain inhibit snowfall there.

The longevity of the snow cover is as important as the amount of snow, because it determines the maintenance of high-albedo versus low-albedo surfaces. As documented by Potter (1965), snow cover persists from 7.5 months in the low subarctic to more than 10 months in the High Arctic. In years of accumulation on ice caps, seasonal snow cover persists all year. Figure 3.10 shows qualitatively the patterns of snow cover from high temperate latitudes to the High Arctic. For the former, snow accumulates quickly to a maximum in midwinter and then melts rapidly. For the latter, it collects rapidly in early winter, when atmospheric moisture is still available. There is little or no accumulation for a long period in midwinter; a second period of accumulation follows in late winter, when moderating temperatures favour more atmospheric moisture. A final rapid melting usually occurs, caused by warm summer air masses. Global warming of the climate may modify the High Arctic's pattern so as to resemble the lower arctic pattern, and the latter in form to resemble the subarctic.

Surface wind flow in the northern latitudes is difficult to characterize. Despite well-developed westerly flow south of the Arctic Circle, poleward of this there is no persistent wind direction. However, some generalizations can be made. In the open arctic tundra, surface wind speeds in midwinter are about one-third greater than those of midsummer; the forested subarctic experiences little seasonal differences. Wind speeds across the tundra are about one-half greater in midwinter than those of the subarctic open forest, but only about one-fifth greater in midsummer. The coniferous trees of the open subarctic woodland allow little seasonal difference in surface roughness. By contrast, both in summer and especially in winter, the subarctic woodland exerts greater frictional drag on the wind than does the tundra. It is on the open tundra that arctic blizzards are frequent and that wind-chill becomes intolerable to humans.

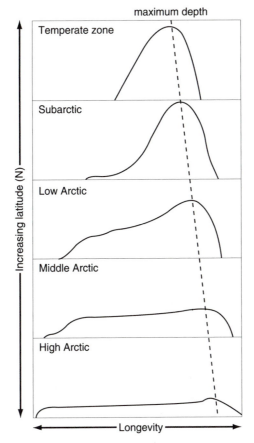

Figure 3.10
Latitudinal variations in seasonal patterns of snow cover.

CLIMATE TYPES

Table 3.5 gives climate data for a number of Canadian localities in the subarctic, the low and middle Arctic, and the High Arctic. They provide representative geographical coverage and include most of the same locations as Table 3.1. The term *subarctic* designates areas within the open coniferous forest belt, where annual temperatures are near or below freezing. The five locations so identified represent a considerable range in latitude. For example, Moosonee, at the southern limit, is at the same latitude as London, England. Its harsh climate, however, results from its continental location and its position at the tip of James Bay. For the five subarctic stations listed in Table 3.5, net radiation, expressed as a percentage of total solar radiation (i.e. $Q*/K{\downarrow}$), varies little. At an average of 25 per cent, this value compares

Table 3.5
Annual climate data for high-latitude stations

Station	Latitude (°N)	Elev. (m)	Q*/K↓ (%)	T (°C)	P (mm)	S (cm)	U (m/s)
Subartic							
Whitehorse, Yukon	60°41'*	324	29	−0.8	260	128	4.2
Norman Wells, NWT	65°19'	64	25	−6.2	328	136	3.0
Moosonee, Ont.	51°18'*	11	31	−0.9	785	280	3.5
Schefferville, Que.	54°50'*	512	28	−4.6	723	320	4.5
Goose Bay, Nfld	53°15'*	44	31	0.2	877	400	4.3
Low and middle Arctic							
Inuvik, NWT	68°30'	9	15	−8.9	236	99	2.0
Baker Lake, NWT	64°45'*	4	9	−11.9	208	58	5.6
Churchill, Man.	58°45'	35	25	−7.3	397	176	6.3
Iqaluit, NWT	63°45'*	207	10	−8.9	415	205	4.9
High Arctic							
Resolute, NWT	74°43'	64	11	−16.4	136	79	6.1
Alert, NWT	82°30'	62	12	−18.0	136	156	3.0

Note: $K\downarrow$ is solar radiation; $Q*$ is net radiation; T is temperature; P is precipitation; S is snowfall; and U is wind speed.

to Montreal and Vancouver, at 35 per cent and 42 per cent, respectively. There is also a pronounced increase in annual rainfall and snowfall eastward across Canada, primarily because of greater available moisture from sources in the Gulf of Mexico, Hudson Bay, and the Atlantic Ocean. Finally, wind speeds are moderate and comparable to those at temperate latitudes.

Except for Iqaluit (Frobisher Bay), the low and middle arctic localities are on the mainland. Both Inuvik and Churchill are located near the forest-tundra boundary, the former just south of it and the latter just north. At latitudes 68°30'N and 58°45'N, respectively, they illustrate the southeastward trend of the treeline across Canada. Although average solar radiation received at the surface ($K\downarrow$) in the low and middle arctic locations is slightly larger than for those of the subarctic, net radiation, expressed as a percentage of total solar radiation (i.e. $Q*/K\downarrow$), averages only 15 per cent. The long, high-albedo snow season inhibits absorption of solar radiation and promotes a very low-energy environment on an annual basis. As in the subarctic, precipitation and snowfall increase eastward, but differences are less, as the smaller water-vapour capacity of the colder air makes itself felt. Average annual precipitation and snowfall are each 88 per cent less for the low and middle arctic localities than for those of the subarctic.

Annual average wind speeds are large, except at Inuvik, where they are measured at the airport, which is located in a flat, lowland area surrounded by forest. In contrast, the anemometer at Churchill airport is fully exposed to prevailing winds from the open tundra and Hudson Bay. In the low and middle arctic, wind-chill becomes an important climatic feature (see chapter 1). For example, annual wind-chill–equivalent temperatures for a clothed human body are -9, -12, -12, and -15 at Inuvik, Churchill, Iqaluit, and Baker Lake, respectively. In like fashion, the frequency of blowing snow between October and March inclusive averages 2, 12, 7, and 19 per cent for the same locations (Winston 1964).

The High Arctic is represented by data (Table 3.5) from Resolute and from Alert, the most northern permanently inhabited place in the world. Both are essentially coastal stations at low elevations and represent tundra rather than ice-cap environments. They have little or no vegetation. The climatological station at Resolute is located on relatively flat terrain, whereas Alert is nestled in hilly terrain, which accounts for its relatively low wind speeds. Although precipitation totals at the two stations are about the same, most of Alert's precipitation arrives as of snow, whereas Resolute gets substantial summer rains and less winter snow. Overall, precipitation in the High Arctic is less than half that of the low and middle Arctic and one-quarter that of the subarctic. For these reasons, much of Canada's High Arctic is classified by soil scientists and botanists as polar desert (see chapter 4).

ENERGY AND WATER BALANCE AT THE SURFACE

The energy balance at the Earth's surface can be expressed in terms of net radiation, Q^*, heat flux into the substrate, Q_S, latent heat of evapotranspiration, Q_E, and sensible heat flux between surface and atmosphere, Q_H, giving

$$Q^* = Q_S + Q_E + Q_H.$$

Surface water balance is given by

$$\Delta M = P - E - R,$$

where ΔM is change in stored water, P is precipitation input, E is evapotranspiration loss, and R is runoff loss. Both surface energy balance and water balance affect distribution of surface moisture and surface vegetation in Canada's north. Energy and water balances, together with vegetation, help determine the distribution, character, and behaviour of permafrost by shaping the ground thermal and moisture regimes.

Most studies in northern energy and water balances have been confined to late winter and summer. Early and midwinter are very low-energy periods, with a simple water cycle consisting of snowfall, its redistribution, and its storage; they are also very difficult periods for carrying out meaningful surface measurement. The classic study of the energy balance in the High Arctic is Ohmura (1982), who investigated conditions on Axel Heiberg Island at an upland tundra site vegetated with hummocky grasses. Ohmura found that during the dry-snow season of late winter net radiation is the main source of energy and sensible heat flux into the atmosphere is the main

Table 3.6
Select energy balance studies

Reference	Site characteristics	PF	Season	Q^*	Q_s	Q_E	Q_H	B_r
High Arctic								
Ohmura (1982)	Upland tundra	Y	Pre-melt	18	2	6	10	1.67
	(Axel Heiberg I., NWT)		Melt	83	28	28	27	0.96
			Snow-free	90	13	46	31	0.67
Low Arctic								
Rouse (1984c)	Upland tundra	Y	Melt	27	10	9	8	0.89
	(Churchill, Man.)		Snow-free	120	20	51	49	0.90
Rouse and Bello	Wet sedge tundra	Y	Melt	108	24	45	39	0.87
(1985)	(Churchill, Man.)		Snow-free	143	21	63	59	0.94
	Upland lichen heath	Y	Melt	122	38	45	39	0.87
	(Marantz Lake, Man.)		Snow-free	138	15	63	60	0.95
Subarctic								
Rouse, Mills, and Stewart	Lichen woodland	N	Midsummer	132	6	83	43	0.52
(1977)	Burned lichen woodland	N	Midsummer	117	4	62	51	0.82
	Shallow tundra lake	N	Midsummer	148	0	109	39	0.39
	(Thor Lake, NWT)							
Lafleur and	Upland sedge	N	Snow-free	112	5	55	46	0.84
Rouse (1988)	Sedge marsh	N	Snow-free	116	11	62	43	0.70
	Woodland swamp	N	Snow-free	124	12	71	41	0.58
	(southern James Bay)							

Note: Q^*, Q_s, Q_E, and Q_H are net radiation, ground heat storage, and sensible and latent heat fluxes, respectively
(W/m²); B_r is the Bowen ratio; *PF* refers to the presence (*Y*) or absence (*N*) of permafrost.

sink (Table 3.6). This gives a large Bowen ratio ($B_r = Q_H/Q_E$) which, because of
the cold atmosphere, produces large vertical temperature gradients and small vapour
pressure gradients.

During melt season, most energy is used in melting snow, but immediately after
snowmelt, in June, Q_H and Q_E increase. On average, for the snow-free period of
July and August, the relatively large Bowen ratio indicates substantial resistance to
evaporation. During the same period, ground heat storage (Q_S) is approximately 9
per cent of net radiation (Q^*), indicating that a substantial portion of energy is used
in thawing the seasonally frozen zone (active layer). In another study, Woo (1976)
determined evaporation as part of a water-balance study near the head of Vendome
Fiord on Ellesmere Island, on terrain similar to that of Ohmura. Woo found that,
for high summer, evaporation (*E*) was 60 per cent greater than that measured by
Ohmura. He concluded that evaporation was the major cause of change in the suprapermafrost water table and thus crucial to summer runoff.

In the low Arctic and subarctic there have been several energy and water-balance
studies (Table 3.6). Roulet and Woo (1986) report on a study carried out west of
Baker Lake in central Keewatin in a wetland underlain by continuous permafrost.

A   B

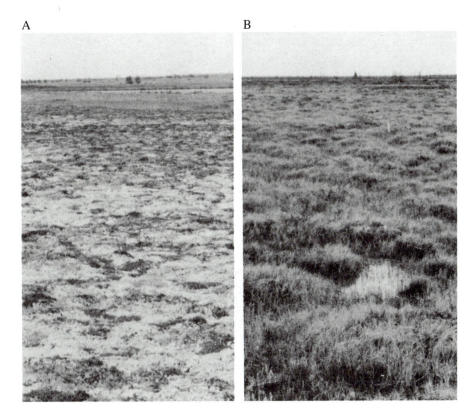

Figure 3.11
Comparative terrain types in coastal tundra: (A) lichen heath; (B) sedge meadow.

They found wetlands as effective as nearby lakes, in evaporation, because capillary water from the water table keeps bog surfaces wet. Rouse (1984a; 1984b; 1984c) reported on radiation and energy-balance studies in the vicinity of the arctic treeline near Churchill. Accumulations of deep winter snow near treeline meant that snowmelt in the forest could lag behind that on the tundra by up to one month. However, in both winter and summer the forest has greater net radiation than open tundra, primarily because of smaller albedos. Rouse found also that in the wet soils of the Hudson Bay lowlands heat flux into the substrate, Q_S, used 16 and 18 per cent of Q^* in tundra are forest, respectively. Most of this energy is used in thawing the active layer.

Despite clear differences between wet and dry tundra in the Hudson Bay lowlands (Figure 3.11), energy flux differs surprisingly little. For example, on dry gravel soils of sedge-covered upland tundra, average Bowen ratios during and after snowmelt were 0.89 and 0.90, respectively. In comparison, on a nearby coastal wet site dominated by sedge grasses, Bowen ratios of 0.87 and 0.94 were obtained (Table 3.6).

Figure 3.12
Instrumental system for determining surface energy balance in wet sedge meadows of southern James Bay.

For the subarctic forest, Rouse, Mills, and Stewart (1977) report on studies of mature lichen woodland and burned woodland sites on dry, non-permafrost upland areas southeast of Yellowknife. For sunny midsummer conditions they found average Bowen ratios of 0.52, 0.82, and 0.39 for mature lichen woodland, burned lichen woodland, and a shallow lake, respectively. For a wetland environment in southern James Bay, Lafleur and Rouse (1988) measured the energy balance of one open forest and two sedge-grass–dominated surfaces. Their instrumentation on one of the sedge-grass sites is shown in Figure 3.12. During the growing season, Bowen ratios of 0.84 and 0.70 were typical of upland sedge and sedge swamp, respectively, and a value of 0.58 typified willow-alder swamp (Table 3.6).

Northern terrain has several significant micro-climatic characteristics. Where permafrost is ice-rich, heat flux into the substrate during the snow-free period is larger than in temperate latitudes. This is promoted by large vertical temperature gradients between the soil surface and the frost table. Where permafrost is absent, Q_S is not greater than in temperature latitudes.

Even when surfaces are very wet, Bowen ratios are substantial, indicating resistance to evapotranspiration. Except for the shallow tundra lake site, Bowen ratios exceed

0.50, and for wet vegetated surfaces, such as wet, sedge-grass tundra, they approach unity during the snow-free period (Table 3.6). This situation indicates a strong control on rates of evapotranspiration.

Indeed, the concept of potential evapotranspiration, as used to describe an un-inhibited vapour flux from vegetated surfaces with unlimited water-supply, does not apply in the higher latitudes, for a number of reasons. First, plant canopy often exerts strong control over transpiration (Lafleur and Rouse 1988), by means of large stomatal resistance or small leaf area, or both. Stomatal and canopy resistance increases with cold soil water in the rooting zone, with warm and dry atmospheric conditions, and with physiological immaturity or senescence of the plant. Most plants in high latitudes grow during a period that does not usually exceed six weeks. Before and after this period, they are relatively inactive. Second, subarctic and arctic species are usually xerophytically adapted to avoid moisture stress. Stomata are often few, sunken or covered in hair or both, and located mainly on lower leaf surfaces. Leaf cuticles are sometimes covered in wax or are very thick. Third, lichens are non-transpiring, and where they are a significant part of the canopy, as in Figure 3.11A, they serve as a mulch to the underlying surface, which inhibits direct evaporation. These various botanical factors make a large canopy resistant to evaporation.

Of equal and often greater importance than surface characteristics is the control exerted by a cold and moist atmosphere. With small vapour-pressure deficits in the atmosphere, vapour-pressure gradients are small, whether between overlying air and saturated vapour pressure at a wet surface or in the substomatal cavity of a leaf. As a result, Q_E is small, which favours a large Bowen ratio. A cold atmosphere is usually associated with small vapour-pressure deficits. At the same time, the vertical temperature gradient between the surface and the cold atmosphere is large. This drives Q_H toward large values, which also helps produce large Bowen ratios. Canopy and atmospheric factors thus inhibit evaporation, even in very wet areas.

Advection – the horizontal transport of air that has humidity and temperature characteristics different from those of the area into which it moves – plays a large role in the surface energy balance, particularly in higher latitudes, where it operates on regional (meso), local, and micro scales.

Regional- or synoptic-scale advection is caused by movement of air masses. We have already seen how it affects the position of treeline (Bryson 1966) and the melting of glaciers (Atl 1978; 1979). In their synoptic classification of the Canadian High Arctic, Bradley and England (1979) observed that, depending on location and season, infrequent synoptic types can substantially influence climate. At Alert, for example, the two types accounting for more than 50 per cent of the precipitation occurred on only 36 per cent of the days. Singh and Taillefer (1986) examined the effect of synoptic-scale advection on evapotranspiration rates from black spruce forest in northern Quebec. With warm air advection over a moist surface, enhancement of evapotranspiration often exceeded 25 per cent.

Mesoscale advection has a pronounced influence on temperature, humidity, and energy balance. In his summertime study of the Lewis Glacier on Baffin Island,

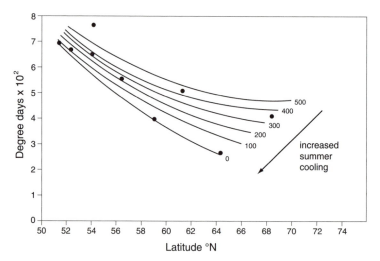

Figure 3.13
Thawing degree–days (1 May–31 Oct.), by latitude and distance from coast; isopleths show distance (km) from Hudson Bay and James Bay coast. *Source*: Rouse (1991).

Rannie (1977) discovered that on warm days with gentle off-glacier winds advection of cold air lowered temperature on average by 2.3°C at the glacier margin and also caused very frequent evening inversions in the adjacent valley. Likewise, Jacobs and Grondin (1988), in examining the influence of large lake systems on mesoclimate in south-central Baffin Island, conclude that the lakes act as heat reservoirs. The latter moderate temperatures locally during cold-air intrusion in summer and retard seasonal cooling in fall and early winter.

Advection is also a major component of the synoptic climatology of the Hudson Bay lowlands. Rouse (1984c), Rouse and Bello (1985), and Rouse, Hardill, and Lafleur (1987), for instance, have extensively studied the effect of advection of cold, moist summer air from Hudson Bay and James Bay on the climate of the adjacent tundra and forest-tundra surfaces. Winds blowing onshore (for about half of the time May–September) strongly enhance sensible heat flux (Q_H) while suppressing ground heat storage (Q_S) and latent heat fluxes (Q_E), compared to offshore winds. This effect extends at least 65 km inland.

The regional impact of advection is illustrated in Figure 3.13. It shows that summer cooling attributable to Hudson Bay and James Bay is substantial to at least 500 km inland and increases with higher latitude. Intensive cooling near the coast shapes the arctic treeline and the southern boundary of continuous permafrost, both of which parallel the west and south coasts of Hudson Bay and intersect James Bay near Ekwan Point. The cold advective source is long-lasting in summer due to sea ice in Hudson Bay and northern James Bay. Rouse and Bello (1985) have calculated that earlier melting of the sea ice would move the treeline and the boundary of continuous permafrost further northwest. Such a situation might occur if global warming were to take place.

Local-scale advection strongly influences the energy balance. Bello and Smith's (1990) study of a shallow tundra lake in the Hudson Bay lowlands found that latent heat flux from the lake exceeds available radiant energy during the summer, with temperature inversions over the lake on the majority of days. Advective enhancement of regional advection is caused by air flow from the warmer and drier upwind terrestrial surfaces and synoptic effects of daily weather variations.

Micro-scale advection can also be large. Halliwell (1989) measured differences in horizontal surface temperatures over a few metres of up to 12C° in hummocky sedge tundra on sunny days. Wet areas are cool, and dry areas are warm. As a result, vertical soil temperature profiles differ greatly in the top layers of ground, over horizontal distances of only one or two metres. The resulting horizontal atmospheric flux of sensible heat enhances evapotranspiration from wet areas.

Unique features of high-latitude vegetation can shift surface energy and water balances. Ground lichen is fully non-transpiring: according to Bello and Arama (1990), even in tundra regions that receive relatively abundant rainfall a lichen canopy can intercept up to 61 per cent of the summer rain. This rain never reaches the soil and subsequently evaporates directly into the atmosphere. Maximum storage capacities of the lichen mat can be more than double those reported for forests. Thus a combination of open forest and ground lichen, such as occurs in many subarctic regions, may facilitate very large canopy storage. Lafleur (1988) notes strong climatic-plant physiological interactions in the subarctic: stomatal resistance to transpiration, from woodland and sedge-grass species, becomes greater with increasing atmospheric vapour-pressure depressions. Because similar relationships are documented for temperate latitudes, one might expect them to exist in low- and middle-arctic regions, where resistance to moisture loss, under conditions of warm dry air and frozen soils, is essential to survival.

REFERENCES

Alt, B.T. 1978. "Investigation of Summer Synoptic Climate Controls on the Mass Balance of Meighen Ice Cap." *Atmosphere-Ocean* 17: 181–99.

– 1979. "Synoptic Climate Controls of Mass Balance Variations on Devon Island Ice Cap." *Arctic and Alpine Research* 10: 61–80.

Barrie, L.A., 1986. "Arctic Air Pollution: An Overview of Current Knowledge." *Atmospheric Environment* 20: 643–63.

Barry, R.G. 1967. "Seasonal Location of the Arctic Front over North America." *Geographical Bulletin* 9: 79–95.

Bello, R.L., and Arama, A. 1989. "The Water Balance of Lichen Canopies." *Climatological Bulletin* 23: 74–8.

Bello, R.L., and Smith, J.D. 1990. "The Effect of Climate Variability on the Energy Balance of a Lake in the Hudson Bay Lowlands." *Arctic and Alpine Research* 22: 98–108.

Bradley, R.S., and England, J. 1979. "Synoptic Climatology of the Canadian High Arctic." *Geografiska Annaler* 61A: 187–201.

Bryson, R.A. 1966. "Air Masses, Streamlines and the Boreal Forest." *Geographical Bulletin* 8: 228–69.

Danielson, E.W., Jr. 1971. "Hudson Bay Ice Conditions." *Arctic* 24: 90–107.

Davies, J.A. 1963. "Albedo Investigations in Labrador-Ungava." *Archiv. Meteorol. Geophys. Bioklim* B13: 137–51.

Findlay, B.F. 1969. "Precipitation in Northern Quebec and Labrador: An Evaluation of Measurement Technique." *Arctic* 22: 140–50.

Halliwell, D. 1989. "A Numerical Model of the Surface Energy Balance and Ground Thermal Regime in Organic Permafrost Terrain." PhD thesis, McMaster University, Hamilton.

Hare, F.K. 1968. "The Arctic." *Quarterly Journal of the Royal Meteorological Society* 94: 439–59.

Hare, F.K., and Hay, J.E. 1974. "The Climate of Canada and Alaska." In R.A. Bryson and F.K. Hare, eds., *World Survey of Climatology: North America*, vol. 2, Amsterdam: Elsevier, 49–129.

– 1971. "Anomalies in the Large-Scale Water Balance over Northern North America." *Can Geographer* 15: 79–94.

Hare, F.K. and Ritchie, J. 1972. "The Boreal Bioclimates." *Geographical Review* 62: 333–64.

Hay, J.E. 1971. "Precipitable Water over Canada: Distribution." *Atmosphere-Ocean* 9: 101–11.

Henderson-Sellers, A., and Robinson, P.J. 1986. *Contemporary Climatology*. UK: Longman Scientific and Technical.

Hengeveldt, H.G. 1986. *Understanding CO_2 and Climate*. Annual Report 1986, Canada Climate Centre, Environment Canada.

Jacobs, J.D., and Grondin, L.D. 1988. "The Influence of an Arctic Large Lakes System on Mesoclimate in South-Central Baffin Island." *Arctic and Alpine Research* 20 no. 2: 212–19.

Lafleur, P.M. 1988. "Field Observation of Stomatal Conductance in Three Wetland Species Growing in a Subarctic Marsh." *Canadian Journal of Botany* 66: 1,367–75.

Lafleur, P.M., and Rouse, W.R. 1988. "The Influence of Surface Cover and Climate on Energy Partitioning and Evaporation in a Subarctic Wetland." *Boundary Layer Meteorology* 44: 327–47.

Leblond, P.H. 1980. "On the Surface Circulation in Some Channels of the Canadian Arctic Archipelago." *Arctic* 33: 189–97.

LeDrew, E.F. 1983. "Arctic Weather." *Geos* 12 no. 2: 6–9.

List, R.J. 1958. *Smithsonian Meteorological Tables*. Smithsonian Miscellaneous Collection, U114, 419.

Maxwell, J.B. 1980. *The Climate of the Canadian Arctic Islands and Adjacent Waters*. Climatological Studies 30, vol. 1, AES, Environment Canada.

Newell, R.E., Vincent, D.G., Dopplick, T.G., Femuza, D., and J.W. Kidson. 1969. "The Global Circulation of the Atmosphere." In G.A. Corby, ed., London: Royal Meteorological Society, 42–90.

Ohmura, A. 1982. "Climate and Energy Balance on the Arctic Tundra." *Journal of Climatology* 2: 65–84.

Petterson, S. 1969. *Introduction to Meteorology*. Toronto: McGraw-Hill.

Potter, J.G. 1965. *Snow Cover*. Climatological Studies 3, Canada Department of Transport, Meteorological Branch.

Rannie, W.F. 1977. "A Note on the Effect of Glacier on the Summer Thermal Climate of an Ice-Marginal Area." *Arctic and Alpine Research* 9: 301–4.

Robinson, D.A., and Kukla, G. 1985. "Maximum Surface Albedo of Seasonally Snow-Covered Lands in the Northern Hemisphere." *Journal of Climate and Applied Meteorology* 24: 402–11.

Roulet, N., and Woo, M.K. 1986. "Wetland and Lake Evaporation in the Low Arctic." *Arctic and Alpine Research* 18: 195–200.

Rouse, W.R. 1976. "Microclimate Changes Accompanying Burning in Subarctic Lichen Woodland." *Arctic and Alpine Research* 8: 357–76.

– 1984a. "Microclimate at Arctic Tree Line. I. Radiation Balance of Tundra and Forest." *Water Resources Research* 20: 57–66.

– 1984b. "Microclimate at Arctic Tree Line. II. Soil Microclimate of Tundra and Forest." *Water Resources Research* 20: 67–73.

– 1984c. "Microclimate at Arctic Tree Line. III. "The Effects of Regional Advection on the Surface Energy Balance of Upland Tundra." *Water Resources Research* 20: 74–8.

– 1991. "Impacts of Hudson Bay on the Energy Balance of the Hudson Bay Lowlands." *Arctic and Alpine Research* 23: 24–30.

Rouse, W.R., and Bello, R.L. 1983. "The Radiation Balance of Typical Terrain Units in the Low Arctic." *Annals American Association of Geographers* 73: 538–49.

– 1985. "Impact of Hudson Bay on the Energy Balance in the Hudson Bay Lowlands and the Potential for Climatic Modification." *Atmosphere-Ocean* 23: 375–92.

Rouse, W.R., Hardill, S.G., and Lafleur, P. 1987. "The Energy Balance in the Coastal Environment of James Bay and Hudson Bay during the Growing Season." *Journal of Climatology* 7: 165–79.

Rouse, W.R., Mills, P.F., and Stewart, R.B. 1977. "Evaporation in High Latitudes." *Water Resources Research* 13: 909–14.

Singh, B., and Taillefer, R. 1986. "The Effect of Synoptic-Scale Advection on the Performance of the Priestly-Taylor Evaporation Formula." *Boundary Layer Meteorology* 36: 267–82.

Tuller, S.E. 1973. "Seasonal and Annual Precipitation Efficiency in Canada." *Atmosphere* 11 no. 2: 52–66.

Vowinckel, E., and Orvig, S. 1970. "The Climate of the North Polar Basin from Orvig, S., Climates of the Polar Regions." In H.E. Landsberg, ed., *World Survey of Climatology*, vol. 14, Amsterdam: Elsevier, 129–225.

– 1984. "Energy Balance of the Arctic. I. Incoming and Absorbed Solar Radiation at the Ground in the Arctic." *Archiv. Meteorol. Geophys. Bioklim.* B13: 352–77.

Winston, W.C. 1964. *A Study of Winds and Blowing Snow in the Canadian Arctic*. Meteorology Branch, Department of Transport, CIR 4162, TEC 548, Toronto, Dec. 1964.

Woo, M.K. 1976. "Evaporation and Water Level in the Active Layer." *Arctic and Alpine Research* 8: 213–17.

Woo, M.K., Heron, R., Marsh, P., and Steer, P. 1983. "Comparison of Weather Station Snowfall with Snow Accumulation in High Arctic Basins." *Atmosphere-Ocean* 21: 312–25.

CHAPTER FOUR

Northern Vegetation

JAMES C. RITCHIE

Let us at the outset clarify our perceptions about the plant life of Canada's northern and polar lands, particularly the ways in which it differs from the flora and vegetation that surround the southern fringe which contains Canada's cities, transportation routes, and recreational centres. A few misrepresentations should be dispelled.

First, it is said that arctic plants show little diversity (numbers of species) because the arctic ecosystems are of recent origin and thus have not had enough time for diversity to evolve. In fact, however, arctic biosystems have existed for at least four million years, roughly the same period as North American deserts, grasslands, and boreal forests.

Second, we often read that arctic plant life is impoverished and sparse because the environment is extremely "harsh," "severe," "difficult," "marginal," or "inhospitable." These metaphors are rooted in our culture, but, as Raup (1969) pointed out, an Inuit ecologist would describe the environment more accurately and less dramatically: arctic plants have long been adapted to life in an environment where the major factors limiting growth are lack of nutrients and an annual radiant-energy balance that is negative or nearly so.

Third and finally, textbooks in ecology often state that arctic plants reproduce mainly by vegetative spread and that normal sexual propagation is rare. Murray (1987) dispels this notion.

In this chapter, I elaborate these three themes – origins and history, ecology, and reproduction – drawing examples from both arctic and subarctic systems. Additional useful references about northern Canadian plant life include those by Chung (1984), an illustrated botanical conspectus of the North American Arctic and Rockies, and two volumes by the author (Ritchie 1984; 1987). In addition, Britton (1957), Beschel (1970), Savile (1972), Bliss (1977), and Tieszen (1978) analyse in detail the biological and ecological characteristics of northern plants and communities, while Aleksandrova (1988) treats circumpolar arctic vegetation.

ORIGINS AND HISTORY

The origins of northern flora and vegetation are still imperfectly understood, but as more fossil information comes to hand the time span lengthens. A recent account of plant macrofossils recovered from the Beaufort Formation in the Canadian Arctic concludes that "a type of nascent tundra [existed] around the Arctic Basin for nearly all of the Pliocene" (Matthews 1987: 85). Fossils of *Saxifraga oppositifolia*, *Dryas integrifolia*, and *Oxyria digyna* have been recorded from the Beaufort Formation on Meighen Island, at latitude 80°N in the Canadian High Arctic. They occur in association with other plant remains that indicate forest-tundra communities (Table 4.1). Funder et al. (1985) have made similar discoveries from northern Greenland. In other words, the arctic flora and vegetation of Canada have existed for about as long – more than four million years – as the regional grasslands and deserts of the Great Plains and southwest of North America. The arctic ecosystems, however, have probably experienced greater spatial displacement in the ensuing millennia of the late Tertiary and especially the Quaternary. During the latest, roughly two million, years of the Quaternary, the Canadian arctic region has been occupied and vacated successively by a series of continental glaciations. The most recent, one of the most extensive, culminated at about 18000 yr BP, when a complex of thick (up to 3 km) ice sheets occupied most of Canada. It is pertinent to note, however, that substantial ice-free areas remained in the Canadian Arctic and adjacent Alaska.

Assemblages of plant fossils containing many arctic species have been found in northern US sites that were beyond the southern margin of the ice sheet (Figure 4.1). Similarly, continuous pollen profiles from glacial refugia in Alaska and Yukon have demonstrated the presence of tundra in the northwest during the latest glacial age (reviewed recently by Barnosky, Anderson, and Bartlein 1987 and by Lamb and Edwards 1988). Examples are the long record by Anderson (1985), from Kaiyak Lake in Alaska and the Hanging Lake site in northern Yukon (Cwynar 1982) (Figures 4.2 and 4.3). The full glacial (15000 to 20000 yr BP) pollen spectra from these and several other sites are dominated by herb taxa. Suggested reconstructions of vegetation have ranged from continuous steppe to polar desert, but most observers agree that the Beringian refugium served as the source area for many of the plants found today in the arctic-subarctic tundras of western and central Canada. Alternations between interglacial conditions, as at the present when most of the Arctic is ice-free, and the several, usually long glacial periods have put arctic plants through cycles of fragmentation and disruption, followed by reunion of divided geographic ranges.

Stebbins (1984; 1985) and Murray (1987) have discussed these issues using a theory of secondary contact to explain the modern evolutionary status of North America's arctic flora. Populations were broken up during glacial periods into segments surviving in various ice-free regions, primarily south or northwest of the continental ice sheets (Figure 4.4). Genus-type diversification occurred during the long glacials (roughly 100,000 years); subsequent secondary contact of these populations after interglacial ice recession produced complex taxonomic races displaying a wide range of reproductive methods.

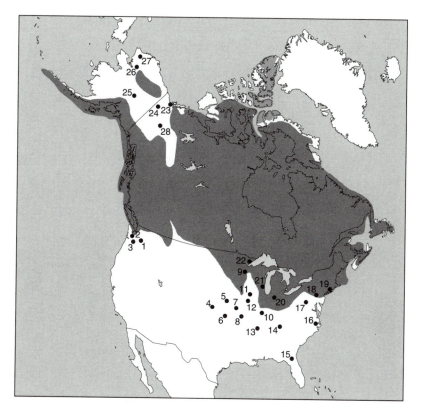

Figure 4.1

Approximate limits of North American ice sheets at latest glacial maximum (18000 yr BP) and locations of sites from which tundra fossil plants have been recovered. *Source*: Ritchie (1987).

As Murray (1987) concludes, "the arctic offers not the simple reproduction system one often hears about, but one complicated by dioecious anemophilous and ento-mophilous species, gynodioecious taxa, self-incompatible diploids, self-compatible popyploids, and apomictic polyploids." In other words, we find a rich diversity of modes of achieving seed-set species, with separate male and female flowers on different plants, some pollinated by wind, others by insect; species with built-in mechanisms either to ensure or to prevent self-fertilization; and genera, like *Taraxacum*, the northern dandelion, with an astonishing array of reproductive meth-ods that include normal sexual crossing and seed formation without the 'normal' fusion of sex cells (apomixis).

Possible causal links between occurrence of ice-free refugia and distribution of locally distinct races of plants (endemic species or subspecies) has long fascinated plant geographers. For example, most observers would explain the appearance of species in such widely separated regions as Fennoscandia, Greenland, and eastern North American Arctic by survival of the species in nunataks (ice-free, often montane

Table 4.1
Summary of plant fossils, grouped by modern habitats, reported from Beaufort Formation sites on Meighen Island, NWT

Aquatic and shoreline

Taxa	Duck Hawk Bluffs Lower	Duck Hawk Bluffs Upper	Ballast Brook	Meighen Island
Selaginella selaginoides				+
Potamogeton sp.	+	+	+	+
Potamogeton epihydrus			+	+
Potamogeton filiformis			+	+
Nuphar sp.			+	+
Nymphaea sp.		+	+	
Ranunculus (Batrachium)		+		+
Callitriche sp.	+		?	
Myriophyllum sp.			+	
Sparganium hyperboreum		+	+	
Triglochin maritimum			+	
Sagittaria sp.	?			
Glyceria sp.			+	+
Carex sect. Acutae				+
Carex sect. Chordorrhiza				+
Dulichium sp.			+	
Eleocharis sp.				+
Scirpus sp.	+		+	
Juncus sp.				+
Potentilla palustris			+	+
Potentilla norvegica			+	+
Hippuris sp.			+ +	+ +
Heath and tundra				
Betula (low shrub type)		+		+

Wet Lowlands (Cont.)

Taxa	Duck Hawk Bluffs Lower	Duck Hawk Bluffs Upper	Ballast Brook	Meighen Island
Ranunculus hyperboreus			+	
Ranunculus lapponicus			+	+
Ranunculus sceleratus				+
Rorippa				+
Sedum sp.	+			
Physocarpus sp.			+	
Hedysarum sp.				?
Viola sp.			+	+
Decodon sp.	+		+ +	
Andromeda polifolia		+	+	+
Menyanthes trifoliata			+ +	+ +
Verbena sp.			?	
Bidens sp.	?			
Forest and woodland				
Abies grandis	+ +			
Abies sp.			?	+
Larix sp.	+		+	+
Pinus sp.			?	+
Pinus 5-needle type				+
Picea sp.	+		+	+
Tsuga sp.	+	+		
Metasequoia sp.	+			
Thuja sp.			+	+
Populus sp.			+	+

Oxyria digyna			+
Saxifraga oppositifolia			+
Dryas integrifolia			+
Empetrum nigrum			+
Ledum palustre			+
Cassiope tetragona	+		+
Andromeda polifolia	+		+
Chamaedaphne sp.	+		
Vaccinium sp.			+
Oxycoccus sp.			+
Wet lowlands and floodplain			
Populus sp.	+		+
Salix sp.	+		+
Myrica gale	+		+
Alnus incana	+	+	+
Chenopodium gigantospermum			+
Chenopodium sp.	+	+	
Claytonia sp.	+		+

Juglans eocineria	+		
Betula arboreal type	+	+	+
Rubus idaeus			+
Aralia sp.	+	?	
Ilex sp.			+
Viola sp.	+		+
Oxycoccus sp.			+
Cornus sp.			?
Teucrium sp.	+		
Sambucus sp.	+		
Open uplands			
Betula (low shrub type)	+		+
Papaver sp.	+		+
Cleome sp.	+ +		
Polanisia sp.	+		
Hedysarum sp.	+		?
Hypericum sp.	+		
Verbena sp.	?		

Source: Matthews (1987).

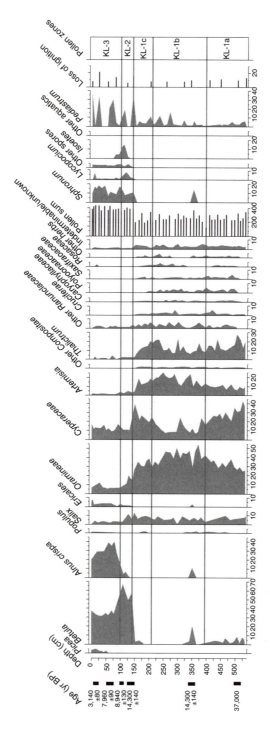

Figure 4.2

Percentage-pollen diagram for Kaiyak Lake (68°7′N; 161°25′W), northwest Alaska, showing presence of treeless plant communities dominated by herbaceous taxa (grasses, sedges, and other taxa) between 37000 and 14000 yr BP. *Source*: Anderson (1985).

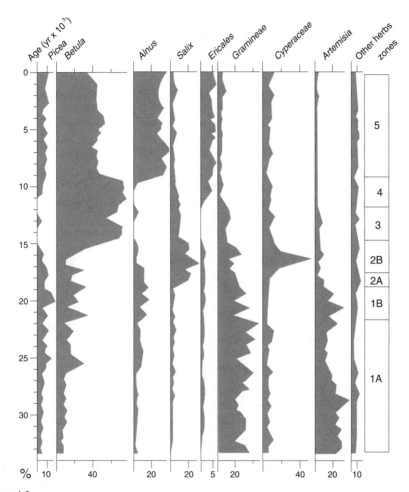

Figure 4.3
Summary percentage-pollen diagram for Hanging Lake, northern Yukon, showing presence of treeless
plant communities dominated by herbaceous taxa during maximum of latest glacial period, centred on
18000 yr BP. *Source*: Cwynar (1982).

habitats) through several glaciations. Nordal (1987) suggests, in the light of current
knowledge about rates of evolution (rapid) and possible long-distance dispersal of
seeds by birds or drifting ice, that survival on refugia "no longer seems to be a
biological necessity."

BIOCLIMATES

The primary source of energy for the growth of green plants is the sun; the annual
amount of solar radiation decreases from the equator to the poles (see chapter 3).

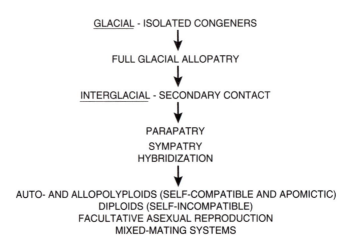

GLACIAL - ISOLATED CONGENERS

↓

FULL GLACIAL ALLOPATRY

↓

INTERGLACIAL - SECONDARY CONTACT

↓

PARAPATRY
SYMPATRY
HYBRIDIZATION

↓

AUTO- AND ALLOPOLYPLOIDS (SELF-COMPATIBLE AND APOMICTIC)
DIPLOIDS (SELF-INCOMPATIBLE)
FACULTATIVE ASEXUAL REPRODUCTION
MIXED-MATING SYSTEMS

Figure 4.4
Secondary-contact hypothesis of Stebbins (1984; 1985) and proliferation of varied arctic breeding systems, including polyploidy.

The simple latitudinal relationship varies chiefly through the effects of large oceans and major orographic features. A glance at a climatic or vegetational map of Canada illustrates clearly these primary causal relationships (Figure 4.5). For example, the treeline, however defined and mapped, is a latitudinal boundary that dips southward from the northwest around Hudson Bay and then resumes its latitudinal course across Nouveau Québec. The departure from a latitudinal course is a response to the effect of the Cordilleran axis on the westerlies and to the influence of Hudson Bay.

We learned in chapter 3 how the pattern of regional radiation balance is arranged spatially across northern Canada (Figure 3.2). A simple correlation with plant cover provides a useful starting-point in understanding factors that might control plant distribution, but researchers have not tested and verified these relationships. We shall begin at the highest latitudes and move southward, describing the climate and associated vegetation of each latitudinal segment.

In subarctic and arctic environments, leaf water potentials which govern leaf abscission (the process of cutting off of the leaf at the end of the growing season) and resulting growth efficiency in plants of temperate climatic regions rarely fall enough to stimulate abscission (Black and Bliss 1980; Woodward 1987). Although high-latitude climates are dry (<150 mm/yr precipitation), permafrost and low evaporation rates maintain high soil-moisture balances. The northern limits of subarctic and arctic plants appear to be controlled by thermal factors. Sakai (1978) has shown that the subarctic conifers (*Picea, Pinus, Larix*) have a variety of physiological devices to prevent formation of ice crystals in cell cytoplasm and, so can tolerate temperatures as low as − 70°C. Their northward extension into arctic regions appears to be limited by the length of the growing season, and a degree-day (above 5°C) of 600 is the approximate minimal value required by northern spruce and larch.

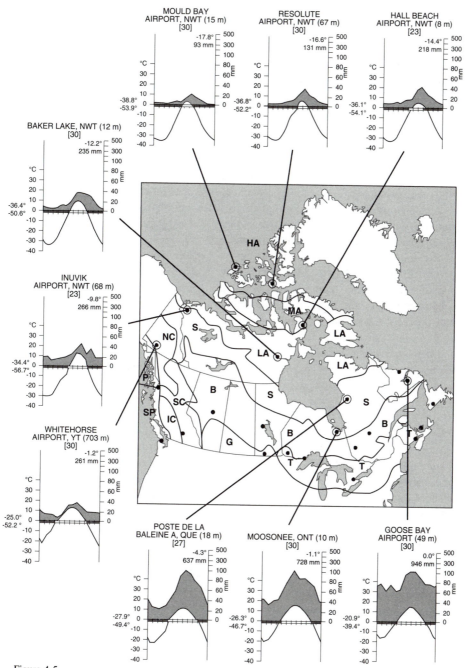

Figure 4.5

Limits of arctic, subarctic, and boreal zones. Bracketed numbers represent number of years of record.

COMMUNITY STRUCTURE AND COMPOSITION

Thorough analysis by Raup (1969) of the ecological tolerances of many High Arctic plants showed that many species occur as relatively isolated individuals found over a wide range of micro-topographic and edaphic conditions. Arctic plants in general have an unusually high proportion (76 per cent) of widespread, common species compared to other ecosystems: there is little interaction between plants (competition), and the evolutionary history of the flora has promoted widespread ecotypic differentiation. As Murray (1987) points out, "species that appear extremely tolerant likely consist of populations of ecotypes" so that the ecological amplitude of species varies widely from one region to another. Distinct, easily classified communities are rare, and systems of classifying and grouping vegetation are based primarily on physical features of the environment rather than floristic differences between communities (Raup 1969; French 1981; and many other authors).

The following précis of arctic-subarctic vegetation draws heavily on Polunin (1948), Bliss (1981), Edlund (1983a; 1983b), and Bliss and Svoboda (1984). Subdivision of Canada's northern vegetation into broadly latitudinal zones has been debated at length, and no attempt is made here to review that large, turgid literature. I shall describe three zones – the High Arctic, characterized by very sparse, discontinuous tundra on uplands; the low Arctic, with continuous shrub tundra on uplands; and the subarctic, a transitional region between the southern boundary of the shrub tundras of the low Arctic and the northern limit of the closed-crown, continuous conifer forests of the boreal zone.

High Arctic

The high arctic vegetation zone occupies the Arctic Archipelago, excluding the southern third of Baffin Island, with its southern boundary across southern Banks and Victoria islands. The vegetation of uplands is highly discontinuous, with total plant cover varying from 5 per cent to 50 per cent (Figure 4.6A), of which bryophytes and lichens make up a large proportion. Vascular plants are tufted or cushion perennials, often widely spaced, giving rise to the long-held assumption (Darwin 1988) that interspecific competition is unimportant. Common vascular plants are *Cerastium*, *Draba*, *Dryas integrifolia*, *Oxyria*, *Oxytropis astragalus*, *Papaver*, *Potentilla*, species of *Saxifraga*, and others, associated variably with sedge and grass species. Extremely xeric sites, such as rocky ridges, promote communities dominated by lichens. Local snowbed habitats are dominated by the heath *Cassiope tetragona* and the arctic willow, *Salix arctica*, assumes local abundance where snow accumulates on shallow upland depressions.

However, zonal groupings are of limited use because community composition and structure appear to be controlled by "maintenance of moist to wet soil surfaces throughout the summer," by soil texture, and by physical and chemical properties

Figure 4.6
Community structure and composition in High Arctic and low Arctic: (A) discontinuous herb tundra, Banks Island, NWT – individual swards dominated by *Dryas integrifolia* and *Salix arctica*; (B) shrub tundra, uplands near Horton River, NWT – continuous cover of dwarf willows, dwarf birch, and ericads.

of the parent rock (Edlund 1983a; Bliss, Svoboda, and Bliss 1984). Bliss, Svoboda, and Bliss (1984: 322) offer the useful concept of "the plant communities of the central and western Queen Elizabeth Islands as a mosaic pattern rather than as zones."

The compositional range, plant cover, and associated habitats of high arctic vegetation can be seen in surveys by Bliss and Svoboda (1984) and by Bliss, Svoboda, and Bliss (1984). They subdivide high arctic vegetation into polar desert and polar semi-desert categories and imply that the major controlling factors are microtopography, rock type, and moisture.

Polar Desert

In the central High Arctic, plant communities on Bathurst, Cornwallis, Devon, King Christian, Prince of Wales, and Somerset islands have been grouped into barren,

Table 4.2

Prominence values (cover X $\sqrt{\text{frequency}}$) and percentages of cover for polar barren community types

Species	Stand no.								
	1D	2D	3D	4D	5D	8C	9C	10C	17C
Saxifraga oppositifolia	–	2.1	0.2	–	0.5	2.2	8.4	0.8	–
Draba corymbosa	–	2.3	0.1	1.3	2.4	2.3	2.3	1.3	3.2
Draba subcapitata	–	–	–	T	–	0.2	–	0.2	–
Papaver radicatum	–	0.9	2.3	1.9	2.7	4.4	6.1	–	2.9
Minuartia rubella	–	0.2	0.2	0.3	0.8	0.4	0.2	T	–
Puccinellia angustata	0.6	0.8	0.2	0.3	0.7	0.1	1.1	0.3	0.4
Cerastium alpinum	–	–	–	–	–	–	–	–	–
Hypnum bambergeri	–	–	0.5	0.2	0.3	1.7	1.2	T	–
Tortula ruralis	–	0.1	–	–	T	0.4	T	T	–
Thamnolia subuliformis	–	0.1	–	–	–	6.9	1.0	0.1	0.4
Dermatocarpon hepaticum	–	–	1.7	–	–	2.3	0.5	–	–
Lecanora epibryon	–	–	–	–	–	0.1	0.3	–	–
Other species	3	3	6	4	3	4	4	3	3
Total species	4	10	13	10	10	15	14	11	7
Total vascular plant cover (%)	0.3	1.1	0.7	0.6	1.0	1.6	2.6	0.5	0.9
Total bryophyte cover (%)	T	T	0.1	0.3	T	0.9	1.0	T	0.5
Total lichen cover (%)	0.0	0.0	0.4	0.0	0.0	1.3	0.4	T	0.1
Total plant cover (%)	0.3	1.1	1.2	0.9	1.0	3.8	4.0	0.5	1.5
Litter cover (%)	0.1	0.6	0.6	0.3	0.4	0.2	0.8	T	0.3
Total bare soil and pebbles (%)	99.6	98.3	98.2	98.8	98.6	96.0	95.2	99.5	98.2

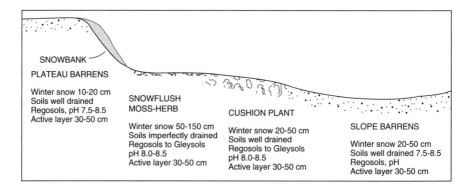

Figure 4.7

Polar desert landscape: generalized profile diagram showing spatial relations of plant communities. *Source*: Bliss, Svoboda, and Bliss (1984).

Table 4.2 (cont.)

Prominence values (cover X $\sqrt{\text{frequency}}$) and percentages of cover for polar barren community types

						Stand no.							
18C	19C	20C	24S	25S	26S	28S	29S	33PW	34PW	37B	38B	40KC	41KC
–	–	4.9	2.3	–	17.1	28.5	–	–	1.9	0.2	–	–	0.5
3.2	8.5	1.8	1.9	1.0	4.6	0.3	–	–	0.7	0.4	0.5	0.4	–
–	–	–	–	–	–	–	1.6	0.2	–	2.9	0.1	0.3	6.0
1.1	2.2	5.7	2.6	0.4	1.0	1.3	1.0	–	3.9	5.6	0.5	T	7.6
–	–	T	T	–	T	0.2	0.5	0.8	T	–	–	–	–
1.0	6.2	0.5	–	–	0.5	0.2	2.7	0.1	–	–	–	T	0.8
–	–	1.7	–	–	2.9	0.6	0.3	T	0.5	0.1	–	–	–
–	–	–	–	–	–	0.8	0.2	0.1	0.1	0.4	T	–	T
–	–	T	–	–	0.6	1.5	–	0.1	–	–	–	–	–
–	–	0.4	–	–	0.4	0.9	–	–	–	–	–	–	–
–	–	–	–	–	0.2	1.9	–	–	–	0.1	–	–	T
–	–	–	–	–	0.2	0.1	–	–	20.1	–	–	–	–
1	3	5	2	4	4	5	3	3	6	5	5	0	3
4	6	13	6	6	14	16	10	9	13	12	9	4	9
0.8	2.1	2.7	1.2	0.3	3.7	3.8	1.2	0.3	1.3	1.3	0.3	0.1	1.9
T	T	T	T	T	1.4	0.9	0.1	0.1	0.1	0.2	T	0.0	0.1
0.0	0.0	T	0.0	T	0.3	0.7	0.0	0.7	2.4	0.2	0.2	0.0	1.2
0.8	2.1	2.7	1.2	0.3	5.4	5.4	1.3	1.1	3.8	1.7	0.5	0.1	3.2
0.5	0.9	1.9	0.2	T	2.0	1.6	0.4	0.0	0.2	0.1	T	T	T
98.7	97.0	95.4	98.6	99.7	92.6	93.0	98.3	98.9	96.0	98.2	99.5	99.9	96.8

Source: Bliss, Svoboda, and Bliss (1984).

Note: Devon Island (D), Cornwallis Island (C), Somerset Island (S), Prince of Wales Island (PW), Bathurst Island (B), and King Christian Island (KC).

snowflush, and cushion plant communities (Figure 4.7). Polar barrens are characterized by snow cover (15–50 cm), shallow soils, and abundant surface permafrost phenomena; plant cover is low (0.3 to 4 per cent) with 4 to 15 species (average) per sampled stand (Table 4.2). Snowflush communities occupy 3 to 5 per cent of the landscape, with maximum species richness (30, mean 13) and total plant cover up to 60 per cent (Table 4.3). Cushion plant communities occupy sites with higher thermal conditions, early snowmelt, and well-drained soils; the preponderant cover consists of cushion perennials with *Salix arctica* and small amounts of mosses and lichens (Table 4.4).

Polar semi-deserts consist of "a complex of moss-graminoid, graminoid steppe and the cryptogam-herb group of communities of the high arctic" (Bliss and Svoboda 1984: 341). Controlling variables appear to be soil texture and moisture. Uplands generally have a discontinuous cover of lichens, mosses, and vascular plants. In the western islands, arctic graminoid species are dominant (*Alopecurus, Luzula, Phippsia,* and *Puccinellia*), associated with cushion perennials (*Cerastium, Draba,*

Table 4.3

Prominence values and cover percentages for snowflush plant communities

Species	11C	12C	13C	14C	15C	16C	21C	22C	23C	27S	30S	35PW
Saxifraga oppositifolia	9.9	3.9	0.1	7.1	0.8	0.5	33.0	47.5	43.2	14.5	–	6.7
Papaver radicatum	7.1	6.1	0.1	5.6	9.5	0.4	0.4	9.5	1.4	2.9	–	12.6
Draba corymbosa	5.2	5.0	–	–	0.3	0.2	12.8	1.2	2.2	–	–	2.2
Minuartia rubella	0.4	0.1	–	–	0.3	–	2.5	1.4	T	4.2	–	–
Saxifraga cernua	1.0	1.3	0.5	2.6	–	0.8	0.6	T	–	0.2	–	0.4
Puccinellia angustata	1.0	–	0.4	–	1.8	–	4.0	1.5	–	–	–	0.1
Cerastium alpinum	1.8	T	–	T	T	–	T	T	2.6	–	–	1.0
Stellaria crassipes	1.1	T	–	0.4	0.4	0.8	2.2	0.9	1.7	1.2	–	0.6
Alopecurus alpinus	–	0.2	–	41.5	–	54.2	–	11.9	0.4	–	1.9	–
Eriophorum triste	–	–	–	–	–	88.0	–	–	–	–	119.7	–
Salix arctica	–	–	–	–	–	–	–	–	–	–	66.9	–
Phippsia algida	0.6	8.8	–	11.6	T	T	–	–	10.3	0.6	–	0.1
Saxifraga caespitosa	–	1.4	–	1.6	–	0.2	0.8	3.2	T	4.1	–	4.5
Oxyria digyna	–	–	–	–	–	–	20.0	10.8	1.3	–	–	–
Orthothecium chryseum	–	10.6	T	186.1	2.0	130.9	–	–	–	–	57.8	13.3
Ditrichum flexicaule	–	3.8	13.8	94.5	2.0	1.1	–	–	12.5	–	22.8	–
Drepanocladus revulens	–	20.5	37.4	3.6	4.8	12.7	–	–	–	–	92.0	–
Bryum pseudotriquetrum	–	24.7	–	0.2	–	–	–	–	–	–	–	–
Hypnum bambergeri	12.3	8.0	–	2.2	–	–	9.3	–	–	14.1	22.7	50.8
Schistidium holmenianum	8.7	–	–	–	–	–	0.6	3.7	15.2	52.1	16.8	5.2
Thammolia subuliformis	8.4	0.5	4.5	2.4	T	0.7	0.3	2.7	5.7	0.4	0.1	28.3
Dermatocarpon hepaticum	5.8	10.2	6.2	120.3	6.0	32.9	8.2	79.1	99.1	53.5	–	138.4
Lecanora epibryum	1.0	0.9	1.3	4.9	T	37.8	–	22.5	9.9	0.5	–	21.0
Lecidea ramulosa	2.1	3.3	78.0	5.7	–	–	–	1.3	11.1	–	–	–
Other species	5	10	3	9	7	12	9	14	16	10	9	12.0
Total species	20	29	22	26	21	27	23	30	32	22	18	27.0
Total vascular plant cover (%)	3.3	4.0	0.3	10.6	2.9	18.1	9.5	10.5	17.2	4.2	21.1	12.0
Total bryophyte cover (%)	6.3	15.4	24.8	34.1	1.9	33.7	4.1	22.6	6.4	11.5	41.5	19.8
Total lichen cover (%)	2.8	2.4	12.0	18.5	0.9	6.6	1.7	12.6	14.1	6.5	T	20.1
Total plant cover (%)	12.4	21.8	37.1	63.2	5.7	58.4	15.3	45.7	37.7	22.2	62.6	51.9
Litter cover (%)	2.4	6.0	2.4	8.3	1.4	6.5	6.1	10.3	8.9	7.4	25.2	13.1
Total bare soil and pebbles (%)	85.2	72.2	60.5	28.5	92.9	35.1	78.6	44.0	53.4	80.4	12.2	35.0

Source: Bliss, Svoboda, and Bliss (1984).

and *Saxifraga*) and relatively rich moss and lichen communities, with total plant cover from 60 to 85 per cent. In the eastern High Arctic, "cushion plant communities ... dominate the polar semi-desert landscapes" (Bliss and Svoboda 1984: 341).

However, several systems of grouping plant communities in the High Arctic are very subjective, often missing biological and physical criteria at the same hierarchical level. Because many areas are difficult of access, distances are vast, and costs of field work are high, level of understanding and density of study sites remain low.

Lowland habitats in the High Arctic occur frequently, characterized by poor drainage and shallow (20 to 50 cm depth) peat soils, dominated by *Carex stans, C. membranacea*, and *Eriophorum*, forming continuous plant cover supporting a relatively varied fauna (Bliss 1975; 1981).

Table 4.4
Prominence values and cover percentages for cushion plant communities

Species	Stand no.					
	6C	7C	34S	32S	36PW	39B
Dryas integrifolia	11.7	7.6	27.7	1.2	–	–
Salis arctica	0.1	0.3	0.1	38.7	–	–
Saxifraga oppositifolia	T	2.5	8.4	3.1	28.4	19.3
Saxifraga caespitosa	–	–	–	0.1	–	3.0
Minuartia rubella	0.1	0.3	–	0.5	–	0.8
Draba corymbosa	0.1	0.8	T	–	0.2	0.9
Draba subcapitata	–	–	–	0.5	2.4	0.2
Papaver radicatum	–	T	–	–	2.1	4.6
Puccinellia angustata	–	–	–	T	0.5	1.3
Hypnum bambergeri	0.3	0.3	–	–	1.1	2.5
Tortula ruralis	–	–	–	–	0.3	1.1
Thamnolia subuliformis	2.8	6.1	–	3.6	5.0	5.1
Dermatocarpon hepaticum	–	5.5	–	0.9	6.6	8.0
Lecanora epibryon	0.1	2.4	22.1	7.1	4.3	0.9
Other species	3	4	2	6	11	9
Total species	11	14	7	16	21	21
Total vascular plant cover (%)	1.6	2.2	4.4	5.4	4.5	4.7
Total bryophyte cover (%)	0.1	0.1	0.0	1.3	0.5	1.1
Total lichen cover (%)	1.5	1.4	3.4	1.9	2.4	7.6
Total plant cover (%)	3.2	3.7	7.8	8.6	7.4	13.4
Total bare soil and pebbles (%)	96.8	96.3	92.2	91.4	92.6	86.6

Source: Bliss, Svoboda, and Bliss (1984).
Note: Cornwallis (C), Somerset (S), Prince of Wales (PW), and Bathurst (B) Islands

Low Arctic

This zone includes northern Yukon, the mainland part of the Northwest Territories, southern portions of Baffin, Banks, and Victoria islands, and northern Nouveau Québec. Upland sites, except rocky or sandy surfaces, support continuous cover of predominantly dwarf shrub vegetation (Figure 4.6B). Common plants are *Betula glandulosa, Salix* species, and species of the ericoid genera *Arctostaphylos, Ledum*, and *Vaccinium*. Tall shrubs of *Alnus* and *Salix* are common on alluvial sites. Poorly drained habitats support deep peats with permafrost and associated patterned ground, dominated by sedges, cottongrasses, shrubs, and mosses.

The Subarctic and Boreal Zone

The transcontinental boreal forest is the largest terrestrial ecosystem in Canada, occupying roughly 40 per cent of total land area. Its great importance ecologically and economically is based on its varied forest products and its vast array of wildlife habitats, including a large fraction of the global total of freshwater lakes and wetlands.

Figure 4.8
(A) Mackenzie delta, a unique wetland, extends northern limit of trees to 69°N; typical pattern of horsetails, sedges, grasses, and willows parallels drainage channels. (B) The arctic treeline is not a precise, linear feature; at Old Man Lake, NWT, white spruce (*Picea glauca*) extends along flanks of an esker several tens of kilometres beyond limit of continuous woodlands.

Forest communities are remarkably uniform, but major regional patterns display distinctive groupings of dominant trees and appear to be correlated with the presence of particular landforms.

A striking example is the large delta of the Mackenzie River, where Canada's longest river discharges into the Arctic Ocean. The long axis of the delta is roughly 180 km, and it reaches almost 100 km at its maximum width. The vast area of alluvial deposits, with its myriad of small, shallow lakes and intricately anastomosing tributary streams and major channels, supports a characteristic pattern of vegetation. The oldest surfaces, particularly in the southern two-thirds of the delta, support fine closed stands of white spruce (*Picea glauca*) – the most productive forests in the northern boreal forest. Younger surfaces – often distinct, long arcuate depositional zones – are dominated by pure stands of balsam poplar (*Populus balsamifera*), while in the most recent alluvium zonally arranged communities form a striking longitudinal pattern from river's edge landward of dense stands of *Equisetum*, swards of sedge and grass, and dense thickets dominated by a few species of willow (*Salix*) (Figure 4.8A).

In this and other large deltas that discharge into the Arctic Ocean, such as those of the Ob and the Lena rivers in Russia, tree species – for example, white spruce in Canada – extend farther north on these relatively deep, fine-grained soils than they do on adjacent uplands, with their extensive, often shallow morainic soils or bedrock. A recent investigation of the Arctic treeline between Hudson Bay and the Mackenzie delta by Timoney (1988) has shown the role of regional and site-specific geological factors in controlling the position of treeline (Figure 4.8B) and, more generally, the structure and composition of boreal woodlands. It can be demonstrated, by thorough mapping from air photographs supported by detailed field analyses, that the western treeline does not exactly follow the trends of regional climatic factors. Instead, it dips farther to the south on the eastern sector, where extensive outcrops of Shield bedrock and thin soils appear to have inhibited development of continuous boreal woodlands. In climatically equivalent parts of the northwestern sector, where sedimentary bedrock and earlier deglaciation have produced relatively deep soils, continuous woodlands occur. Also, black spruce (*Picea mariana*) is the dominant species in the eastern sector, between Great Bear Lake and Hudson Bay, while white spruce (*Picea glauca*) is more common to the west.

The apparent influence of bedrock on boreal and arctic plant communities, in both structure and composition, varies widely. It is expressed most strongly in areas, confined to the far northwest of Canada, that have not been covered by continental glaciers and that therefore lack an overburden of glacial debris. Striking examples can be seen in the Richardson Mountains of Yukon. In some localities a shale-sandstone bedrock yields coarse, slightly acidic soils and supports dense shrub tundra dominated by dwarf birch and heaths, with local stunted spruce trees. Elsewhere, limestone bedrock produces shallow, calcareous soils bearing sparse tundra of an entirely different floristic composition and community structure, dominated by Arctic avens, sedges, and various herbs (Ritchie 1984).

Another abiotic factor controlling spatial patterns and species composition of bo-real, subarctic, and arctic vegetation is permafrost (discussed more fully in chapter 6). The interrelationships of peat growth (often referred to as paludification of land-scapes), and subsequent aggradation of permafrost are not fully understood, but recent investigations in the northwest (e.g. by Zoltai and Tarnocai 1974; Ovenden 1982; and Gorham 1988) indicate several apparent trends of co-variation between com-munity patterns and permafrost.

FIRE

Sites in the boreal subarctic forest that have been free of fire for a century or more are generally dominated by conifers. The two spruces, *Picea mariana* and *P. glauca*, fir, *Abies balsamea*, and jackpine, *Pinus banksiana*, are the most abundant. Short-lived, rapid-growing trees with facility for both sexual and vegetative reproduction often appear in sites that are opened up immediately after fire; the chief species are white birch (*Betula papyrifera*) and poplar (*Populus* species) in these secondary communities, though the pattern of post-fire succession varies with the fire rotation cycle.

In dry, interior regions, such as the northern prairies and southern Northwest Territories, fire cycles of 80 to 150 years are common and the boreal forest is a mosaic of conifer and deciduous tree patches that reflect largely post-fire succession patterns. In eastern, maritime regions, fire cycles are longer, and southeast Labrador has the longest rotations, accompanied by the most humid ground cover and soils, with the result that the spruce-fir forests regenerate directly without an intermediate stage dominated by the short-lived deciduous taxa (Foster 1985).

Sirois (1988) has shown that exceptionally widespread fires in northern Quebec in the 1950s helped deforest the forest-tundra there. Almost 50 per cent of upland sites in the shrub sub-zone were deforested by fire, and regeneration of black spruce in these sites was minimal. An age structure analysis (Figure 4.9) shows that roughly four years elapsed after the fires (of the 1950s) before successful regeneration, which was maintained for only about a decade.

In addition, Payette et al. (1988) have demonstrated a south-to-north gradient of increasing fire-rotation period, reaching 1,460 years in a shrubby forest-tundra zone. They suggest that the very weak post-fire regeneration potential of black spruce populations in the northern part of the forest-tundra zone could affect the length of the period of fire rotation. These investigations suggest that treeline communities dominated by arboreal spruce are replaced, after fire, by Krummholtz forms, because of such changes in microhabitat as diminished snow cover. Sexual reproduction of Krummholtz forms is less than that of arboreal forms of black spruce, and subsequent fires can cause regional extinction of black spruce. Sirois (1988) has also shown that repeated natural fires at intervals shorter than the length of time to fecundity of regenerating black spruce speeds deforestation.

In the boreal-subarctic forests and woodlands of Nouveau Québec, few tree species participate in post-fire successions. The communities are essentially monospecific,

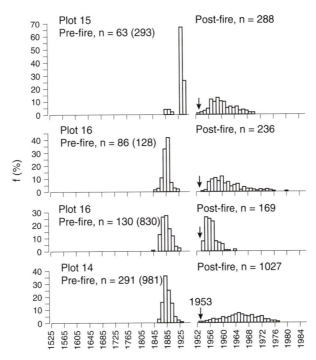

Figure 4.9

Age structure of pre- and post-fire black spruce (*Picea mariana*) populations in the boreal forest of northern Quebec. Numbers is parentheses
refer to entire pre- and post-fire populations. Analysis was made of three
plots in upland sites (the upper three) and one in a depression, in an area
burned in 1953. Pre-fire populations refer to living stems present in stand
and killed by fire in 1950s. *Source*: Sirois (1988).

dominated by *Picea mariana*. As a result, rapid-growing, actively vegetative trees
such as white birch and poplar that dominate other regions of the boreal forest
following fire, and the highly fire-resistant jackpine, are unavailable to fill this ecological niche. Communities remain open, and fruticose lichens regenerate rapidly
and create seedbed conditions inimical to regeneration of black spruce. As a result,
spruce regeneration declines rapidly after thirty years from the time of fire. In summary, despite fire's small role within the tundra zone, it has probably been influential
in forest-tundra areas, causing increases in low-arctic tundra vegetation at the expense
of northern boreal and forest-tundra vegetation, summarized diagrammatically by
Payette and Gagnon (1985) (Figure 4.10).

Paleoecological evidence from both the northern and southern limits of the boreal
region shows that these ecotones are sensitive to climatic change. For example, the
northern limit extended north beyond its present position during the summer thermal
maximum at the beginning of the Holocene period, centred on 9000 yr BP (Ritchie,

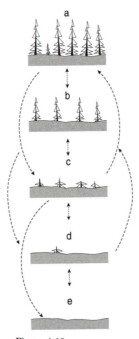

Figure 4.10
Reforestation-deforestation process in Quebec's northern forest-tundra
zone. "Fire recurrence in spruce forest stands (a) during the Holocene
cooling (downward arrows) caused a shift towards depauperate coniferous populations (b. c. d); subsequent fires culminated in total deforestation indicated by treeless communities (e). Short warming periods over
the last 1000 yr were responsible for local reforestation (upward arrows)." *Source*: Payette and Gagnon (1985).

Cwynar, and Spear 1983). By contrast, the southern limit, in the prairie provinces,
was reached relatively late in the Holocene (roughly 2,000 years ago), in response
to a slight climatic change toward cooler and more humid summers (Ritchie 1987).

These findings suggest that the boreal forest of the continent's interior will be
sensitive to any climatic warming that might follow the global increase in anthropogenic gases and a "greenhouse effect" on climate. Several commentators have
pointed out that if summer drought continues to increase over one or more decades,
major ecological changes can be expected in the boreal ecosystems. More frequent
fires (Harrington, Flannigan, and Van Wagner 1983), greater concentrations of certain
chemicals in water bodies (Schindler 1988), and release of acidic anions from
desiccating wetlands will aggravate acidification (Bayley and Schindler 1987; Schindler 1988). Preliminary predictions from climatic modelling identify the boreal regions of Canada's western interior as among the world's most sensitive to warming
responses to the "greenhouse" gases (Manabe and Wetherald 1986).

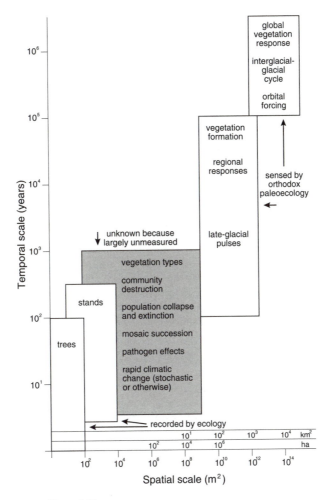

Figure 4.11
Spatio-temporal gap (shaded): critical biotic responses from environmental changes missed by ecology (lower left) and paleoecology (upper right). *Source*: Birks (1986).

Global warming may change the northern ecotone of the boreal forest. For example, implications for permafrost are outlined in chapter 12. Increased thicknesses of the active layer in mires would probably significantly lower regional water tables and increase thermokarst erosion.

Pathogenic insects, notably spruce budworm, may also affect the boreal forest and its productivity. Rate of spread and total area of infestation are related in complex ways to both anthropogenic influences and natural fire frequency (Blais 1983; Baskerville 1988).

The ecological problems in Canada's northern vegetation, particularly the boreal zone, which have just been described have brought into sharp focus a serious deficiency in our knowledge. Most available data on biotic responses to environmental change are at spatial and temporal scales either too small or too large to give us a sense of the potential effects, direct and indirect, of change. The gap is in the spatio-temporal scales that lie between those of traditional ecological investigations and those of paleoecology. The biotic responses in this "window" (Figure 4.11) have gone largely unmeasured. However, recent innovative approaches using an array of tools (pollen analysis, charcoal analysis, stable isotope measurement, fire-scar analysis, process modelling, and remode sensing) hold out promise for the future.

REFERENCES

Anderson, P. 1985. "Late Quaternary Vegetational Change in the Kotzebue Sound Area, Northwestern Alaska." *Quaternary Research* 24: 307–21.

Aleksandrova, V.D. 1988. *Vegetation of Soviet Polar Deserts: Studies in Polar Research.* Translated by D. Love. First published in Russian, 1983. New York: Cambridge University Press.

Barnosky, C.W., Anderson, P.M., and Bartlein, P.J. 1987. "The Northwestern U.S. during Deglaciation: Vegetational History and Paleoclimatic Implications." In W.F. Ruddiman and H.E. Wright, Jr., eds., *North America and Adjacent Oceans during the Last Deglaciation: The Geology of North America*, Vol. K-3, Boulder, Colo.: Geological Society of America. 289–321.

Baskerville, G.L. 1988. "Redevelopment of a Degrading Forest System." *Ambio* 17 no. 5: 314–22.

Bayley, S.E., and Schindler, D.W. 1987. "Sources of Aklalinity in Precambrian Shield Watershed Streams under Natural Conditions and after Fire or Acidification: Effects of Atmospheric Pollutants on Forests, Wetlands and Agriculture Ecosystems." In T.C. Hutchinson and M. Hayes, eds., *Proceedings of the Workshop*, 12–17 May 1985, Toronto, Ont. New York: Springer-Verlag, 531–47.

Beschel. R.L. 1970. "The Diversity of Tundra Vegetation." In W.A. Fuller and P.G. Kevan, eds. *Productivity and Conservation in Northern Circumpolar Lands*, International Union for Conservation of Nature and Natural Resources Publications, Morges, New Series 16: 85–92.

Birks, H.J.B. 1986. "Late Quaternary Biotic Changes in Terrestrial and Lacustrine Environments, with Particular Reference to North-west Europe." In B.E. Berglund. ed., *Handbook of Holocene Palaeoecology and Palaeohydrology*, New York: Wiley, 3–65.

Black, R.A., and Bliss, L.C. 1980. "Reproductive Ecology of *Picea Mariana* (Mill.) BSP, at Tree Line near Inuvik, Northwest Territories, Canada." *Ecological Monographs* 50: 331–54.

Blais, J.R. 1983. "Trends in the Frequency, Extent and Severity of Spruce Budworm Outbreaks in Eastern Canada." *Canadian Journal of Forest Research* 13: 539–47.

Bliss, L.C. 1975. "Tundra Grasslands, Herblands, and Shrublands and the Role of Herbivores." *Geoscience and Man* 10: 51–79.

– 1977. "General Summary, Truelove Lowland Ecosystem." In L.C. Bliss, ed., *Truelove Lowland – A High Arctic Ecosystem*, Edmonton: University of Alberta Press, 657–75.

– 1981. "North American and Scandinavian Tundras and Polar Deserts." In L.C. Bliss, O.W. Heal, and J.J. Moore, eds., *Tundra Ecosystems: A Comparative Analysis*, International Biological Program 25, Cambridge: Cambridge University Press, 8–24.

Bliss, L.C., and Svoboda, J. 1984. "Plant Communities and Plant Production in the Western Queen Elizabeth Islands." *Holarctic Ecology* 7: 325–44.

Bliss, L.C., Svoboda, J., and Bliss, D.I. 1984. "Polar Deserts, Their Plant Cover and Plant Production in the Canadian High Arctic." *Holarctic Ecology* 7: 305–24.

Britton, M.E. 1957. "Vegetation of the Arctic Tundra." In H.P. Hansen, ed., *Arctic Biology*, Corvallis: Oregon State University Press.

Chung, I.-C. 1984. *The Arctic and the Rockies as Seen by a Botanist Pictorial*. Seoul: Samhwa Printing Co.

Cwynar, L.C. 1982. "A Late-Quaternary Vegetation History from Hanging Lake, Northern Yukon." *Ecological Monographs* 52: 1–24.

Edlund, S.A. 1983a. "Bioclimatic Zonation in a High Arctic Region: Central Queen Elizabeth Islands." Current Research, Part A, Geological Survey of Canada, Paper 83–1A, 381–90.

– 1983b. "Reconnaissance Vegetation Studies on Western Victoria Island, Canadian Arctic Archipelago." Current Research, Part B, Geological Survey of Canada, Paper 83–1B, 75–81.

Foster, D.R. 1985. "Vegetation Development following Fire in *Picea Mariana* (Black Spruce)–*Plerozium* Forests of South-eastern Labrador, Canada." *Journal of Ecology* 73: 517–34.

French, D.D. 1981. "Multivariate Comparisons of IBP Tundra Biome Site Characteristics." In L.C. Bliss, O.W. Heal, and J.J. Moore, eds., *Tundra Ecosystems: A Comparative Analysis*, International Biological Program 25, Cambridge: Cambridge University Press, 124–39.

Funder, S., Abrahamsen, H., Bennike, O., and Feyling-Hanssen, R.W. 1985. "Forested Arctic: Evidence from North Greenland." *Geology* 13: 542–6.

Gorham, E. 1988. "Canada's Peatlands: Their Importance for the Global Carbon Cycle and Possible Effects of 'Greenhouse' Climatic Warming." *Transactions, Royal Society of Canada* 67.

Harrington, J.B., Flannigan, M.D., and Van Wagner, C.E. 1983. Information Report PI-X-25, Petawawa National Forest Institute, Canadian Forestry Service.

Lamb, Henry F., and Edwards, Mary E. 1988. "The Arctic." In B. Huntley and T. Webb III, eds., *Vegetation History*, New York: Kluwer Academic Publishers, 519–55.

Manabe, S. and Wetherald, R.T. 1986. "Reduction in Summer Soil Wetness Induced by an Increase in Atmospheric Carbon Dioxide." *Science* 236: 626–8.

Matthews, J.V., Jr. 1987. "Plant Macrofossils from the Neogene Beaufort Formation on Banks and Meighen islands, District of Franklin." In Current Research, Part A, Geological Survey of Canada, Paper 87–1A, 73–87.

Murray, D.F. 1987. "Breeding Systems in the Vascular Flora of Arctic North America." In K.M. Urbanska, ed., *Differentiation Patterns in Higher Plants*, London and New York: Academic Press, 239–62.

Nordal, I. 1987. "Tabula Rasa after All? Botanical Evidence for Ice-free Refugia in Scandinavia Reviewed." *Journal of Biogeography* 14: 377–88.

Ovenden, L.E. 1982. "Vegetation History of a Polygonal Peatland, Northern Yukon." *Boreas* 11: 209–24.

Payette, S., and Gagnon, R. 1985. "Late Holocene Deforestation and Tree Regeneration in the Forest-Tundra of Québec." *Nature* 313: 570–2.

Payette, S., Morneau, C., Sirois, L., and Desponts, M. 1988. "Recent Fire History of the Northern Québec Biomes." *Ecology* 70: 656–73.

Polunin, N. 1948. *Botany of the Canadian Eastern Arctic. III. Vegetation and Ecology.* Bulletin 104, National Museum of Canada, Ottawa.

Raup, H.M. 1969. "Studies of Vegetation in Northwest Greenland." In K.H. Greenridge, ed., *Essays in Plant Geography and Ecology*, Nova Scotia Museum, 45–62.

Ritchie, J. 1984. *Past and Present Vegetation of the Far Northwest of Canada.* Toronto: University of Toronto Press.

– 1987. *Postglacial Vegetation of Canada.* Cambridge: Cambridge University Press.

Ritchie, J.C., Cwynar, L.C., and Spear, R.W. 1983. "Evidence from Northwest Canada for an Early Holocene Milankovitch Thermal Maximum." *Nature* 305: 126–8.

Sakai, A. 1978. "Freezing Tolerance of Evergreen and Deciduous Broad-leaved Trees in Japan with Reference to Tree Regions." *Low Temperature Science* B36: 1–19.

Savile, D.B.O. 1972. *Arctic Adaptations in Plants.* Monograph 6, Canadian Department of Agriculture, Ottawa.

Schindler, D.W. 1988. *The Effects of Global Change on the Boreal Landscape.* Winnipeg: Freshwater Institute.

Sirois, L. 1988. "La déforestation subarctique du Québec. Une analyse écologique et démographique." PhD thesis, Université Laval, Quebec.

Stebbins, G.L. 1984. "Polyploidy and the Distribution of the Arctic-Alpine Flora: New Evidence and a New Approach." *Botanica Helvetica* 94: 1–13.

– 1985. "Polyploidy, Hybridization, and the Invasion of New Habitats." *Annals, Missouri Botanical Gardens* 72: 824–32.

Tieszen, L.L., ed. 1978. *Vegetation and Production Ecology of an Alaskan Arctic Tundra.* New York: Springer.

Timoney, K.P. 1988. "A Geobotanical Investigation of the Subarctic Forest-Tundra of the Northwest Territories." PhD thesis, University of Alberta, Edmonton.

Woodward, F.I. 1987. *Climate and Plant Distribution.* Cambridge: Cambridge University Press.

Zoltai, S.C., and C. Tarnocai. 1974. *Soils and Vegetation of Hummocky Terrain.* Environmental-Social Program, Northern Pipelines, Report 74–5.

CHAPTER FIVE

Northern Hydrology

MING-KO WOO

The magnitude and timing of hydrological processes vary spatially within the vast regions of northern Canada according to climate, geology, and vegetation. Yet these northern regions all experience a low level of solar radiation and tremendous seasonal contrast in radiation regime compared with southerly latitudes. With an extended period of radiative loss each year, winters are long and cold. This distinctive period of prolonged cold exerts an overwhelming influence on hydrological activities.

The main effect is that cold-climate phenomena such as snow, ice, and frost play a major part in the hydrological cycle. Thawing of ice and snow involves latent heat, and thus heat and water fluxes are closely linked in the north. Moreover, snow and ice exert a strong feedback on the receipt of radiation through high reflectivity, or albedo. This effect reduces the amount of radiation received in spring and early summer, and, as a consequence, cold conditions remain even longer. Another effect is retention of water in storage for a protracted period in the form of semi-permanent snowbanks, icings, and multi-annual lake ice cover. In addition, in the eastern Arctic, vestiges of the former glacial period remain, with present-day glaciers representing very long-term storage of water. A further effect relates to the short period each year when temperatures rise above freezing. Thus the activity of most hydrological processes is concentrated within a much shorter period than in temperate regions.

A combination of these climatic and hydrological characteristics influences not only vegetation growth and soil genesis but also overall evolution of the northern landscape. Moreover, snow, ice, and frozen ground determine storage and circulation of water in the arctic and subarctic regions of Canada.

In this chapter we consider snow, glacier, and permafrost hydrology. This is followed by a discussion of surface hydrological processes, lake hydrology, and stream flow.

SNOW HYDROLOGY

Snow is a major form of water storage. It withholds six to nine months of precipitation accumulation and releases it quickly, over several weeks. Because wind redistributes snow, the spatial pattern of water release is uneven. The mean duration of snow cover ranges from 180 days in the subarctic to over 270 days in the arctic islands (*Hydrological Atlas of Canada* 1978). As such, snow is an integral part of the northern landscape for most of the year. In this section we shall consider the properties, accumulation, and melting of snow, the movement and runoff of meltwater, and late-lying snowbanks.

Snow Properties

Extreme coldness distinguishes northern snow cover from that of temperate latitudes. Winter snow temperatures may drop to as low as $-15°C$ in the Arctic and occasionally below $-10°C$ in the subarctic.

Williams (1957) compared the snow cover of Aklavik in the subarctic (68°N) with that of Resolute in the High Arctic (75°N) and found that the latter is always denser and harder. This difference may be attributed to the absence of forest cover in the Arctic which enables wind scouring and redeposition to compact the snow.

In detail, local variation in snow depth, density, and hardness depends on the intensity of several processes. These include (1) accumulation from individual snowfall events, (2) redistribution and compaction through drifting, often forming a hardened layer called wind slab, (3) compaction caused by internal pressure of the snow, (4) formation of depth hoar because of vapour flux within the snow, (5) addition of rainwater, including freezing rain, in early winter or early spring and refreezing into a hard layer, and (6) occasional early winter snowmelt, followed by refreezing of meltwater to produce an ice layer. As a result of these processes, typical snow cover exhibits a combination of strata with distinctive properties (Figure 5.1).

Snow Accumulation

Many northern weather stations underestimate snowfall because conventional precipitation gauges tend to undercatch blowing snow (Goodison 1978) and many trace events are not recorded by the gauges. The amounts are significant. For example, Findlay (1969) analysed data for subarctic Quebec and suggested annual underestimation of 150 mm, while Woo et al. (1983) concluded that actual snow accumulation in the arctic islands is 130 to 300 per cent higher than that recorded by weather stations.

Vegetation in the subarctic acts to intercept snowfall. While some intercepted snow may be lost to sublimation, the remainder may be either blown off the vegetation or shed to underlying snow cover. In more exposed areas, wind drifting redeposits the snow on more sheltered terrain. These processes invariably make snow distribution move uneven.

In the High Arctic, where vegetation is minimal, topography largely controls snow distribution. At McMaster River basin, Cornwallis Island, for instance, snow surveys

Figure 5.1
Dye injected into cold snow cover near Resolute, NWT, showing development of flow fingers and horizontal deflection of flow paths as meltwater encounters changes in snow strata. *Photo*: P. Marsh.

show a consistent pattern of low snow accumulation on the hilltops, followed by flat areas and slopes, with gullies and valleys holding the deepest snow. Windswept areas generally have little snow, while topographical depressions gather the maximum amount at the end of winter.

Snowmelt

The energy available for snowmelt, Q_m, can be considered as:

$$Q_m = Q^* + Q_H + Q_{LE} + Q_R + Q_G,$$

where Q^* is net radiation, Q_H sensible heat flux, Q_{LE} latent heat flux, Q_R melt energy produced by rain on snow, and Q_G ground heat flux. Q_G is seldom important because the ground is colder than the snow at the time of melt.

Radiation melt is related to short- and long-wave receipt and loss at the snow surface:

$$Q^* = (1 - \alpha) K{\downarrow} + L{\downarrow} - L{\uparrow},$$

where α is snow albedo, $K{\downarrow}$ is incoming short-wave radiation, and $L{\downarrow}$ and $L{\uparrow}$ are incoming and outgoing long-wave radiation. When snowmelt begins in northern Canada, days are already long and short-wave radiation is a major consideration. High albedo noticeably reduces net receipt of radiation. As the melt season advances, albedo decreases, enhancing radiation snowmelt. Within and near subarctic and arctic settlements, the snow is usually dirtier and the albedo is lower than the surrounding areas (Drake 1981; Woo and Dubreil 1985). This locally accelerates melt.

Sublimation (a form of latent heat loss in terms of melt energy) depends on the dryness of air relative to the snow surface. In most parts of the Arctic and the subarctic, sublimation is not a serious water loss. There are exceptions, however. At Truelove Inlet, Devon Island, Ryden (1977) estimated sublimation loss at about 20 per cent of annual precipitation, and Ohmura (1982) showed that half of the heat gained by radiation at Expedition Fiord, Axel Heiberg Island, was spent by latent heat loss.

Air temperature commonly increases on clear days, making Q_H a major source of melt energy. Toward the end of the melt season, snow cover usually fragments into patches as thinner snowpacks ablate. Then, advection of sensible heat from bare ground quickens the melt further. Rain-on-snow melt is less important in the Arctic, where precipitation is low, but may be significant occasionally in the subarctic.

Meltwater Movement through Snow

When meltwater descends the snow cover, a melting front marks the abrupt transition from the dry zone to the zone saturated with meltwater. Downward advance of the wetting front is governed by the rate at which meltwater is supplied to the front, the requirement to fill the liquid storage of the dry snow, and the amount of refreezing needed to warm the snow to 0°C (Colbeck 1976).

When the wetting front encounters a sudden change in snow strata, water is ponded at the interface. Lateral flow is facilitated along the boundary until flow fingers develop (see Figure 5.1). These fingers penetrate to the lower stratum to form a network of flow routes (Marsh and Woo 1984). Refreezing of meltwater along this network produces ice lenses and ice columns, measuring several millimetres to several centimetres thick.

Some meltwater infiltrates the frozen soil. As this water freezes within the soil, latent heat is released and the soil temperature rises quickly. Figure 5.2 shows that the frozen soil at Resolute was warmed up rapidly in this manner between 2 and 6 July 1978 (as is shown by closeness of the soil isotherms). Infiltration ceases as soil pores are sealed by ice. Then, any meltwater percolating to the ground surface

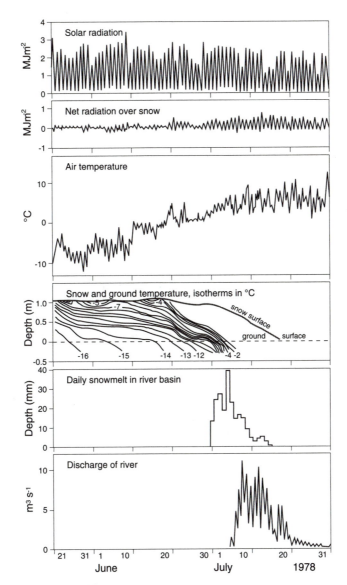

Figure 5.2

Generation of meltwater, McMaster River, near Resolute, NWT, May–
July 1978. Note delay of stream-flow and snowmelt responses after air
temperature has risen above freezing point and after radiation has reached
a positive balance. River has basin area of 32 km^2.

freezes to form a basal ice layer (Woo and Heron 1981), while the remaining melt-water yields runoff.

Meltwater Runoff

Uneven snow distribution means that part of the snow cover may be totally ablated while remaining snow patches are still sustaining runoff (Figure 5.3A). Thus runoff from a slope may move alternately across snow and bare ground. Runoff over the snow takes several forms. On the surface, unconcentrated flow includes spreading of water over snow (sheet flow) or downslope movement of water and snow mixture (slush flow). Sometimes water is concentrated into rills carved in the snow or may flow along tunnels and tubes developed within the snow cover. Water flow on bare areas takes the form of either surface or subsurface flow. Surface flow is particularly common in early spring because the frozen ground does not favour infiltration and much of the meltwater runs off as sheet flow on the slopes.

Streamflow initiation may be delayed for several days after snowmelt has commenced, because of retention of meltwater by deep snow. This is often the case in the Arctic, where much snow has accumulated in depressions and valleys. Typically, initial saturation of the valley snow cover is followed by water movement either within or over the valley snow (Woo and Sauriol 1980). Several processes then occur, including formation of channels in the snow, vertical incision and lateral shifting of channels in the snow (Figure 5.3B), formation and collapse of tunnels carved in deep snowdrifts, and ponding and subsequent release of water behind snow dams formed by drifts. Stream channels open up in segments along different parts of the valley. Until the channels coalesce into an integrated drainage network, water will not be conveyed to the outlet of the basins. In this case, streamflow response to snowmelt is much delayed (see Figure 5.2).

Late-lying Snowbanks

Large snowdrifts are common in the Arctic. They linger for weeks in summer, and some may not melt completely except in very warm years. These snowbanks differ from seasonal snow cover in several respects. First, because most of the snow is composed of basal ice, densities often reach 700 kg/m^3. Second, surfaces are often covered by wind-blown dust which both reduces the high albedo and increases surface roughness. Third, although areally not extensive, snowbanks keep the ground beneath frozen. This affects the pattern of water movement on the slope (Young and Lewkowicz 1988).

These snowbanks provide continuous runoff throughout the summer, sometimes maintaining patchy wetlands immediately downslope of the snowbank (Roulet and Woo 1986). In the polar deserts, such snow is a steady source of water to an otherwise arid environment. For some basins, snowbanks provide long-term storage of precipitation over a period of years. Runoff is withheld in cooler years but is released by

A

B

Figure 5.3
(A) Fragmentation of snow cover in small arctic basin produces combination of bare and snow-covered surfaces. At centre is lake (area 0.1 km^2) impounded by snow dam which prevented drainage of meltwater in spring 1978. *Photo*: R. Heron. (B) Stream carving channel in snow that has infilled an arctic valley.

more intense melting in warmer years. In this regard, the hydrology of snowbanks represents a transition from seasonal cover to glaciers.

GLACIER HYDROLOGY

Many northern glaciers blanket the landscape to form ice caps or ice sheets, sometimes sending tongues along their peripheries to produce valley glaciers (Figure 5.4A). Their distinct hydrological characteristics can be summarized conveniently by examining glacier mass balance, associated runoff, and complicating factor – the presence of glacier-dammed lakes.

Glacier Mass Balance

A glacier mass balance is derived from annual accumulation, ablation, and change in mass of ice held in storage. The difference between accumulation and ablation, if positive, indicates a net gain of snow, ice, and water to the glacier. A negative balance indicates the reverse.

Glacier accumulation results from snowfall, avalanche, drifting of snow onto the glacier, and/or formation of superimposed ice. Compacted snow, known as firn, has a density that exceeds 500 kg/m^3. It turns into glacier ice when most of the voids between snow crystals are sealed. Superimposed ice, analogous to basal ice in snow cover, is formed when snow meltwater refreezes on contact with the cold glacier surface (Koerner 1970). Glacier ablation is caused primarily by melt (Müller and Keeler 1969), but ice may be lost to sublimation, erosion by running water on and in the ice, and by calving of ice blocks along the glacier margins (Figure 5.4B).

Figure 5.5 illustrates annual mass balance of the Devon Island Ice Cap as determined along two transects (Koerner 1970). Higher elevations experience net gain in mass, and lower elevations show a loss. Gains or losses vary from year to year. When the net balances for various elevation zones are weighted by the area occupied by the respective elevation bands, overall mass balance can be computed. For the Devon Island Ice Cap, data show a positive balance for 1963–65 and a negative one for 1961–62 and 1965–66. Considering that mean annual precipitation for the ice cap is about 300 mm (Koerner 1979), net change in storage from year to year may be considerable.

The trend toward net accumulation at higher zones and net ablation at lower areas is compensated for by glacier flow which transfers mass from the zone of accumulation to the zone of ablation. Normally, glacier flow rates are in the order of 0.1 m per day, but occasionally a glacier surges at a speed exceeding 15 m per day. In this case, there will be a large loss of mass as ice blocks break off at the glacier front (e.g. Stanley 1969).

Runoff from Glaciers

Glacier runoff responds mainly to the energy budget that causes the snow, firn, and

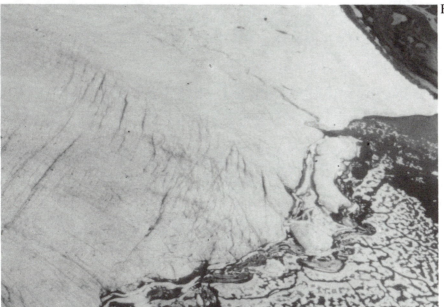

Figure 5.4
Northern glaciers: (A) Ellesmere Ice Cap sending glacier tongues down valley at head of Mackinson Inlet, still covered by sea ice: (B) crevasses and supraglacial drainage channels on tongue of Ellesmere Ice Cap – ice calves from ice front to produce iceberg.

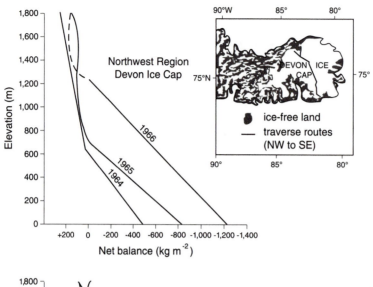

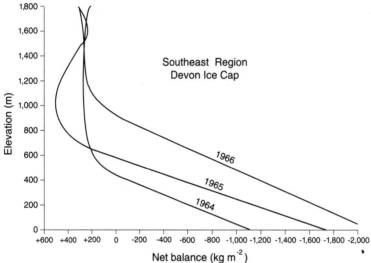

Figure 5.5

Annual mass balance (1964–66) along two transects of Devon Ice Cap, showing gain of mass at high elevations and loss at lower altitudes; inset shows ice caps on eastern Devon Island. *Source*: Koemer (1970).

glacier ice to melt. Rain falling on the glacier generates much less runoff. Heat supply on rainy or cloudy days is often less than on clear days, and melt is retarded accordingly. Glacier runoff tends therefore to be low during cool years and higher in warm, dry years.

Runoff occurs on the glacier surface (supraglacial), within the glacier (englacial), or along the base of the glacier (subglacial). Several distinguishing characteristics can be identified. First, within the snow and firn zone of arctic glaciers, slush flow (a mixture of snow and water) is common (Adams 1966). Sometimes snow impounds supraglacial lakes which drain rapidly when the dams are breached. Second, the flow of water melts and erodes the snow and ice that form streambeds and thus modifies channel morphology. Third, glacier movement and development of crevasses gradually alter drainage morphology on or in the glacier (see Figure 5.4B). Finally, because ice is impermeable, there is negligible seepage loss. However, fissures and moulins (or vertical shafts) allow supraglacial and englacial passages to link and produce a three-dimensional drainage network.

Runoff associated with arctic glaciers ceases as winter commences and begins again with the melt season. Müller and Iken (1973) provide examples of hydrographs for a supraglacial stream (drainage area about 0.25 km^2) and for drainage from englacial and subglacial sources (Figure 5.6). The supraglacial stream formed on 15 June, and its hydrograph showed prominent diurnal cycles in response to daily melt, with peak flows in early afternoons and low flows in early mornings. Englacial and subglacial flows followed similar patterns (see Figure 5.6), but these cycles were interrupted during cold spells when melt was reduced. During these cold spells supraglacial flow diminished and englacial and subglacial discharge ceased. Rainfall increased runoff, but, depending on temperature, rain turned into snow at higher elevations and the snow-covered area generated little flow.

Glacier-Dammed Lakes

At glacier margins, abutment of ice against land frequently dams up lakes of various sizes (Maag 1969). Small lakes built up by runoff of meltwater may drain rapidly as the ice dam is breached by overflow or by subglacial drainage. Where large lakes have been impounded, opening of englacial or subglacial passages through the dam can release large volumes of lake water in a short time, until such passages are again closed by ice. This episodic bursting of ice-dammed lakes may produce dramatic floods, called jökulhlaups.

One example of a jökulhlaup has been described by Blachut and McCann (1981) from Ellesmere Island, where two ice-dammed lakes along the western margin of the Ellesmere Ice Cap formed as ice blocked the exit of a structural depression. An upper lake drained via a surface stream to lower lakes. The water level rose through the summer, as snowmelt and inflow added water to the lower lake (Figure 5.7). On 13 August, lake level dropped suddenly when a conduit was opened in the ice dam. The several manual discharge measurement made during the glacier outburst flood compare reasonably well with the rate of lake volume loss. Maximum discharge was estimated to be 73 m^3/s, and total volume of the flood was 3.4×10^7 m^3. It appears that jökulhlaups from this lake occur regularly: such floods were observed three years in a row. For many other ice-dammed lakes, however, frequency and magnitude of jökulhlaups may differ.

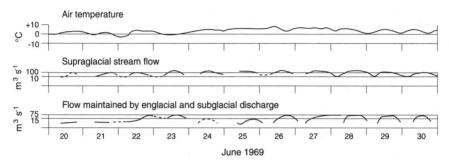

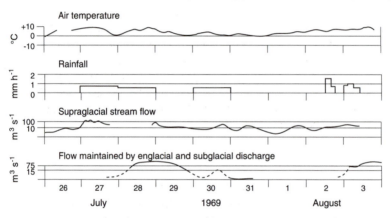

Figure 5.6
Runoff from an arctic glacier (daily variations) at two sites, White Glacier, Axel Heiberg Island, NWT; discharges are plotted on logarithmic scale. *Source*: Müller and Iken (1973).

PERMAFROST HYDROLOGY

Permafrost hydrology studies distribution, movement, and storage of water as it is directly or indirectly influenced by the presence of perennially frozen ground (Woo 1986). Analysis of the hydrology of the north must consider both heat and water because freeze-thaw activities are also involved (see chapter 6).

Hydrological Characteristics of Frozen Soil

The hydrological behaviour of seasonal and perennially frozen materials differs mainly in the length of time that the material is below 0°C; otherwise they share similar properties. Sub-zero temperatures may cause freezing, but not all water freezes at 0°C. For instance, water with impurities freezes at a lower ice-nucleation tem-

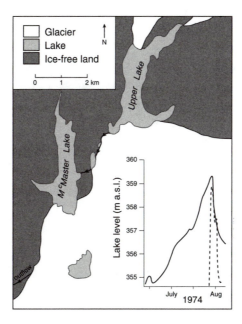

Figure 5.7
"McMaster Lake," Ellesmere Island, NWT; inset: water level July–August 1974 and computed outflow (dashed line) during jökulhlaup. *Source*: Blachut and McCann (1981).

perature than does water held in the interstices of soil. Within soil, water that is strongly bonded, or adsorbed, to soil particles may not freeze at temperatures even below $-50°C$. The finer the particles, the larger is the surface area available for the absorbed water film. Thus the amount of unfrozen water in the soil will decrease with falling temperature.

During freezing, water moves from warmer to colder zones in the soil. Winter freezing progresses downward, and water in the warmer soil moves upward toward the freezing front. Similarly, permafrost is colder than soil in the active layer during the thawed season, and water also migrates toward the permafrost table. The results are significant redistributions of moisture as soil freezes.

Conversion of water into ice may seal off flow passages in the soil, considerably reducing hydraulic conductivity. When ice content is high, soil will become less permeable. Although migration of water can occur at temperatures much below $0°C$, frozen soils may be considered impermeable at the scale of most hydrological investigations.

Formation and degradation of ground ice are discussed in more detail in chapter 6. From the hydrologic viewpoint, these processes can radically change the thermal property of the soil because the thermal conductivity of ice is about ten times larger than that of water. The higher the ice content, the more notable is the thermal effect

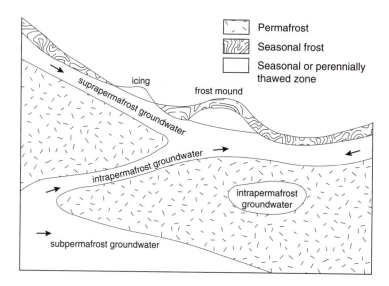

Figure 5.8
Occurrence of groundwater in permafrost regions.

of ice growth and decay. Thus, while heat flux alters hydrological characteristics of soil, water and ice content strongly affects the soil temperature regime.

Subsurface Hydrological Processes

Permafrost is not completely impermeable (van Everdingen 1987), but its permeability decreases tremendously relative to the same soil in a thawed state. Permafrost behaves as an aquiclude, limiting groundwater storage and movement. Depending on whether water is held in the thawed zones above, within, or beneath the permafrost, it exists as suprapermafrost, intrapermafrost, or subpermafrost groundwater, respectively (Tolstikhin and Tolstikhin 1976). Suprapermafrost groundwater is found either in the active layer or in isolated taliks below some river and lake beds. The storage capacity of its aquifer changes during the year because the active layer undergoes freeze and thaw (Woo and Steer 1983). Intrapermafrost groundwater occurs in taliks kept unfrozen by the flow of warm water or in transient taliks that have not yet chilled below 0°C (Figure 5.8). Subpermafrost groundwater seldom reaches the surface in continuous permafrost areas because few interconnecting flow passages cross through the very thick permafrost.

Discharge of groundwater to the surface produces springs, seepage through river and lake beds, or saturation overland flow. Discharge of suprapermafrost groundwater is seasonal and ceases when water stored in the active layer is frozen or depleted. In contrast, subpermafrost groundwater can sustain perennial flow, unless conduits to the surface are sealed by freezing. During winter, water discharged by subsurface

Figure 5.9
Icings: (A) Babbage River, northern Yukon: large icings form each year, 4–5 m thick and several km^2 in extent; (B) Engineer Creek, Yukon: Dempster Highway crossing ripped by freshet because culverts were blocked by icings, July 1983.

Figure 5.10
Measuring surface flow: (A) north-central Banks Island, NWT, using a portable weir and recording device constructed by A.G. Lewkowicz; (B) Vendom Fiord, Ellesmere Island, NWT, measuring surface-water level at a McMaster University experimental site.

flow freezes on reaching the ground surface, producing sheet-like masses of layered ice, called icings (Carey 1973). In the subarctic, icings (Figure 5.9A) can block entire sections of a river, leaving little room to convey snowmelt freshet in the spring. A highway crossing the river at such a point can be severely damaged by overflow (Figure 5.9B). Extensive icing is a steady source of water for local runoff long after snow has been depleted in summer.

SURFACE HYDROLOGICAL PROCESSES

Other than snow accumulation, most surface hydrological processes come to a halt in winter. Spring comes late in northern Canada, when more hours of daylight provide ample energy to melt snow rapidly, releasing months of accumulated snow within several weeks. As meltwater is released from the snow, some will infiltrate into the frozen soil (Marsh 1988), as noted earlier. Much water is discharged as flow either through snow or as surface runoff (e.g. Lewkowicz and French 1982) (figure 5.10A).

Before surface flow begins, the many depressions created by microtopography have to be filled. Then water flows over the frozen ground in sheets or rills (Figure 5.10B). Surface runoff follows pronounced diurnal cycles, reflecting daily variation of snowmelt contribution. The abundance of water at the surface, combined with large amounts of energy available, enables high evaporation in the spring.

As summer advances, surface flow declines, because thawing of the active layer provides a thicker zone where suprapermafrost groundwater can be stored, and the water table drops below the surface. Also, most of the snow has been depleted and summer rainfall is the only major source of water-supply. Finally, evaporation and lateral flow continue to withdraw water from the active layer, leaving far less water to sustain surface flow than in the spring. Evaporation also decreases in the Arctic during summer as the surface dries out, while in the subarctic, transpiration of vascular plants speeds evaporation.

Surface flow occurs where the water table intersects the ground surface. Given that most slopes are topographically irregular, depressions may intersect the shallow water table, and these depressions are local zones with surface runoff. On different segments of a slope surface and subsurface flows can therefore take place together or alternately, depending on whether water emerges or submerges in the active layer (Figure 5.11).

A shallow, seasonally thawed zone cannot retain much meltwater or rainwater input. Water table usually rises rapidly, and the zone with surface flow will expand quickly once the water table is at the surface. Since surface flow is orders of magnitude faster than subsurface flow, water is delivered quickly to the lower slopes or stream-banks. Flow recession is also rapid once the water table falls below the slope surface. It is mainly because of this fast response of surface flow to both snowmelt and rainfall that water is delivered quickly to rivers.

LAKE HYDROLOGY

Northern Canada has numerous lakes, many of them concentrated on the Canadian Shield. Low air temperatures favour an extended duration of ice cover, and most hydrological activities are confined to the season with open water on the lakes. The presence of permafrost beneath many arctic lakes may isolate them from the regional groundwater system. However, large bodies of water commonly impede permafrost development, and a talik is often found beneath them.

Ice Cover

Three broad categories of ice are found on northern lakes. First, clear, transparent ice, known also as "black ice" because one can see through it to the darker lake bottom, is formed by the freezing of water under tranquil conditions. Second, white ice, which appears white because it reflects light, is caused by incorporation of snow or ice particles (frazil ice) into the ice cover. In some arctic lakes, remnants of ice from previous years can remain to produce – third – multi-year ice that shows evidence of partial summer thawing and fracturing, followed by refreezing.

In northern Canada, ice formation is facilitated by large and rapid heat loss and the low level of solar radiation receipt. Ice grows mainly through heat conduction from the lake water to the cold atmosphere. This may be minimized if early snowfall produces slush on the lake surface. If an insulating snow cover is subsequently

Figure 5.11
Patchy surface and flow (darker zones) on an arctic slope near Resolute, Cornwallis Island, NWT, depending on emergence of suprapermafrost water table at ground surface.

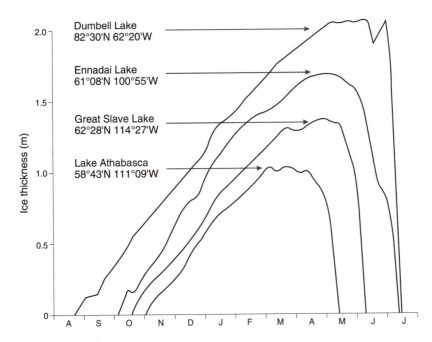

Figure 5.12
Seasonal growth and decay (August–July) of ice on subarctic and arctic lakes, showing poleward increase in maximum ice thickness and duration of ice cover.

established over the ice, conductive heat loss is definitely retarded. During winter, the ice surface may be flooded by inflow from the shore area or by upward seepage of lake water through large cracks in the ice. Flood water will slush the snow on the ice and subsequently freeze to produce a new white ice layer. This mechanism is more common in the subarctic.

The seasonal ice regime of several northern lakes along approximately 100°W is shown in Figure 5.12. Ice-growth season and maximum ice thickness tend to increase poleward. In the subarctic, ice grows to 1.0–1.5 m, while in the arctic islands maximum thickness can reach 2.7 m (Billelo 1980). Ice decay is usually insignificant during the northern winter and seldom begins until either snowmelt has depleted most of the snow on the ice or snow meltwater from the shore has reached the lake. Ice decay involves both ice melt and mechanical disintegration; the former can be further partitioned into surface, bottom, internal, and lateral melt.

The energy flux involved in surface melt is similar to melting of snow cover, except that less short-wave radiation is consumed because part of this radiation penetrates into the ice to cause internal melt. Where the ice is flooded, the water is kept warm by radiative heating, and the warm water will further melt surface ice. Bottom melt is produced by contact of lake water with the base of the ice. Lateral melt occurs where fractured ice is in contact with warm lake water.

Fracturing may be initiated when the addition of meltwater from the shores raises the lake level and arches the ice. Once fractured ice floats free on the lake, it is subject to fragmentation and attrition. This process is facilitated by internal melt which etches microscale channels along individual ice crystals to produce a loose agglomeration of elongated crystals called "candle" ice. Such ice crumbles easily when attacked by wave action and when driven against the shore or against other ice floes. Ice decay eventually destroys the ice cover unless the summer is brief and cold. Then the residual ice will be incorporated into the new ice formed in the fall.

The Lake and Its Drainage Basin

The exchange of flow between the lake and its basin is minimal during the long winters, when a lake's ice cover inhibits evaporation and when snow accumulation and ground freezing reduce slope runoff. Only in the subarctic might there be sub-surface flow to the lakes, but the magnitude is low. For example, in the Knob Lake basin near Schefferville, Quebec, FitzGibbon and Dunne (1981) estimated the rate to be 0.19 $m^3/s/km^2$. Lake outflow may continue throughout winter in the subarctic but is rare for small arctic lakes. For larger lakes such as Lake Hazen, Ellesmere Island (82°N), winter discharge can produce icing mounds (Harington 1960; 33).

Snowmelt produces a large influx of water to the lakes from their basins. However, increase of lake storage often lags behind snowmelt on the basin slopes, because of retention of considerable meltwater within the snow cover and in the microtopographical depressions on the land surface (FitzGibbon and Dunne 1981).

The outlets of many small lakes are blocked by thick snowdrifts accumulated in winter. As lake storage increases, water level rises until the snow dam is breached. This usually yields the peak annual outflow, accompanied by rapid depletion of lake storage accumulated during melt (FitzGibbon and Dunne 1981; Woo, Heron, and Steer 1981). Afterward, ice decay enlarges open water areas on the lake where evaporation is effective. Then the slopes gradually become free of snow, exposed ground thaws, and surface runoff diminishes as meltwater supply declines and evaporation increases.

STREAM FLOW

Runoff from small basins in northern Canada is often influenced by either snowmelt or glacier ablation. As basin size increases, however, storage dampens stream-flow response to water-supply, and flow tends to be more influenced by a range of processes operating at different intensities, through different times, over various parts of the basin. The resultant flow pattern reflects regional rather than local controls of runoff. In this discussion, small basins are considered to be in the order of 10^2 km^2 or less, and medium to large basins between 10^2 and 10^5 km^2.

Regimes of Small Basins

Most northern rivers do not provide a sufficiently long record to allow meaningful

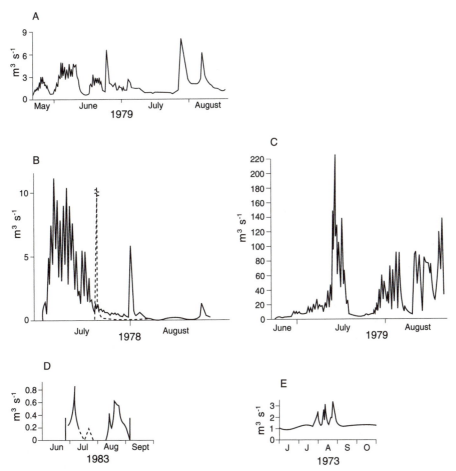

Figure 5.13
Hydrographs illustrating runoff regimes of small subarctic and arctic rivers: (A) subarctic nival regime;
(B) arctic nival regime (dashed line shows lake outflow modified by the breaching of a snow dam); (C)
proglacial regime; (D) wetland regime; and (E) spring-fed regime.

determination of average flow regimes. I use the term *regime* loosely to describe
flow characteristics revealed by selected samples of available data.

Nival regimes are dominated by snowmelt floods in spring. Church (1974) distinguished between subarctic and arctic nival regimes. In the subarctic, streams may
or may not have winter runoff, depending on the supply of groundwater. In spring,
snowmelt produces high runoff which typically shows diurnal cycles in response to
melt (Figure 5.13A). High flows may be enhanced by the breaking up of the river
ice cover to form ice jams (Gerard 1979). Formation of ice jams causes upstream
water levels to rise, and failure of the jams creates high velocities in unexpected
places downstream. As summer evaporation increases, runoff declines, unless rainstorms replenish the water loss (Onesti and Walti 1983). The arctic nival regime

shows one major flood period in spring, followed by rapid recession to base flow, interrupted occasionally by rainstorm-generated peaks (Figure 5.13B). Winter flow is absent because the suprapermafrost groundwater reservoir is too limited to maintain flow (Woo 1983).

Proglacial regimes tend to have spring snowmelt runoff superseded by peak flow from glacier melt (Figure 5.13C). Glacier melt runoff responds primarily to summer energy supply, at a time when most nival regime streams have little remaining snow to sustain runoff. During cool years, there may be little distinction in flow pattern between a nival and a proglacial stream if late-lying snowbanks can maintain high flow in the former type of basin through the summer (Marsh and Woo 1984). For some glacierized basins with ice-dammed lakes, jökulhlaups can produce high flows when the lakes drain.

Wetland-regime streams are fed by poorly drained areas which normally have considerable water-retention capacity to attenuate high flows. This does not happen in spring when the wetland is frozen and has little ability to absorb the snow meltwater (Woo 1988). A pronounced spring freshet is generated (Brown, Dingman, and Lewellin 1968), though often not as severe as the spring floods of nival regime streams (Figure 5.13D). Only in summer, when the active layer has thawed, will the wetland reduce rainfall-generated floods and sustain moderate recession flow.

Spring-fed regimes were recognized by Craig and McCart (1975). Often associated with carbonate rocks in the discontinuous permafrost zone, such streams have relatively stable discharges because the main source is groundwater (Figure 5.13E). One study near Great Bear Lake by van Everdingen (1981) showed tht subsurface flow can account for an equivalent of 15 per cent of annual precipitation. Springs continue to flow in winter, and snowmelt runoff usually produces peak flows.

Floods are of particular concern to northern development and can disrupt river and road transport, inundate settlements along river banks, and alter riverine habitats. Church (1988) classified floods in cold regions under three categories. First, hydrometeorological floods may be caused by snowmelt, rainstorms, and mixed rain-on-snow events. Second, a group of floods is created by channel blockage by snow and ice, including river ice and snow in channels, icings, and glacier ice (to generate jökulhlaups). Third, azonal floods, such as those resulting from landslide failures, can be found in the north.

Regional Stream Flow

It is generally accepted that the mean flow of a basin increases logarithmically with drainage basin area, but this relationship varies regionally. In northern Canada each physical region shows a distinctive runoff regime, as evidenced by the percentage distribution of runoff for different months (Figure 5.14). The Cordilleran region, receiving heavy precipitation, yields high runoff per unit area. Glacierized mountainous basins there have the largest unit-area annual flow. Late summer is the period of maximum runoff (e.g. White River; see Figure 5.14A). Glacier melt effect is less

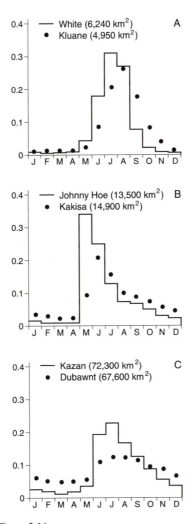

Figure 5.14
Monthly distribution of runoff for rivers in three physiographic regions, with and without large lake immediately upstream of gauging site. Dots represent data for stations lying below major lakes. (A) Glacierized basins in Cordilleran region; (B) northern plains basins; and (C) continuous permafrost basins in the shield.

apparent where glacier coverage is small, becoming indistinguishable from non-glacierized mountain basins. There, peak flow occurs in May or June, in response to snowmelt. The plateau basins have the least runoff in the Cordilleran region, and peak flow is concentrated within a shorter snowmelt season than in mountainous basins.

Rivers on the northern plains yield the lowest flow per unit area, probably because abundant wetlands enhance evaporation. Flow regime is similar to that of the plateau rivers, but the peak arrives earlier because of lower altitudes (e.g. Johnny Hoe River; see Figure 5.14B). Temperatures are lower in the Canadian Shield than in the northern plains, with most precipitation falling as snow. Runoff per unit area for rivers in the discontinuous permafrost zone of the Shield is comparable to that for northern plains rivers; but rivers in continuous permafrost areas have higher runoff because of lower evaporation losses (Figure 5.14C).

REFERENCES

Adams, W.P. 1966. *Ablation and Runoff on the White Glacier, Axel Heiberg Island, Canadian Arctic Archipelago.* Axel Heiberg Island Research Reports Glaciology No. 1, McGill University, Montreal.

Billelo, M.A. 1980. *Maximum Thickness and Subsequent Delay of Lake, River and Fast Sea Ice in Canada and Alaska.* US Army, Cold Regions Research and Engineering Laboratory Report 80–6.

Blachut, S.P., and McCann, S.B. 1981. "The Behavior of a Polar Ice-dammed Lake, Ellesmere Island, Canada." *Arctic and Alpine Research* 13: 63–74.

Brown, J., Dingman, S., and Lewellen, R.I. 1986. *Hydrology of a Drainage Basin on the Alaskan Coastal Plain.* US Army, Cold Regions Research and Engineering Laboratory Research Report 240.

Carey, K.L. 1973. *Icings Developed from Surface Water and Ground Water.* US Army, Cold Regions Science and Engineering Monograph III D-3.

Church, M. 1974. "Hydrology and Permafrost with Reference to Northern North America." In *Proceedings, Workshop Seminar on Permafrost Hydrology*, Canadian National Committee, International Hydrological Decade (IHD), Ottawa, 7–20.

– 1988. "Floods in Cold Climates." In V.R. Baker, R.C. Kochel, and P.C. Patton eds., *Flood Geomorphology*, New York: John Wiley & Sons, 205–29.

Colbeck, S.C. 1976. "An Analysis of Water Flow in Dry Snow." *Water Resources Research* 12: 523–7.

Craig, P.C., and McCart, P.J. 1975. "Classification of Stream Types in Beaufort Sea Drainages between Prudhoe Bay, Alaska, and the Mackenzie Delta, N.W.T., Canada." *Arctic and Alpine Research* 7: 183–98.

Drake, J.J. 1981. "The Effects of Surface Dust on Snowmelt Rates." *Arctic and Alpine Research* 13: 219–23.

Findlay, B.F. 1969. "Precipitation in Northern Quebec and Labrador: An Evaluation of Measurement Technique." *Arctic* 22: 140–50.

FitzGibbon, J.E., and Dunne, T. 1981. "Land Surface and Lake Storage during Snowmelt Runoff in a Subarctic Drainage System." *Arctic and Alpine Research* 13: 277–85.

Gerard, R. 1979. "River Ice in Hydrotechnical Engineering: A Review of Selected Topics." *Proceedings, Canadian Hydrology Symposium 79*, Cold Climate Hydrology, Vancouver, BC, 1–29.

Goodison, B.E. 1978. "Accuracy of Canadian Snow Gauge Measurements." *Journal of Applied Meteorology* 17: 1,541–8.

Harington, C.R., 1960. *A Short Report of Snow and Ice Conditions on Lake Hazen, Winter 1957–58*. Defence Research Board, Department of National Defence, Canada.

Hydrological Atlas of Canada, 1978. Ministry of Supply and Services, Ottawa.

Koerner, R.M. 1970. "The Mass Balance of the Devon Island Ice Cap, Northwest Territories, Canada, 1961–66." *Journal of Glaciology* 9: 325–36.

– 1979. "Accumulation, Ablation, and Oxygen Isotope Variations on the Queen Elizabeth Islands Ice Caps, Canada." *Journal of Glaciology* 22: 25–41.

Lewkowicz, A.G., and French, H.M. 1982. "The Hydrology of Small Runoff Plots in an Area of Continuous Permafrost, Banks Island, N.W.T." *Proceedings, 4th Canadian Permafrost Conference*, Calgary, 2–6 March 1981, National Research Council of Canada, 151–62.

Maag, H. 1969. *Ice-dammed Lakes and Marginal Glacial Drainage on Axel Heiberg Island, Canadian Arctic Archipelago*. Axel Heiberg Island Research Reports, McGill University, Montreal.

Marsh, P. 1988. "Soil Infiltration and Snow-melt Run-off in the Mackenzie Delta, N.W.T." *Proceedings, 5th International Conference on Permafrost*, Trondheim, Norway, 618–21.

Marsh, P., and Woo, M.K. 1984. "Wetting Front Advance and Freezing of Meltwater within a Snow Cover. 1. Observations in the Canadian Arctic." *Water Resources Research* 16: 1,853–64.

Müller, F., and Iken, A. 1973. "Velocity Fluctuations and Water Regime of Arctic Valley Glaciers." *Symposium on the Hydrology of Glaciers*, International Association of Hydrological Sciences (IAHS) Publication 95, 165–82.

Müller, F., and Keeler, C.M. 1969. "Errors in Short-term Ablation Measurements on Melting Ice Surfaces." *Journal of Glaciology* 8: 91–105.

Ohmura, A. 1982. "Regional Water Balance on the Arctic Tundra in Summer." *Water Resources Research* 18: 301–5.

Onesti, L.J., and Walti, S.A. 1983. "Hydrologic Characteristics of Small Arctic-Alpine Watersheds, Central Brooks Range, Alaska." *Proceedings, Fourth International Conference on Permafrost*, Fairbanks, Alaska, 957–61.

Price, A.J., and Dunne, T. 1976. "Energy Balance Computations on Snowmelt in a Subarctic Area." *Water Resources Research* 12: 686–94.

Roulet, N.T., and Woo, M.K. 1986. "Low Arctic Wetland Hydrology." *Canadian Water Resources Journal* 11: 69–75.

Ryden, B.E. 1977. "Hydrology of Truelove Lowland." In L.C. Bliss, ed., *Truelove Lowland, Devon Island, Canada: A High Arctic Ecosystem*, Edmonton: University of Alberta Press, 107–36.

Stanley, A.D. 1969. "Observations of the Surge of Steele Glacier, Yukon Territory, Canada." *Canadian Journal of Earth Sciences* 6: 819–30.

Tolstikhin, N.I., and Tolstikhin, O.N. 1976. *Groundwater and Surface Water in Permafrost Region (Translation)*. Inland Water Directorate, Technical Bulletin No. 97, Ottawa.

van Everdingen, R.O. 1981. *Morphology, Hydrology and Hydrochemistry of Karst in Perma-*

frost Terrain near Great Bear Lake, Northwest Territories. National Hydrology Research Institute Paper No. 11, Inland Waters Directorate, Calgary, Alberta.

– 1987. "The Importance of Permafrost in the Hydrological Regime." In M.C. Healey and R.R. Wallace, eds., *Canadian Aquatic Resources*, Fisheries and Oceans Canada, 243–76.

Williams, G.P. 1957. *An Analysis of Snow Cover Characteristics at Aklavik and Resolute, Northwest Territories.* National Research Council, Division of Building Research, Research Paper 40, Ottawa.

Woo, M.K. 1983. "Hydrology of a Drainage Basin in the Canadian High Arctic." *Annals of the Association of American Geographers* 73: 577–96.

– 1986. "Permafrost Hydrology in North America." *Atmosphere-Ocean* 24: 201–34.

– 1988. "Wetland Runoff Regime in Northern Canada." *Proceedings, Fifth International Permafrost Conference*, Trondheim, Norway, 644–9.

Woo, M.K., and Dubreil, M.A. 1985. "Empirical Relationship between Dust Content and Arctic Snow Albedo." *Cold Regions Science and Technology* 10: 125–32.

Woo, M.K., and Heron, R. 1981. "Occurrence of Ice Layers at the Base of High Arctic Snowpacks." *Arctic and Alpine Research* 13: 225–30.

Woo, M.K., Heron, R., Marsh, P., and Steer, P. 1983. "Comparison of Weather Station Snowfall with Winter Snow Accumulation in High Arctic Basins." *Atmosphere-Ocean* 21: 312–25.

Woo, M.K., Heron, R., and Steer, P. 1981. "Catchment Hydrology of a High Arctic Lake." *Cold Regions Science and Technology* 5: 29–41.

Woo, M.K., and Sauriol, J. 1980. "Channel Development in Snow-Filled Valleys, Resolute, N.W.T., Canada." *Geografiska Annaler* 62A: 37–56.

Woo, M.K., and Steer, P. 1983. "Slope Hydrology as Influenced by Thawing of the Active Layer, Resolute, N.W.T." *Canadian Journal of Earth Sciences* 20: 978–86.

Young, K.L., and Lewkowicz, A.G. 1988. "Measurement of Outflow from a Snowbank with Basal Ice." *Journal of Glaciology* 34: 358–62.

CHAPTER SIX

Cold-Climate Processes and Landforms

HUGH M. FRENCH

A number of distinct and sometimes unique processes and landforms characterize cold environments. All are associated with freezing and sub-freezing temperatures, together with the presence of moisture in its various forms. They may be subdivided further by being associated with either glacial or non-glacial conditions. Because the extent of present-day glaciers is localized, it is clear that most of Canada's landmass is cold, non-glacial. Such a condition is commonly called "periglacial" (e.g. French 1976; Clark 1988; Washburn 1980; Williams and Smith 1989).

Cold-climate processes are intimately linked with the action of intense frost, often combined with the presence of permafrost. On this basis, at least 50 per cent of Canada's land surface currently experiences periglacial conditions. There are gradations between environments in which frost action processes dominate, shaping the whole or a major part of the landscape, and those in which they are subservient to others. However, lithologies vary in susceptibility to frost action, and there is no perfect correlation between areas of intense frost action and areas underlain by permafrost. Moreover, since extensive areas of northern Canada have emerged only recently from beneath Late Wisconsinan ice sheets and glaciers, periglacial processes serve only to modify these glaciated landscapes. However, in those areas of Canada that were either outside the Pleistocene glacial limits or have experienced protracted non-glacial histories, such as northern interior Yukon and parts of the high arctic islands, landscapes are probably in equilibrium with current processes.

FREEZING AND THAWING

Central to the operation of most cold-climate processes are freezing and thawing of the ground surface. These may occur either diurnally, as in many temperate and subtropical regions, or seasonally, as in much of northern Canada. The depth of frost penetration depends mainly on the intensity of cold, its duration, thermal and physical properties of the soil and rock, and overlying vegetation. Where the depth of seasonal frost exceeds that of thaw during the summer following, a zone of frozen (i.e.

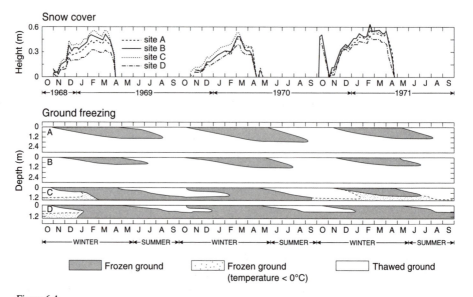

Figure 6.1

Ground thermal regimes and associated snow cover conditions for four sites at Thompson, Man., 1968–71. Only two sites are underlain by permafrost. At site C permafrost degraded during measurement; permafrost can be extremely sensitive to disturbance. *Source*: Data from Brown and Williams, (1972).

temperature <0°C) ground persists throughout the year and is commonly referred to as permafrost, or perennially cryotic ground. All three conditions – diurnal frost, seasonal frost, and permafrost – influence the nature and extent of cold-climate processes.

The seasonal (i.e. annual) rhythm of ground freezing and thawing dominates much of northern Canada where long, cold winters are typical. From a geomorphic view-point, the nature and rate of both spring thaw and autumn freeze-back are of interest. Spring thaw influences the nature of spring runoff, while autumn freeze-back controls frost heaving and ice segregation in the soil. Usually, spring thaw occurs quickly and over three-quarters of the soil thaws during the first four to five weeks in which air temperatures are above 0°C. Ground thermal regimes are closely related to snow thickness and density. At Thompson, Man., for example, located in the discontinuous permafrost zone, years of heavy snowfall retard thaw yet prevent extremes of ground freezing (Figure 6.1). In such regions, extremely subtle site differences related to drainage, soil type, or micro-climate can determine whether or not permafrost exists.

Autumn freeze-back is equally complex. In regions underlain by continuous perma-frost, freezing is two-sided, occurring both downward from the surface and upward from the perennially frozen ground beneath. Moreover, the freezing period is much longer and may persist for six to eight weeks. At Inuvik, NWT, for example, freezing begins in late September and finishes in mid-December. During most of this period the soil remains in a near-isothermal condition, sometimes referred to as the "zero curtain." This phenomenon results from the release of latent heat on freezing and retards the drop in temperature. Initially, freezing progresses slowly from the surface downward, but it dramatically speeds up at depth, because of upward freezing from the permafrost beneath and of moisture decrease with depth, since soil water is initially drawn upward to the freezing plane, preferentially increasing the latent heat effects in the surface layers.

Freeze-thaw cycles in the soil are surprisingly infrequent. One of the earliest field studies to document this fact was at Resolute Bay, NWT (Cook and Raiche 1962). At depth (i.e. >10 cm) usually just the annual cycle occurs. Only at the ground surface are freeze-thaw cycles at all frequent, and even these fluctuations may be stopped by a sizeable snowfall.

FROST HEAVE AND ICE SEGREGATION

Intimately associated with ground freezing are the phenomena of frost heaving and ice segregation, which take place wherever moisture is present within the soil. Frost heaving caused by ice segregation occurs throughout much of Canada. Annual ground displacements of several centimetres are common, with cyclic differential ground pressures of many kilopascals per square centimetre. Engineering hazards caused by these displacements and pressures, together with the adverse effects of accumulations of segregated ice in freezing soil, are widespread and costly.

Ice segregation involves complex interrelationships among the ice, an unfrozen liquid phase, and the bulk porewater. Complex latent heats of phase change as well as variable interfacial energies between phases are involved. Freezing air temperatures create a thermal gradient that induces upward heat flow. Initially, soil water freezes near the ground surface; as freezing progresses, segregated ice crystals form, grow, and ultimately coalesce into lamellar lenses (ice segregation) (Figure 6.2A). Ice lens growth is thus a consequence of continuous upward flow of water from below; a balance is ultimately achieved between dissipation of the latent heat of freezing and the upward flow of soil water. The soil surface is displaced upward (frost heaving) by the growth of a series of ice lenses (Figure 6.2B). Restriction of displacement can cause significant "heaving pressures."

Heave can be primary (i.e. capillary) or secondary (e.g. see Smith 1985a; 1986). In primary heave, the critical conditions for the growth of segregated ice are:

$$P_i - P_w = \frac{2}{r_{iw}} < \frac{2_o}{r} , \tag{1}$$

where P_i is pressure of ice, P_w is pressure of water, o is surface tension between ice and water, r_{iw} is the radius of the ice water interface, and r is the radius of the largest continuous pore openings. Secondary heave is not well understood but may take place at temperatures below 0°C and at some distance behind the freezing front. Porewater expulsion from an advancing freezing front is another mechanism of ice

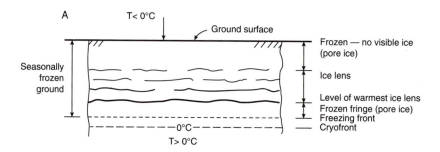

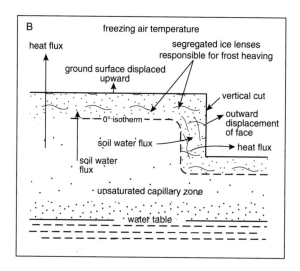

Figure 6.2

Ice segregation: (A) ice lenses growing at some distance behind frost line, separated by frozen fringe; *source*: Associate Committee on Geotechnical Research (1988); (B) cross-section showing how ice lenses form normal to direction of heat flow and how heaving processes and displacements are always in direction of heat flow; *source*: Smith (1986).

segregation, especially for massive ice bodies, provided that porewater pressures are adequate to replenish groundwater that is transformed to ice.

Repeated frost heaving is usually associated with the seasonally frozen layer. Field studies in the Mackenzie delta region indicate that heave occurs not only during autumn freeze-back but also during winter when ground temperatures are below 0°C (Mackay et al. 1979; Smith 1985b). Geomorphic indicators of frost heaving include upheaval of bedrock blocks (Figure 6.3) (Dyke 1984; 1986), upfreezing of objects and tilting of stones (e.g. Mackay 1984b), and sorting and migration of soil particles.

A Expulsion of water

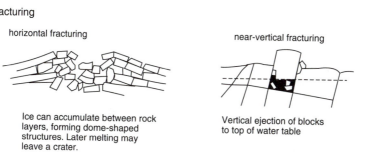

B Fracturing

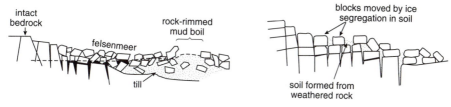

C Displacement

Figure 6.3
Frost heaving of bedrock: (A) water-expulsion mechanism in well-indurated rock with relatively thick
active layer (mainland Canadian Shield); (B) influence of fracture fabric on heaving style; and (C) ice
segregation mechanism displacing rock either by freeze-thaw creep of rock transported to soil surface or
by direct application of presure exerted by ice segregation in weathering products. *Source*: Dyke (1986).

Frost heaving presents numerous geotechnical problems in the construction of roads,
buildings, and pipelines in cold environments (e.g. Carlson et al. 1982; Hayley 1982;
Smith, Dallimore, and Kettle 1985) and in locally induced artificial freezing, such
as beneath ice arenas (Leonoff and Lo 1982).

   Diurnal one-sided freezing at or just beneath the ground surface can result in
formation of needle ice. Ice crystals grow upward in the direction of heat loss and
can range in length from a few millimetres to several centimetres. Occasionally, they
may lift small pebbles or, more commonly, soil particles. Formation of needle ice
is particularly common in the Cordillera wherever wet, silty soils are present. The

thawing and collapse of needle ice are thought to affect frost sorting, frost creep, the differential downslope movement of fine and coarse material, and the creation of certain micro-patterned ground forms. The importance of needle ice as a disruptive agent in the soil has probably been underestimated, especially in exposing soils to wind action, deflation, and cryoturbation activity. In some areas, it may cause damage to plants when freezing produces vertical mechanical stresses within the root zone (e.g. Brink et al. 1967). In the BC Coast Mountains, needle ice occurs in oriented stripes (Mackay and Mathews 1974), and both wind direction and solar radiation have been suggested as explanations; it is not clear whether oriented needle ice is primarily a shadow effect developed by thawing or a freezing effect.

CRYOGENIC WEATHERING

The weathering of bedrock under cold climates is not fully understood. It is currently believed that a group of weathering processes, both physical and chemical, operate either independently or in combination.

The most influential bedrock-weathering process has generally been assumed to be frost wedging, a physical process relying on repeated freeze-thaw activity. Features attributed to frost wedging in northern Canada include extensive areas of angular bedrock fragments (blockfields), scree and talus deposits, and irregular bedrock outcrops termed tors. Porous and well-bedded sedimentary rocks, such as shales, sandstones, and limestones, are regarded as especially susceptible to frost wedging (Figure 6.4A). However, numerous field studies from northern latitudes and mid-latitude alpine sites now demonstrate the limited frequency of freeze-thaw cycles (see above) and the required combination of freezing intensity and moisture availability is rarely encountered under natural field conditions (e.g. Thorn 1979; 1988; Hall 1980). Serious doubt must be cast, therefore, on the efficacy of freeze-thaw rock shattering for much of northern Canada (see French 1981: 267–9).

Several studies have estimated the rate of rockwall recession apparently resulting from frost wedging. For example, in Longyeardalen, Spitsbergen, over the last fifty years frost weathering has caused steep rock faces to retreat at a rate of 0.3 mm/ yr (Jahn 1976). In northern Canada, rates of cliff recession vary with lithology, but figures as high as 0.6 to 1.0 mm/yr have been suggested (see French 1976: 148, Table 7.4).

A recent trend has been to regard hydration weathering – in which the pressure of absorbed water generates forces sufficient to free grains and disintegrate rock – as significant in cold regions. Salt weathering must also be considered. Their overall significance to cryogenic weathering, however, is still to be agreed on (Washburn 1980: 73–4; French 1981). These processes focus attention on moisture rather than temperature requirements for weathering and may be effective without the freezing process being involved. Physical and chemical effects may combine in these processes. For example, field studies on coarse-grained granitic rocks on Ellesmere Island indicate that repeated salt hydration and crystallization following summer precipi-

A

B

Figure 6.4
Cryogenic weathering of bedrock: (A) Paleozoic sandstones, Rea Point, eastern Melville Island, NWT; (B) granite terrain, Mt Fitton, Barn Ranges, northern Yukon.

tation favour microfracturing and formation of weathering pits and associated "grus" (Watts 1983). A second example known to the author is in never-glaciated northern Yukon, where granite weathering phenomena include both angular and rounded tors, block fields, and grus formation (Figure 6.4B). Elsewhere, other studies suggest that limestone solutional effects are prominent in cold regions (e.g. Ford 1984; 1987; van Everdingen 1981) and that permafrost does not necessarily inhibit development of extensive karst topography (e.g. Sloan and van Everdingen, 1988). Finally, experimental studies in Russia indicate that many widely held assumptions concerning cold-climate weathering may need to be re-evaluated. For example, Konischev (1982) concludes that quartz has lower resistance than feldspar when subject to freeze and thaw. According to these studies, the ultimate distribution of grain size resulting from cryogenic weathering is the reverse to that produced under "normal" (i.e. non-cryogenic) conditions.

It is therefore not surprising that cold-climate weathering processes and their associated landforms are currently subject to renewed scrutiny. For example, the efficacy of snow, or nivation, traditionally recognized as a potent geomorphic agent, is now in question (e.g. Thorn 1988). Nivation relies on the assumption that freeze-thaw processes promote physical disintegration of bedrock beneath the snowbank, while the hydration hypothesis emphasizes rock porosity. Many so-called nivation features, such as tors, may be structurally controlled rather than erosional in origin (e.g. Dyke 1983). In the case of the cryoplanation terraces of the Yukon Plateau (e.g. Hughes, Rampton, and Rutter 1972; French, Harris, and van Everdingen 1983), where no obvious structural control exists, their distribution may reflect subtle interactions among rock porosity, humidity, and prevailing winds.

PERMAFROST AND GROUND ICE

A long period of winter cold and a relatively short period of summer thaw may lead to formation of a layer of ground that remains frozen (i.e. below 0°C) throughout the year. This perennially cryotic ground is termed permafrost. Although essentially a thermal condition, the presence of moisture within permafrost, in either a frozen or an unfrozen state, presents problems, as evidenced by frost heave, described above. Soil and rock do not automatically freeze at 0°C, especially if percolating groundwater is highly mineralized or under pressure. As a result, significant quantities of unfrozen porewater may continue to exist at temperatures below zero.

There is a broad zonation of permafrost conditions in Canada according to climate (see Figure 1.8). Zones of either continuous or discontinuous permafrost are recognized, in addition to alpine permafrost and subsea permafrost. In total, approximately 50 per cent of Canada's land surface is underlain by permafrost of one sort or another. The southern limit of the zone of continuous permafrost correlates well with the approximate position of the −6-to-−8°C mean annual air temperature isotherm, and this relates to the −5°C isotherm of mean annual ground temperatures. The discontinuous zone is further subdivided into areas of widespread permafrost and scattered permafrost; at its extreme southern fringes, permafrost exists as isolated "islands" beneath peat and other organic sediments.

Ground ice is a major component of permafrost. In certain areas of the western Canadian Arctic underlain by unconsolidated sediments, it may comprise at least 50 per cent by volume of the upper 1–5 m of permafrost (Table 6.1). Although many types of ground ice can be recognized, pore ice, segregated ice, and wedge ice are the most significant in terms of volume and widespread occurrence (Harry 1988). In general, the volume of ground ice is greatest near to the surface of permafrost where

Table 6.1
Ground ice volumes in upper 5 m of permafrost in selected areas, western Canadian Arctic

Area	Pore/segregated ice (%)		Wedge ice (%)		Total ice (%)	
King Point, Yukon	43.5	(79.2)	11.4	(20.8)	54.9	(100)
Richards Island, NWT	28.3	(79.3)	7.5	(21.0)	35.7	(100)
Southwest Banks Island, NWT	38.7	(66.2)	19.8	(33.8)	58.5	(100)

Sources: Harry, French, and Pollard (1985); Richards Island: Pollard and French (1980); Banks Island: French and Harry (1983).
Note: Values in parentheses indicate percentage contribution of ground ice to total ice volume.

ice-wedges, reticulate ice veins (Mackay 1974), and aggradational ice (Burn 1988) are all present and where the ground is subject to repeated annual thermal changes (e.g. Mackay 1983). The latter induces moisture migration downward from the active layer above and upward from the permafrost below. Burn and Michel (1988), for example, have used the tritium content of ground ice to deduce recent temperature-induced water migration into permafrost near Mayo in central Yukon.

In certain areas of the western arctic coastal lowlands, icy sediments and massive bodies of ice several tens of metres thick are known to exist (Figure 6.5). Because the source of water for such ground ice is usually subterranean rather than subaerial, growth of many massive ground-ice bodies and much segregated ice is associated with aggradation of permafrost and the Quaternary history of the area (e.g. Rampton 1983; French, Harry, and Clark 1982; Burn, Michel, and Smith 1986). In certain areas, cryostratigraphic and petrofabric observations suggest that some massive ice is of buried glacier origin (French and Harry 1988; 1990; Dallimore and Wolfe 1988); in other areas, a segregation-intrusion origin is thought more likely (e.g. Mackay 1973; Harry French, and Pollard 1988). A classification of massive ground ice (Figure 6.6) illustrates some of the possible origins of massive ice.

PERMAFROST-RELATED PROCESSES AND LANDFORMS

A number of processes are directly related to the presence of permafrost and give rise to unique landforms. The most widespread is large-scale thermal-contraction cracking of the ground. This leads to formation of a polygonal pattern of fissures (Figure 6.7A). The fissures may be as much as 4–5 m deep, and the polygons may be 15–30 m in dimensions. If water from melting snow at the ground surface trickles down the fissure and if cracking is repeated at the same locality, a wedge-shaped ice body (ice-wedge) forms (Figure 6.7B).

In the Mackenzie delta, detailed field observations indicate that active ice-wedge cracking is concentrated between mid-January and mid-March. However, controls over cracking are not well established (e.g. see Burn 1990). For example, wedges

Figure 6.5
Massive ground ice of probable segregated origin, Mackenzie delta region, NWT: (A) near Nicholson Point and (B) near Tuktoyaktuk. Deformations in ice result from either glacier ice-thrusting (at Nicholson) or unloading (at Tuktoyaktuk).

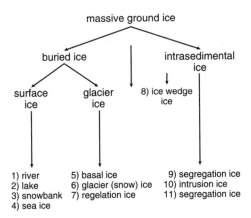

Figure 6.6
Classification of massive ground ice.

do not crack each year; instead, probably less than half of the wedges in any given area crack annually (Mackay 1975; Harry, French, and Pollard 1985). As the ice-wedges grow wider, peaty ridges usually develop parallel to the fissures, and a polygonal, saucer-shaped depression is formed (low centre polygons). The ice-wedges in such polygons are usually active (i.e. the wedges still crack). Many polygons, however, are bun-shaped (high centred), and in these features cracking of the wedge is infrequent. Cracking is not instantaneous but propagates laterally from discrete points, and directional cracking, both upward and downward, apparently occurs (Mackay 1984a). Thermal-contraction cracks have been observed to form within a few years of the ground being exposed to subfreezing temperatures (Figure 6.7C).

Ice-wedges may accelerate other geomorphic processes by promoting differential thaw and erosion along the wedge. For example, the catastrophic drainage of a lake may take place via erosion of ice-wedges at its outlet (Harry and French 1983; Mackay 1988), while rapid retreat of coastal cliffs by block slumping in areas with numerous ice-wedge polygons is well known. In the western Canadian Arctic adjacent to the Beaufort Sea, retreat rates of 1.0–5.0 m per year have been recorded (e.g. Harry, French, and Clark 1983; Mackay 1986a).

Several distinctive landforms result from aggradation of permafrost and growth of ground ice. The best known are perennially frozen peaty mounds (palsas) and perennial ice-cored hills (pingos). Both are types of frost mounds (Figure 6.8). Palsas form in wetlands primarily by ice segregation beneath a peaty organic layer. Typically they occur in the discontinuous permafrost zone and are rarely more than 5.0 m high. Seasonal frost mounds (Pollard and French 1983; Pollard 1988) are sometimes confused with palsas, which can be morphologically similar. However, they are not the result of ice segregation and, instead, possess a core of injection ice. Finally, the variety of small-scale frost mounds commonly observed in tundra regions are usually regarded as being different to palsas, pingos, or seasonal frost mounds.

Figure 6.7
Large-scale thermal-contraction cracking: (A) depressed centred polygons, southern Banks Island; (B) ice-wedge in lacustrine silts near Sachs Harbour, southern Banks Island; (C) thermal-contraction crack developed on recently exposed lake bottom, Illisarvik, Mackenzie delta.

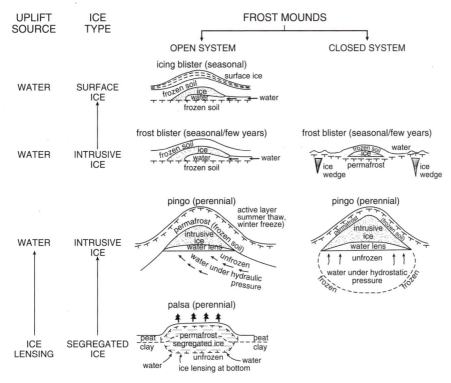

Figure 6.8
Relationship among uplift source, ice type, and frost mounds. *Source*: Mackay (1986b).

Pingos can grow to sizeable dimensions, occasionally exceeding 50 m in height and 300 m in diameter (Figure 6.9A). The ice core of a pingo (Figure 6.9B) may be produced by injection of groundwater under pressure from below, by segregation, or by water freezing in dilation cracks (Mackay 1979; 1985). Hydrostatic (closed-system) pingos derive their water pressure from porewater expelled by downward permafrost growth; hydraulic (open-system) pingos by gravity flow from an upslope source. One of the largest groupings of closed-system pingos, approximately 1,350 in number, is along the Pleistocene Coastal Plain of the Mackenzie delta; smaller groups occur on Banks Island and elsewhere in the arctic islands. The site of pingo growth is usually a shallow residual pond left by rapid lake drainage or an abandoned river channel or terrace. In all cases permafrost has aggraded into unfrozen and saturated sediments (taliks), causing expulsion of porewater.

The largest concentration of open-system pingos is in unglaciated interior Yukon, where over 700 have been identified (Hughes 1969). The major requirement is the presence of thin and/or discontinuous permafrost and the confining of subpermafrost waters. Most open-system pingos therefore develop in distinct topographic situations, such as valley bottoms, or on lower valleyside slopes.

A

B

Figure 6.9
Pingos: (A) Ibyuk Pingo, Pleistocene Mackenzie delta, 40 m high and 1,000 years old, the largest in Canada; (B) ice exposed in core of small pingo which has grown since 1935 in a drained lake basin approximately 15 km west of Tuktoyaktuk.

Degradation of permafrost often involves melting of ground ice accompanied by local collapse and subsidence of the ground. These processes are termed thermokarst, a physical (i.e. thermal) process peculiar to permafrost regions (French 1976: 104–33; 1987). Since thermokarst merely reflects a disruption in the thermal equilibrium of the permafrost, a range of conditions can initiate it, including changes in regional climate, localized slope instability and erosion, drainage alteration, and either natural (i.e. fire) or man-induced disruptions to surface vegetation cover. In the boreal forest, fire frequently initiates permafrost degradation and slope failure (e.g. Harry and MacInnes 1988). Along the western arctic coastal plain, where alluvial sediments with high ice contents are widespread, thermokarst is believed to be one of the principal processes fashioning the landscape (e.g. Rampton 1982; Harry, French, and Pollard 1988). Elsewhere, large-scale thermokarst phenomena include ground-ice slumps (e.g., Lewkowicz 1986; Burn and Freile 1989) and thaw lakes (Bird 1967: 212–16; Harry and French 1983; Burn and Smith 1988). Some types of thermokarst, such as ground-ice slumps, are probably the most rapid erosional agents currently operating in tundra and arctic regions. Where large ice-wedges are present within ice-rich silts, striking badland thermokarst topography can result. Typical man-induced thermokarst terrain consists of an irregular hummocky and unstable topography of enclosed depressions and standing bodies of water (e.g. French 1975; 1987: 245–9).

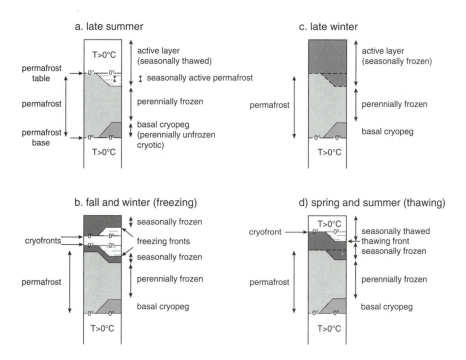

Figure 6.10
Seasonal changes (A–D) in active layer; temperature relative to 0°C and state of water also indicated.
Source: Associate Committee on Geotechnical Research (1988).

ACTIVE LAYER PROCESSES

Between the upper surface of permafrost and the ground surface lies the active layer, a zone that thaws each summer and refreezes each autumn. In thermal terms, it is the layer that fluctuates above and below 0°C during the year (French 1988). The seasonal dynamics of the active layer are illustrated in Figure 6.10A–D. Its thickness varies from as little as 15–30 cm in the High Arctic to over 1.5 m in the Canadian subarctic. Thickness depends on many factors, including ambient air temperatures, angle of slope and orientation, vegetation cover, thickness (depth and density) and duration of snow cover, soil and rock type, and ground moisture conditions.

Mass Wasting

Distinct mass wasting processes operate in the active layer in northern Canada – permafrost creep, frost creep, gelifluction (solifluction), and certain forms of rapid mass movement – and give rise to characteristic landform features (Figure 6.11).
   Permafrost creep results from a variety of time-dependent stresses and strains in ice-rich frozen soil. It is dominated by the rheological properties of ice. Glen's flow

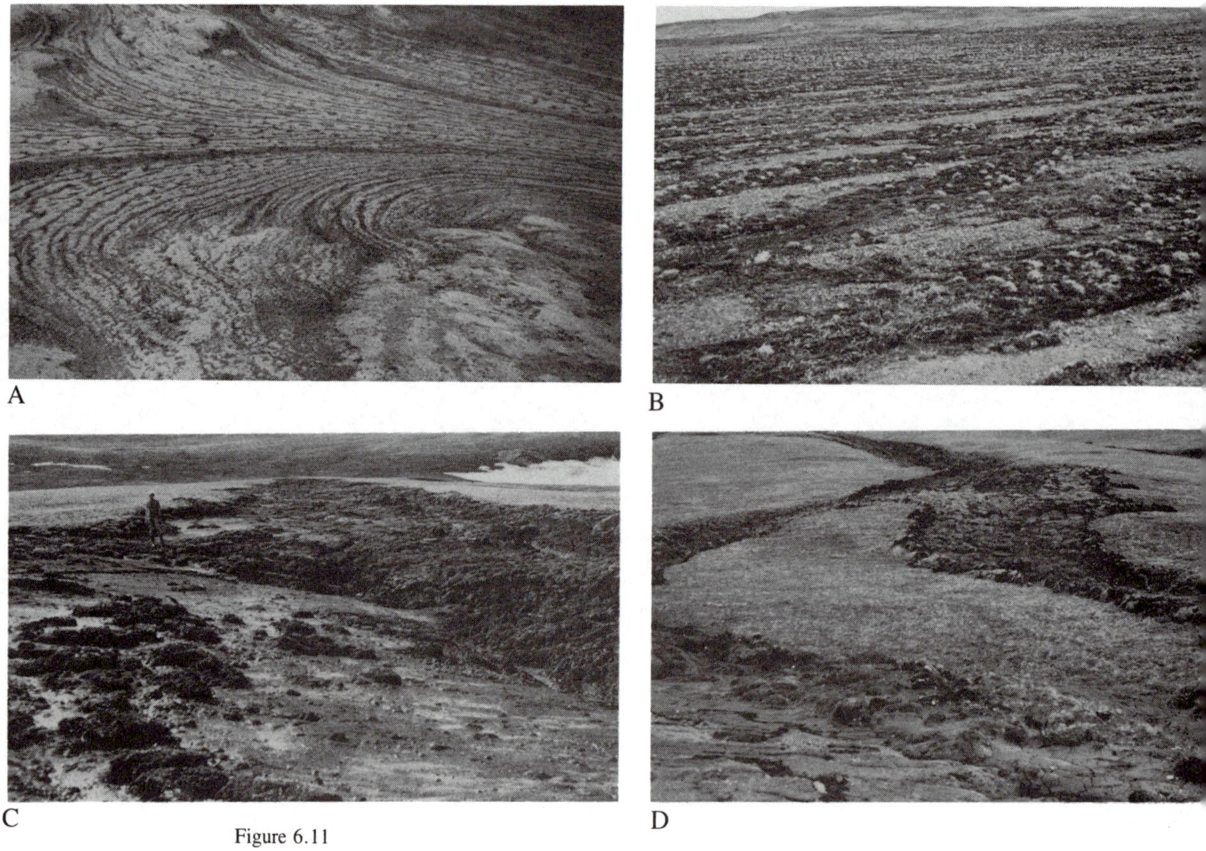

Figure 6.11

Mass wasting phenomena, Banks Island (71–74°N), western Arctic: (A) gelifluction stripes; (B) gelifluction sheet; (C) active layer failure on ice-rich shale of Christopher Formation, Thomsen River, and (D) skin flows and multiple mudflows, Masik Valley.

law suggests that at a temperature of −22°C, the strain rate produced by a given stress is only one-tenth of its value at 0°C, and the strain rate almost doubles between −5°C and 0°C. In relatively warm permafrost in the Mackenzie valley, rate of creep was measured at between 0.15 and 0.30 cm/yr and varied according to ground ice conditions (Savigny and Morgenstern 1986). On Melville Island, however, with mean annual ground temperature of −17°C, rates of creep were substantially less and restricted to the upper 1.0–1.5 m of permafrost. Rates of 0.03–0.05 cm/yr were typical of four sites on a low-angle (5°), west-facing slope near Rea Point (Bennett and French 1988).

Frost creep is the ratchet-like downslope movement of particles produced by frost heaving of the ground and subsequent settling upon thawing – the heaving being predominantly normal to the slope and the settling more nearly vertical. Movement associated with frost creep decreases from the surface downward and depends on the frequency of freeze-thaw cycles, angle of slope, moisture available for heave, and frost susceptibility of soil.

Gelifluction is a form of solifluction in areas underlain by permafrost; *solifluction* is a more general term for the mass wasting typical of any cold region not necessarily underlain by permafrost. Both processes, however, are faster than soil creep. Gelifluction is most likely where the permafrost table limits downward percolation of water through the soil and where melting of segregated ice lenses provides excess water which reduces internal friction and cohesion in the soil. Particularly favoured sites include areas beneath or below late-lying snowbanks. In northern Canada, rates of movement of the order of 0.5–10.0 cm/yr are typical (e.g. Dyke 1981; Mackay 1981; Egginton and French 1985). Features produced by both solifluction and gelifluction include uniform sheets of locally derived surficial materials, tongue-shaped lobes, and alternating stripes of coarse and fine sediment (Figure 6.11A, B).

Rapid mass movements may also occur in the active layer, with the permafrost table acting as a lubricated slip plane and controlling the depth of the failure plane (Figure 6.11C, D). Such failures, termed "skin flows," "detachment failures," or "bimodal flows," are usually attributed to local conditions of soil moisture saturation and high porewater pressure, often the result of thaw consolidation (see French 1988: 168–9). Failure may take the form of a mudflow or of a distinct slump scar or hollow. In several arctic islands, where the active layer is very thin and fine-grained, ice-rich sediments are widespread, active layer failures are both common and difficult to predict. For example, the trigger mechanism is often a period of exceptionally hot or humid weather which thaws the ice-rich layer at the base of the active layer, as happened in the summer of 1988 on Ellesmere Island (Edlund, Taylor Alt, and Young 1989). In the boreal forest regions of northern Canada, failures often follow destruction of vegetation by forest fire (e.g. Zoltai and Pettapiece 1973; Harry and MacInnes 1988).

Cryoturbation

Cryoturbation is the lateral or vertical displacement of soil which accompanies seasonal and/or diurnal freezing and thawing. This process is usually associated with formation of patterned ground phenomena.

The most common type of patterned ground is the non-sorted circle, or "hummock" (Figure 6.12A). This form is ubiquitous wherever fine-grained sediments are present, particularly in the Mackenzie valley and Keewatin. At Inuvik, NWT, for example, hummocks are composed of fine-grained frost-sensitive soils and are typically 1–2 m in diameter and 30–50 cm high. Beneath the hummock, the late-summer frost table is bowl-shaped, and the hummocks grade from completely vegetated (earth hummocks) to bare centred (mud hummocks). On sloping terrain, non-sorted stripes commonly develop (Figure 6.12B).

The mound, or circle, form has traditionally been attributed to upward displacement of material resulting from cryostatic (i.e. freeze-back) pressures generated in a con-

Figure 6.12
Cold-climate patterned ground phenomena. Northwest Territories; (A) mud hummocks (non-sorted circles), boreal forest near Inuvik; (B) non-sorted stripes, Banks Island; (C) section through earth hummock (shovel leaning against top portion) showing cryoturbated organic material and decaying wood, Hume River near Fort Good Hope, Mackenzie valley – *photo*: S.C. Zoltai; and (D) sorted circle, Resolute Bay, site of experimental studies by A.L. Washburn.

fined, wet, unfrozen pocket in the active layer. However, existence of substantial cryostatic pressures in the field has yet to be convincingly demonstrated. Presence of voids in soil, occurrence of frost cracks in winter, and weakness of confining soil layers lying above appear to block pressures of any magnitude. Moreover, on theoretical grounds, cryostatic pressures should not develop in frost-sensitive hummock soil because ice lensing at the top and/or bottom of the active layer will desiccate the last unfrozen pocket so that porewater is under tension, not under pressure. Mackay (1980) concluded that upward displacement of material is caused by freeze and thaw of ice lenses at the top and bottom of the active layer, with a gravity-induced, cell-like movement (Figure 6.13A). The latter occurs because the top and bottom of the freeze-thaw zones have opposite curvatures. Evidence of cell-like

A

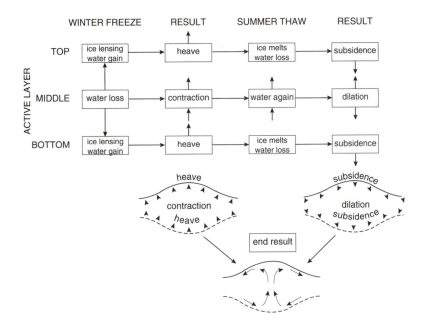

WINTER FREEZE   RESULT   SUMMER THAW   RESULT

ACTIVE LAYER

TOP — ice lensing / water gain → heave → ice melts / water loss → subsidence

MIDDLE — water loss → contraction → water again → dilation

BOTTOM — ice lensing / water gain → heave → ice melts / water loss → subsidence

heave
contraction
heave

subsidence
dilation
subsidence

end result

B

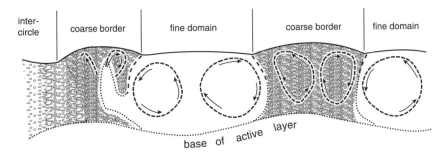

inter-circle   coarse border   fine domain   coarse border   fine domain

base of active layer

Figure 6.13
Models of soil displacement within active layer: (A) equilibrium model for hummocks, western arctic coast (after Mackay 1980); (B) soil circulation patterns in non-sorted circles, western Spitsbergen (after Hallet et al. 1988). Convection of both fine-grained soil and gravel-rich border material is simplest pattern of soil motion compatible with measured surface displacements, microrelief, subduction of surface organics, and indications of occasional diapirisms under circle borders.

circulation is also deduced from the grain-size distribution of the hummocky soils, from radiocarbon dating of organic materials intruded into the hummock centres from the sides (e.g. see Figure 6.12C), and from upward-moving tongues of saturated soil observable in late summer.

Recent quantitative field experiments (Figure 6.12D) support the idea of a cell-like soil circulation in the formation of patterned ground. For example, at Resolute Bay in the High Arctic, movement of dowels indicates radial-outward soil displacement in the order of millimetres per year in the fine-grained central areas of sorted circles which increases to a maximum near the stony borders (Washburn 1989), and on western Spitsbergen, progressive circulatory motion of the soil has been measured, with periods of activity in the summer separated by long quiescent periods (Hallet and Prestrud 1986; Hallet et al. 1988) (Figure 6.13B). In addition, percolative porewater convection through thawed soil has been suggested as a mechanism for displacement (e.g. Ray et al. 1983; Gleason et al. 1986). It might operate during summer when temperature is a few degrees above zero near the ground surface but approximately zero at the base of the layer of thawed soil. It is argued that the density difference of water between 0 and 4°C may be sufficient to induce free convection of water through the soil-pore network. This would lead to spatially non-uniform thawing at the top of frozen ground caused by convective heat transfer. Such a process many explain the concave upward curvature of the frost table beneath circles, a phenomenon not adequately explained in Mackay's (1980) equilibrium model.

Even though soil circulation can be demonstrated, it still does not completely explain the origin of circles. Washburn (1989: 953-4) suggests that the circulation pattern may have evolved subsequent to a protoform, such as a diapir, which has reached a stage of development conducive to initiation of soil circulation. In this context, mudboils are known to form by diapirism resulting from hydrostatic or artesian pore pressures and are widespread in certain areas of northern Canada (e.g. Dyke and Zoltai 1980; Shilts 1978). Finally, the traditional concepts of differential heave and ice segregation have been combined recently with density instability in thawed soils to explain cryoturbation structures both in Spitsbergen and in Pleistocene deposits in Europe (van Vliet-Lanoe 1988). Clearly, therefore, the problematic phenomenon of cold-climate patterned ground, an enigma ever since the excursion to Spitsbergen in 1910 by the International Geological Congress, is complex.

REFERENCES

Associate Committee on Geotechnical Research (ACGR). 1988. *Glossary of Permafrost and Related Ground Ice Terms*. Permafrost Subcommittee, National Research Council of Canada, Technical Memorandum 142, Ottawa.

Bennett, L.P., and French, H.M. 1988. "Observations on Near-Surface Creep in Permafrost, Eastern Melville Island, Arctic Canada." In *Proceedings, Fifth International Permafrost Conference*, Trondheim, Norway, vol. 1, Trondheim: Tapir Publishers, 683-8.

Bird, J.B. 1967. *The Physiography of Arctic Canada*. Baltimore: Johns Hopkins Press.

Brink, V.C., Mackay, J.R., Freyman, S., and Pearce, D.G. 1967. "Needle Ice and Seedling Establishment in Southwestern British Columbia." *Canadian Journal of Plant Science* 47: 135–9.

Brown, R.J.E., and Williams, G.P. 1972. *The Freezing of Peatlands*. Division of Building Research, National Research Council of Canada, Technical Paper No. 381, Ottawa.

Burn, C.R. 1988. "The Development of Near-Surface Ground Ice during the Holocene at Sites near Mayo, Yukon Territory, Canada." *Journal of Quaternary Science* 3: 31–8.

– 1990. "Implications for Palaeoenvironmental Reconstruction of Recent Ice Wedge Development at Mayo, Yukon Territory." *Permafrost and Periglacial Processes* 1 no. 1: 3–14.

Burn, C.R., and Friele, P.A. 1989. "Geomorphology, Vegetation Succession, Soil Characteristics and Permafrost in Retrogressive Thaw Slumps near Mayo, Yukon Territory." *Arctic* 42: 31–40.

Burn, C.R., and Michel, F. 1988. "Evidence for Recent Temperature-Induced Water Migration into Permafrost from the Tritium Content of Ground Ice near Mayo, Yukon Territory, Canada." *Canadian Journal of Earth Sciences* 25: 909–15.

Burn, C.R., Michel, F.A., and Smith, M.W. 1986. "Stratigraphic, Isotopic and Mineralogical Evidence for an Early Holocene Thaw Unconformity at Mayo, Yukon Territory." *Canadian Journal of Earth Sciences* 23: 794–803.

Burn, C.R., and Smith, M.W. 1988. "Thermokarst Lakes at Mayo, Yukon Territory." In *Proceedings, Fifth International Permafrost Conference*, Trondheim, Norway, vol. 1, Trondheim: Tapir Publishers, 700–5.

Carlson, L.E., Ellwood, J.R., Nixon, J.F., and Slusarchuk, W.A. 1982. "Field Test Results of Operating a Chilled, Buried Pipeline in Unfrozen Ground." In *Proceedings, Fourth Canadian Permafrost Conference*, National Research Council of Canada, Ottawa, 475–80.

Clark, M.J., ed. 1988. *Advances in Periglacial Geomorphology*. New York: John Wiley and Sons.

Cook, F.A., and Raiche, V.G. 1962. "Freeze-Thaw Cycles at Resolute, NWT." *Geographical Bulletin*, 18: 64–78.

Dallimore, S.A., and Wolfe, S. 1988. "Massive Ground Ice Associated with Glacio-Fluvial Sediments, Richards Island, N.W.T., Canada." In *Proceedings, Fifth International Permafrost Conference*, Trondheim, Norway, vol. 1, Trondheim: Tapir Publishers, 132–7.

Dyke, A.S. 1981. "Late Holocene Solifluction Rates and Radiocarbon Soil Ages, Central Canadian Arctic." Geological Survey of Canada, Paper 81–1C, 17–22.

– 1983. *Quaternary Geology of Somerset Island, District of Franklin*. Geological Survey of Canada, Memoir 403.

Dyke, A.S., and Zoltai, S.C. 1980. "Radiocarbon-dated Mudboils, Central Canadian Arctic." Geological Survey of Canada, Paper 80–1B, 271–5.

Dyke, L.S. 1984. "Frost Heaving of Bedrock in Permafrost Regions." *Bulletin, Association of Engineering Geologists* 21 no. 4: 389–405.

– 1986. "Frost Heaving of Bedrock." In H.M. French, ed., "Focus, Permafrost Geomorphology." *Canadian Geographer* 30: 358–66.

Edlund, S.A., Taylor Alt, B., and Young, K.L. 1989. "Interaction of Climate, Vegetation and Soil Hydrology at Hot Weather Creek, Fosheim Peninsula, Ellesmere Island, Northwest

Territories." In Current Research, Part D, Geological Survey of Canada, Paper 89–1D, 125–33.

Egginton, P.A., and French, H.M. 1985. "Solifluction and Related Processes, Eastern Banks Island, N.W.T." *Canadian Journal of Earth Sciences* 22: 1,671–8.

Ford, D.C. 1984. "Karst Groundwater Activity and Landform Genesis in Modern Permafrost Regions of Canada." In R.G. Lafleur, ed., *Groundwater as a Geomorphic Agent*, Allen and Unwin, 340–50.

– 1987. "Effects of Glaciations and Permafrost upon the Development of Karst in Canada." *Earth Surface Processes and Landforms* 12: 507–21.

French, H.M. 1975. "Man-induced Thermokarst, Sachs Harbour Airstrip, Banks Island, N.W.T." *Canadian Journal of Earth Sciences* 12: 132–44.

– 1976. *The Periglacial Environment*. London and New York: Longman.

– 1981. "Periglacial Geomorphology and Permafrost." *Progress in Physical Geography* 5: 267–73.

– 1987. "Permafrost and Ground Ice." In K.J. Gregory and D.E. Walling, eds., *Human Activity and Environmental Processes*, New York: J. Wiley and Sons, 237–69.

– 1988. "Active Layer Processes." In M.J. Clark, ed., *Advances in Periglacial Geomorphology*, New York: J. Wiley and Sons, 151–77.

French, H.M., Harris, S.A., and van Everdingen, R.O. 1983. "The Klondike and Dawson City." In H.M. French and J.A. Heginbottom, eds., *Guidebook 3: Permafrost and Related Features of the Northern Yukon Territory and Mackenzie Delta, Canada*, Fourth International Conference on Permafrost and IGU Commission on the Significance of Periglacial Phenomena, Alaska, Division of Geological and Geophysical Surveys, Fairbanks, Alaska, 35–63.

French, H.M., and Harry, D.G. 1983. "Ground Ice Conditions and Thaw Lake Evolution, Sachs River Lowlands, Banks Island." In H. Poser and E. Schunke, eds., *Mesoformen des Reliefs in heutigen Periglacial Raumes*, Abhandlungen der Akademie der Wissenschaften in Göttingen, 35, 70–81.

– 1988. "Nature and Origin of Ground Ice, Sandhills Moraine, Southwest Banks Island, Western Canadian Arctic." *Journal of Quaternary Science* 3: 19–30.

– 1990. "Observations on Buried Glacier Ice and Massive Segregated Ice, Western Arctic Coast, Canada." *Permafrost and Periglacial Processes* 1: 31–43.

French, H.M., Harry, D.G. and Clark, M.J. 1982. "Ground Ice Stratigraphy and Late-Quaternary Events, South-West Banks Island, Canadian Arctic." In *Proceedings, Fourth Canadian Permafrost Conference*, National Research Council of Canada, Ottawa, 81–90.

Gleason, K.J., Krantz, W.B., Caine, N., George, J.H., and Gunn, R.D. 1986. "Geometrical Aspects of Sorted Patterned Ground in Recurrently Frozen Soil." *Science* 232: 216–20.

Hall. K. 1980. "Freeze-Thaw Activity at a Nivation Site in Northern Norway." *Arctic and Alpine Research* 12: 183–94.

Hallet, B., Anderson, S.P., Stubbs, C.W., and Gregory E.C. 1988. "Surface Soil Displacements in Sorted Circles, Western Spitzbergen." In *Proceedings, Fifth International Permafrost Conference*, Trondheim, Norway, vol. 1, Trondheim: Tapir Publishers, 770–5.

Hallet, B., and Prestrud, S. 1986. "Dynamics of Periglacial Sorted Circles in Western Spitzbergen." *Quaternary Research* 26: 81–99.

Harry, D.G. 1988. "Ground Ice and Permafrost." In M.J. Clark, ed., *Advances in Periglacial Geomorphology*, New York: John Wiley and Sons, 113–49.

Harry, D.G., and French, H.M. 1983. "The Orientation of Thaw Lakes, Southwest Banks Island, Arctic Canada." In *Proceedings, Fourth International Conference on Permafrost*, Washington, DC: National Academy Press, 456–61.

Harry, D.G., French, H.M., and Clark, M.J., 1983. "Coastal Conditions and Processes, Sachs Harbour, Southwest Banks Island, Western Canadian Arctic." *Zeitschrift für Geomorphologie* Supplement 47: 1–26.

Harry, D.G., French, H.M., and Pollard, W.H. 1985. "Ice Wedges and Permafrost Conditions near King Point, Beaufort Sea Coast, Yukon Territory." In Current Research, Part A, Geological Survey of Canada, Paper 85–1A, 111–16.

– 1988. "Massive Ground Ice and Ice-Cored Terrain near Sabine Point, Yukon Coastal Plain." *Canadian Journal of Earth Sciences* 25: 1,846–56.

Harry, D.G. and MacInnes, K. 1988. "The Effect of Forest Fires on Permafrost Terrain Stability, Little Chicago–Travaillant Lake Area, Mackenzie Valley, N.W.T." In Current Research, Part D, Geological Survey of Canada, Paper 88–10, 91–4.

Hayley, D.W. 1982. "Application of Heat Pipes to Design of Shallow Foundations on Permafrost." In *Proceedings, Fourth Canadian Permafrost Conference*, National Research Council of Canada, Ottawa, 535–44.

Hughes, O.L. 1969. *Distribution of Open System Pingos in Central Yukon Territory with Respect to Glacial Limits*. Geological Survey of Canada, Paper 69–34.

Hughes, O.L., Rampton, V.N., and Rutter, N.W. 1972. *Quaternary Geology and Geomorphology, Southern and Central Yukon*. 24th International Geological Congress, Montreal, Guidebook A-11.

Jahn, A. 1976. "Contemporaneous Geomorphological Processes in Longyeardalen, Vestspitsbergen (Svalbard)." *Biuletyn Peryglacjalny* 26: 253–68.

Konischev, V.N. 1982. "Characteristics of Cryogenic Weathering in the Permafrost Zone of the European U.S.S.R." *Arctic and Alpine Research* 14: 261–5.

Leonoff, C.E., and Lo, R.C. 1982. "Solution to Frost Heave of Ice Arenas." In *Proceedings, Fourth Canadian Permafrost Conference*, National Research Council of Canada, Ottawa, 481–6.

Lewkowicz, A.G. 1986. "Nature and Importance of Thermokarst Processes, Sandhills Moraine, Banks Island, Canada." *Geografiska Annaler* 69A: 321–7.

– 1988. "Slope Processes." In M.J. Clark, ed., *Advances in Periglacial Geomorphology*, New York: John Wiley and Sons, 325–68.

Mackay, J.R. 1973. "Problems in the Origin of Massive Icy Beds, Western Arctic, Canada." In *Permafrost, North American Contribution*, Second International Conference, Yakutsk, USSR, National Academy of Sciences Publication 2115, Washington, DC, 223–8.

– 1974. "Reticulate Ice Veins in Permafrost, northern Canada." *Canadian Geotechnical Journal* 11: 230–7.

– 1975. "The Closing of Ice-wedge Cracks in Permafrost, Garry Island, Northwest Territories." *Canadian Journal of Earth Sciences* 12: 1668–74.

– 1979. "Pingos of the Tuktoyaktuk Peninsula Area, Northwest Territories. *Géographie physique et quaternaire* 33: 3–61.

– 1980. "The Origin of Hummocks, Western Arctic Coast." *Canadian Journal of Earth Sciences* 17: 996–1,006.

– 1981. "Active Layer Slope Movement in a Continuous Permafrost Environment, Garry Island, Northwest Territories, Canada." *Candian Journal of Earth Sciences* 18: 1,666–80.

– 1983. "Downward Water Movement into Frozen Ground, Western Arctic Coast, Canada." *Canadian Journal of Earth Sciences* 20: 120–34.

– 1984a. "The Direction of Ice-Wedge Cracking in Permafrost: Downward or Upward?" *Canadian Journal of Earth Sciences* 21: 516–24.

– 1984b. "The Frost Heave of Stones in the Active Layer above Permafrost with Downward and Upward Freezing." *Arctic and Alpine Research* 16: 439–46.

– 1985. "Pingo Ice of the Western Arctic Coast, Canada." *Canadian Journal of Earth Sciences* 22: 1,452–64.

– 1986a. "Fifty Years (1935–85) of Coastal Retreat West of Tuktoyaktuk, District of Mackenzie." In Current Research, Part A, Geological Survey of Canada, Paper 86–1A, 727–35.

– 1986b. "Frost Mounds." In H.M. French, ed., "Focus, Permafrost Geomorphology," *Canadian Geographer* 30: 363–4.

– 1988. "Catastrophic Lake Drainage, Tuktoyaktuk Peninsula Area, District of Mackenzie." In Current Research, Part D, Geological Survey of Canada, Paper 88–1D, 83–90.

Mackay, J.R., and Mathews, W.H. 1974. "Needle Ice Striped Ground." *Arctic and Alpine Research* 6: 79–84.

Mackay, J.R., Ostrick, J., Lewis, C.P., and Mackay, D.K. 1979. "Frost Heave at Ground Temperatures below 0°C, Inuvik, Northwest Territories." Geological Survey of Canada, Paper 79–1A, 403–6.

Pollard, W.H. 1988. "Seasonal Frost Mounds." In M.J. Clark, ed., *Advances in Periglacial Geomorphology*, New York: J. Wiley, 201–29.

Pollard, W.H., and French, H.M. 1980. "A First Approximation of the Volume of Ground Ice, Richards Island, Pleistocene Mackenzie Delta, Northwest Territories, Canada." *Canadian Geotechnical Journal* 17: 509–16.

– 1983. "Seasonal Frost Mound Occurrence, North Fork Pass, Ogilvie Mountains, Yukon." *Proceedings, Fourth International Conference on Permafrost*, Washington, DC: National Academy Press, 1,000–4.

Rampton, V.N. 1973. "The Influence of Ground Ice and Thermokarst upon the Geomorphology of the Mackenzie-Beaufort Region." In B.D. Fahey and R.D. Thompson, eds., *Research in Polar and Alpine Geomorphology*, Proceedings, 3rd Guelph Symposium on Geomorphology, Norwich: Geo Abstracts, 43–59.

– 1982. *Quaternary Geology of the Yukon Coastal Plain*. Geological Survey of Canada, Bulletin 317.

Ray, R.J., Krantz, W.B., Caine, T.N., and Gunn, R.D. 1983. "A Model for Sorted Patterned Ground Regularity." *Journal of Glaciology* 29: 317–37.

Savigny, W., and Morgenstern, N.R. 1986. "In-situ Creep Properties of Ice-rich Permafrost Soil." *Canadian Geotechnical Journal* 23: 504–14.

Shilts, W.W. 1978. "Nature and Genesis of Mudboils, Central Keewatin, Canada." *Canadian Journal of Earth Sciences* 15: 1,053–68.

Sloan, C.E., and van Everdingen, R.O. 1988. "Region 28, Permafrost Region," In W. Back, J.S. Rosenstein, and P.R. Seaber, eds., *Hydrogeology*, Geological Society of America, The Geology of North America, Vol. 0-2, 263-70.

Smith, M.W. 1985a. "Models of Soil Freezing." In M. Church and O. Slaymaker, eds., *Field and Theory: Lectures in Geocryology*, Vancouver: University of British Columbia Press, 96-120.

– 1985b. "Observations of Soil Freezing and Frost Heave at Inuvik, Northwest Territories, Canada." *Canadian Journal of Earth Sciences* 22: 283-90.

– 1986. "Frost Action and Soil Freezing." In H.M. French, ed., "Focus, Permafrost Geomorphology," *Canadian Geographer* 30: 358-64.

Smith, M.W., Dallimore, S.R., and Kettle, R.J. 1985. "Observations and Prediction of Frost Heave of an Experimental Pipeline." In S. Kinosita and M. Fukuda, eds., *Proceedings, Fourth International Symposium on Ground Freezing*, 5-7 Aug. 1975, Sapporo, Japan, and Rotterdam: Balkena, 297-304.

Thorn, C.E., 1979. "Bedrock Freeze-Thaw Weathering Regime in an Alpine Environment, Colorado Front Range." *Earth Surface Processes* 4: 211-28.

– 1988. "Nivation: A Geomorphic Chimera." In M.J. Clark, ed., *Advances in Periglacial Geomorphology*, New York: J. Wiley and Sons Ltd., 3-31.

van Everdingen, R.O. 1981. *Morphology, Hydrology and Hydrochemistry of Karst in Permafrost Terrain near Great Bear Lake, Northwest Territories*. National Hydrology Research Institute, Calgary, Paper 11, I.W.D. Scientific Series No. 114.

Van Vliet-Lanoe, B. 1988. "The Significance of Cryoturbation Phenomena in Environmental Reconstruction." *Journal of Quaternary Science* 3: 85-96.

Washburn, A.L. 1980. *Geocryology: A Survey of Periglacial Processes and Environments*. New York: J. Wiley.

– 1989. "Near-Surface Soil Displacement in Sorted Circles, Resolute Area, Cornwallis Island, Canadian High Arctic." *Canadian Journal of Earth Sciences*, 25: 941-55.

Watts, S.H. 1983. "Weathering Pit Formation in Bedrock near Cory Glacier, Southeastern Ellesmere Island, Northwest Territories." In Current Research, Part A, Geological Survey of Canada, Paper 83-1A, 487-91.

Williams, P.J., and Smith, M.W. 1989. *The Frozen Earth: Fundamentals of Geocryology*. Cambridge: Cambridge University Press.

Zoltai, S.C., and Pettapiece, W.W. 1973. *Terrain, Vegetation and Permafrost Relationships in the Northern Part of the Mackenzie Valley and Northern Yukon*. Report 73-4, Environmental-Social Committee Northern Pipelines, Task Force on Northern Oil Development, Ottawa: Information Canada.

PART THREE

Mountain Environments

CHAPTER SEVEN

Cold Mountains of Western Canada

OLAV SLAYMAKER

DISTINCTIVE CHARACTERISTICS

The most obvious and distinctive attribute of mountain environments is rugged topography or, more precisely, great local relief – large difference in elevation between the highest and lowest points within a small area. Great local relief gives rise to steep slopes and high-elevation areas. These three topographic attributes – great relief, steep slopes, and high elevations – have further implications, as recognized by Hewitt (1972). Local relief gives rise to variations in energy and moisture conditions over short distances. Steep slopes result in high "geopotential energy" and high elevations, in turn, give rise to cold climatic conditions, more analogous to those of the Arctic than the temperate climates of the valleys.

In this chapter our starting point is the assertion that "regionally characteristic mountain environments ... exist throughout the mountains of the world" (Thompson 1960). Coldness is central to such a discussion. Here the distinctiveness of western Canada's mountain environments is used to identify unifying characteristics. Three themes can be recognized: the roles of snow, ice, and water in Canada's Cordilleran drainage basins; the distinctive nature of sediment and solute movement in such basins; and mountain zonation as a conceptual framework for integration of the morphology and mass fluxes typical of these cold mountainous environments.

Whereas processes associated with Canada's northern environments are distinctively cold-temperature–controlled, those connected with Canada's mountain environments are more distinctively controlled by water, snow, and glacier flux. The so-called alpine mosaic of land-cover types emphasizes mesoscale variability (Table 7.1). More general discussions of mountain geomorphology and hydrology can be found in Slaymaker and McPherson (1972), Slaymaker (1974), Barsch and Caine (1984), Alford (1985), and Slaymaker (1987a).

WATER FLUXES

Availability of water, whether gaseous, liquid, or solid, and the ways in which it

Table 7.1
Major processes in mesoscale alpine systems

Environmental system	$>10^3$ yr	$10–10^3$ yr	$10^{-1}–10$ yr	$<10^{-1}$ yr (c. 1 month)
1 Glaciers	Water storage term in hydrologic cycle	Glacier retreat and advance	Mass balance	Physics of glacier motion
2 Snowpacks	–	–	Snow metamorphism	Water balance
3 Alpine lakes	Glacier scour Lake formation Sedimentation	Sedimentation	Lake draining events Ice formation and breakup	Water balance Sediment dynamics
4 Mountain streams	Glacial erosion	Channel evolution	Extreme events	Runoff concentration "Tumbling" flow regime Sediment dynamics
5 Morainic mounds	Glacial deposition	Degradation of morainic slopes	Chemical weathering Frost action Vegetational change	Water balance Sediment and solute sources
6 Alpine and sub-alpine meadows, valley bottom	Sedimentation in glacial lakes	Dissection by streams Floodplain development Soil formation	Periodic inundation Vegetational change	Water balance Sediment and solute sources
7 Alpine and sub-alpine meadows, adret slope	Postglacial soil development	Slope degradation Channel dissection	Vegetational change Mass movement	Water balance Sediment and solute sources
8 Alpine and sub-alpine treed slopes, adret	Postglacial soil development Soil creep Gullying	Slope degradation Channel dissection	Frost action Mass movement	Water balance Sediment and solute sources

9 Alpine and sub-alpine treed slopes, ubac and ridge top	Postglacial soil development Slides and flows of earth and mud	Slope degradation	Frost action Mass movement	Water balance Sediment and solute sources
10 Alpine barren	Glacial erosion Postglacial rockfall and talus Slope formation	Slope degradation	Frost action Mass movement	Water balance Sediment and solute sources

Source: Slaymaker (1974).

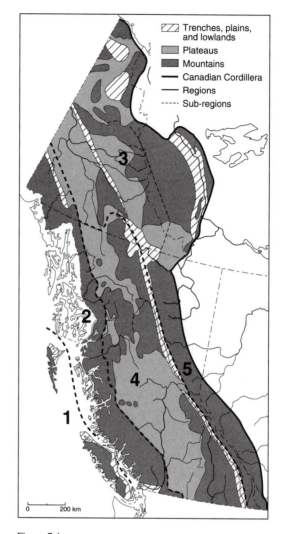

Figure 7.1
Hydrologic regions and climate stations in Cordillera: (1) Insular Mountains, (2) St Elias and Coast Mountains, (3) Subarctic Dry Interior Mountains, (4) Temperate Interior Alpine Mountains, and (5) Temperate Dry Interior Mountains. Climate station data in Table 7.2

transports and transforms elements of its environment help explain the distinctiveness of contemporary mountain environments. Because mountains enhance precipitation totals and variability, receive much of their precipitation as snow, and contain the greatest variety of mechanisms for removing water from mountain peak to valley bottom, water, sediment, and solute fluxes are relatively more important in mountains

Table 7.2
Climatic data for Canadian Cordillera (1951–80)

	Elevation (m)	R	S	Days P	P_{24} max.	T (°C)	T_{min} (°C)	Frost-free days
Insular Mountains								
Bear Creek	351	3,266	247	191	363	7.0	−17.2	165
Masset	12	1,353	88	211	76	7.6	−18.9	160
Port Alice	15	3,181	58	212	234	9.3	−12.8	222
Tofino	20	3,226	53	199	174	8.9	−15.0	203
Coast Mountains								
Allison Pass	1,341	473	1,432	174	81	1.8	−42.8	32
Alta Lake	668	801	657	185	80	5.7	−30.6	116
Ocean Falls	5	4,232	155	218	234	8.1	−18.3	199
Prince Rupert	82	2,966	181	246	175	6.7	−24.4	156
Subarctic Dry Interior Mountains								
Aishihik	966	162	105	109	45	−4.4	−56.7	47
Elsa	814	220	203	122	46	−4.4	−51.7	86
Snag	587	210	155	122	53	−6.1	−62.8	51
Watson Lake	689	239	229	153	46	−3.3	−58.9	93
Temperate Interior Alpine Mountains								
Barkerville	1,265	506	538	177	71	1.4	−46.7	48
Germansen Landing	747	275	269	157	54	0.3	−48.8	55
Glacier (Mt Fidelity)	1,875	474	1,975	169	72	0.2	−30.5	65
Old Glory Mountain	2,347	179	551	183	46	−1.8	−37.8	21
Temperate Dry Interior								
Banff	1,397	263	251	136	54	2.5	−51.1	89
Columbia Icefield	1,981	239	643	157	62	−2.1	−41.1	16
Grande Cache	1,250	369	278	140	89	2.1	−40.0	86
Red Pass Junction	1,059	355	405	154	71	1.7	−39.4	59

Source: AES (1982b).

Note: R = rainfall (mm), S = snowfall (water equivalent, mm), *Days P* = number of days with measurable precipitation, P_{24} *max* = maximum rainfall recorded in 24 hours (mm), T = mean daily temperature and T_{min} = minimum temperature recorded.

than in other environments. For water fluxes a traditional water balance framework is used. The water balance is defined as:

$$P - ET = Q + \Delta s$$

where P is precipitation, ET evapotranspiration, Q runoff, and S storage (whether ice, snow, soil moisture, groundwater, or lakes) over a specified period (t). In mountain regions, available climatic data are confined largely to low-elevation sites, so interpolation and extrapolation from standard meteorological stations of Environment Canada's Atmospheric Environment Service (AES) are often necessary (Figure 7.1 and Table 7.2).

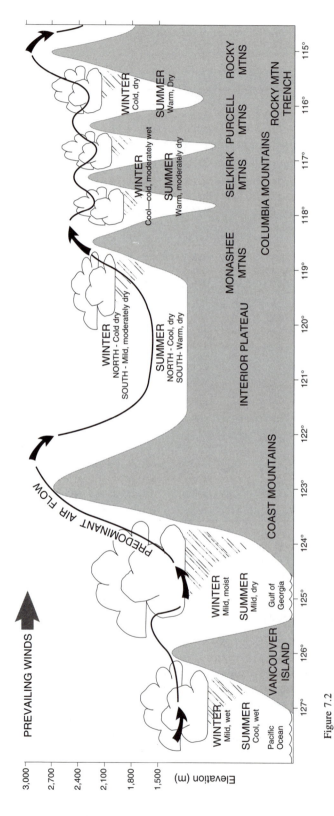

Figure 7.2
West-east profile of Canada's Cordillera at 50°N.

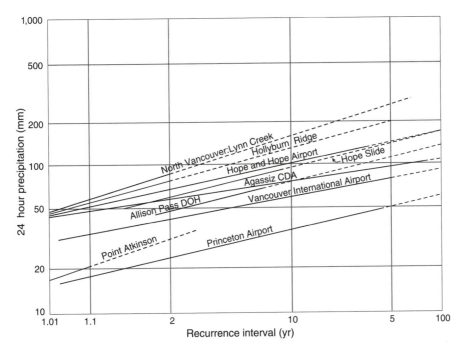

Figure 7.3
Magnitude and frequency of 24-hour precipitation in the Coast Mountains. *Source*: AES (1982b).

Precipitation

Over windward slopes in British Columbia and southwest Yukon (Outer and Coast Mountains), mean annual precipitation exceeds 3,000 mm over large areas. Leeward slopes and interior valleys of the region have annual totals of 400 mm or less (Figure 7.2). The lowest values are 200 mm, around Old Crow in interior northern Yukon, and the highest values recorded are 7,600 mm, on the Queen Charlotte Islands. Local variations in magnitude and frequency of precipitation are great (Figure 7.3).

Annual snowfall is greatest in the St Elias Mountains, the northern and central Coast Mountains, and the Columbia and Rocky Mountains, each with over 600 cm (or 600-mm water equivalent). Windward slopes on Vancouver Island and the Queen Charlotte Islands receive less than 100 cm of snowfall. Indeed, on the latter, in eight years out of ten there is no continuous snow cover.

Variation of ground-snow load with elevation in southern British Columbia (Claus et al. 1984) is shown in figure 7.4A–D. The Coast Mountains (B) are relatively mild and humid and receive extremely heavy precipitation; snow loads increase most rapidly with elevation. The Columbia Mountains and the Rockies have moderate

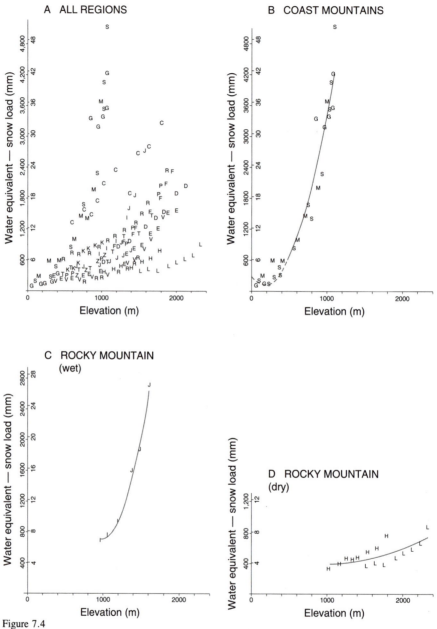

Figure 7.4
Elevation v. snow load for (A) five mountain regions in southern British Columbia, (B) Coast Mountain region (C) Rocky Mountain (wet) region, and (D) Rocky Mountain (dry) region. *Source*: Claus et al (1984).

Figure 7.5
Glacier retreat, Coast Mountains: Oval, Scimitar, Chaos, and Radiant glaciers, typical glacier retreat of past century, with valley glaciers penetrating well below timberline. *Source*: Province of British Columbia air photo BC 274–47.

precipitation on west-facing slopes (C), while valleys and east-facing slopes are semi-arid (D). Consequently, on the Alberta side of the Rocky Mountains, snow loads increase more slowly with elevation.

Glaciers

Data on mass balance of glaciers in Canada's western mountains are extensive, but only on Berendon, Place, and Sentinel (Coast Mountains) and Peyto and Ram (Rocky Mountains) glaciers have complementary runoff measurements been made (Figure 7.5). Peyto, Blue, and South Cascade glaciers (the latter two in adjacent Washington state) are three of only eighteen glacier basins in the world with unbroken records longer than fifteen years. Comparatively little monitoring of cold or dry-based glaciers, such as exist in the Mackenzie Mountains, has been carried out in

Canada's Cordillera. For temperate or wet-based glaciers, there is clear differentiation between the mass balance of continental glaciers (greatly affected by energy inputs) and maritime glaciers (which respond more to precipitation inputs). Also, the occasional and exceptional year is critical in the overall glacier/climate reaction (Young and Ommanney 1984).

Yarnal (1984) compared the mass balance, summer balance, winter balance, and equilibrium-line altitude of three Canadian glaciers (Peyto, Place, and Sentinel) with the meteorological records of neighbouring stations for the period 1966–84. While Peyto Glacier's mass balance is almost entirely related to summer temperature, that of Sentinel is controlled mostly by winter precipitation, and Place is influenced by both elements. The net effect of the past 100 years of climatic amelioration on Coast Mountains glaciers is illustrated in Figure 7.5.

Groundwater and Soil Moisture

There are few studies on groundwater recharge and storage in Canada's coastal mountain regions. There is huge local variation in hydraulic conductivities of complex geologic structures such as Quaternary volcanic piles in the Coast Mountains, with local relief of up to 2,300 m a.s.l. By contrast, groundwater systems in the limestone massifs of the Eastern Cordillera have been explored at regional and local scales (see chapter 8).

Storage of moisture in Canada's western mountains is influenced by the presence or absence of permafrost (Woo 1986). South of the discontinuous permafrost limit, distribution of alpine permafrost is complex: it is affected not only by geothermal heat flux, periodic temperature variations, and diffusivity of earth materials, but also by snow depth and density distribution. In general, the great depth of snow cover in the Western Cordillera inhibits development of permafrost at altitudes where lapse rates in the atmosphere predict frozen ground. As a result, alpine permafrost is more common in the northern interior and Eastern Cordillera; it is absent from the Queen Charlotte Islands and Vancouver Island and occurs only above 2,300 m a.s.l. in the Pacific Ranges of the Coast Mountains and the Cascade Mountains.

Soils of mountain regions are generally shallow and coarse. Hydraulic conductivities are therefore high, and water moves rapidly into and through the regolith. Where present, highly indurated basal till with low hydraulic conductivity increases surface runoff and encourages local gullying. A further factor is that of water repellency. Although it has been understood for some time that slash burning affects the generation of water repellency for up to two years in the perhumid Coast Mountains, extensive natural water repellency has been documented recently in the alpine regions of BC mountains (Barrett and Slaymaker 1989). Its source has not been identified, but it appears to be common in patchy soils above timberline in the Coast, Selkirk, and Rocky Mountains. It leads to bypassing of the soil matrix, encourages gullying, and reduces soil moisture storage even more than hydraulic conductivity estimates would predict.

Runoff

The Coast Mountains and Outer Mountains can be viewed as a single region in terms of precipitation regime, flood regime, and regional analysis of flood extremes. Floods induced by rainfall greatly exceed those brought on by snowmelt, and almost all floods occur in the six months of fall and winter.

By contrast with the coastal region, the remainder of western Canada's mountains are dominated by nival floods and floods produced by either blockage by snow and ice or by land failure. Church (1988) discusses these flood-producing mechanisms and the runoff regime and associated geomorphic work. He illustrates (Figure 7.6A) nival runoff regimes from Fraser River (Southern Rocky Mountains, Coast Mountains, and Interior Plateau), Beatton River (Northern Rocky Mountains), Boundary Creek (Southern Rocky Mountains), Jamieson Creek (Coast Mountains), and Babbage River (Brooks Range); a glacier-influenced regime of Lillooet River (Coast Mountains); and a glacier-burst flood (Coast Mountains).

The Cordilleran region yields high runoff per unit area. Glacierized mountainous basins there have the largest unit-area annual flow, and late summer is the period of maximum runoff. In non-glacierized mountain basins, peak flow occurs in May or June, in response to snowmelt. The plateau basins have the least runoff in the Cordilleran region, and peak flow is concentrated within a snowmelt season shorter than that in the mountainous basins (Figure 7.6B).

When plotted against basin area, mean annual runoff clusters along several lines, each of which may be related to the hydrological characteristics of the region (Figure 7.7A). For example, analysis of records for ninety-seven BC gauging stations (1965–84) yields clear differentiation in mean annual flood characteristics among exposed coast and fjords, the Coast Mountains, and the Columbia and Rocky Mountains (Figure 7.7B).

Hydrologic Regions

At least five hydrologically distinct regions can be defined for Canada's western mountains (Figure 7.8). Only the first is not dominated by snow or permafrost:

(1) perhumid insular mountains, rainfall-dominated and with extremely intense runoff;
(2) moist maritime mountains, snow-dominated and glacier-influenced, with intense runoff;
(3) subarctic dry interior mountains, permafrost-dominated and with low runoff intensity;
(4) temperate interior alpine mountains, snow-dominated, and with average runoff intensity, and
(5) temperate dry interior mountains, snow-dominated, with sporadic permafrost and low runoff intensity.

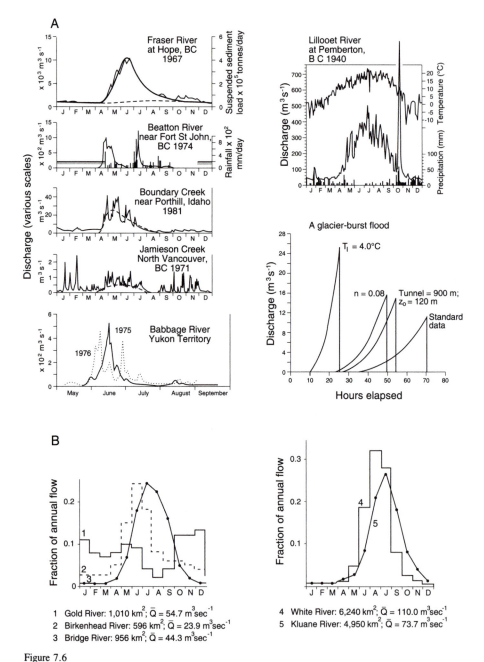

Figure 7.6

(A) Runoff regimes in the Cordillera; *Source*: Church (1988). (B) Monthly distribution of runoff for five Cordilleran basins; *Sources*: Water Survey of Canada (1989a; 1989b).

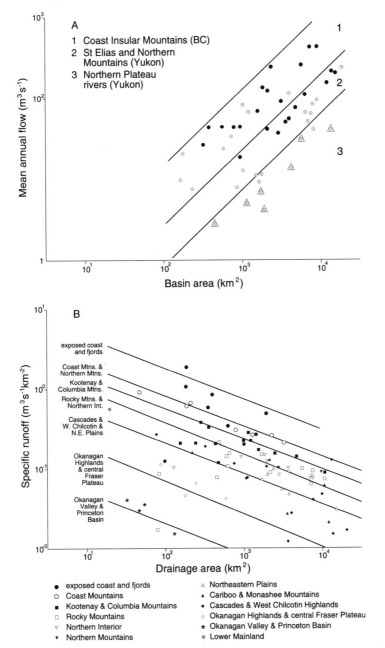

Figure 7.7
(A) Mean annual flow v. basin area for non-glacierized and glacierized Cordillera mountain basins and Interior Plateau; (B) mean annual flood (1965–84) in British Columbia, by region. *Sources*: Water Survey of Canada (1989a; 1989b).

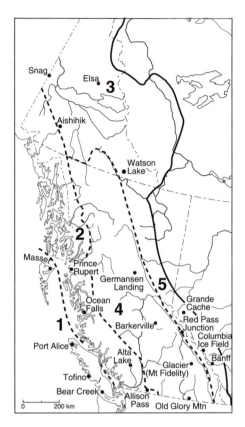

Figure 7.8
Hydrologic and physiographic regions of Canada's Cordillera.

There is a gradation in permafrost conditions between regions 3 and 4. The northern part of region 3 experiences continuous permafrost down to sea level. In central Yukon, widespread discontinuous permafrost at lower altitudes is associated with continuous permafrost in the mountains; in southern Yukon and northern British Columbia, a zone of sporadic discontinuous permafrost occurs at low levels, with continuous permafrost in the mountains. In central British Columbia, permafrost is restricted to the alpine zone, and in the southernmost part of the province, sporadic permafrost occurs in the Monashee Mountains.

SEDIMENT AND SOLUTE FLUXES

Sediment Fluxes

With respect to clastic sediment production in cold mountain environments, ongoing field experiments concern frequency, magnitude, and spatial distribution of rockfalls (e.g. Gardner 1983). Sediment sources on slopes have been described by Schreier and Lavkulich (1985), and sediment accumulations in small ponds have provided rates of sediment production over longer periods (Souch and Slaymaker 1986). A quantitative statement of the rates of detachment, transport, and discharge of rock and soil materials from a basin is as follows:

$$(e_s + e_f + m) - (S_s + S_t) = S_o,$$

where e_s is slope erosion, e_f is fluvial erosion, m is mass movement, S_c is sediment stored, S_t is sediment in transit within the basin, and S_o is sediment leaving the basin (Slaymaker 1988b).

Roberts and Church (1986), applied a sediment budget approach superbly in the Queen Charlotte Ranges. They assessed the impact of logging on sediment production and delivery to stream channels by comparing pre-logging and post-logging sediment budgets. In two of four mountainous basins, clear-cutting produced more sediment to channels than forested areas, but in the other two basins the reverse was the case. They identified the buildup of coarse sediment storages in the stream channels of each basin and related sediment transport out of the reach to in-channel sediment storage. This finding demonstrates the difficulty of differentiating anthropogenic effects from "natural" landslides. The sediment budget approach permits ranking of the importance of sediment producing, transporting, and accumulating processes and provides increased understanding for watershed managers.

Attempts to establish indices of freeze-thaw frequency based on mean temperatures from standard sources (Fraser 1959; Williams 1964) have led to a useful regionalization scheme for the Canadian Cordillera in terms of duration, severity, and frequency of freeze-thaw events. Three zones are commonly identified: (1) the BC coast; (2) Yukon, Northwest Territories, and the BC northern interior; and (3) the rest of the BC interior and Alberta. Zone 1 experiences fewer than 25 freeze-thaw events per year; zone 2, 25 to 50; and zone 3, more than 100, extending over nine to ten months. Landform evidence lends credence to these three zones from the viewpoint of mechanical weathering. For example, rock glaciers are not found on the western slopes of the BC Coast Mountains and Vancouver Island but are present on the eastern slopes of the Cascade and Coast Mountains and in the BC northern interior and Yukon (Johnson 1984). Deep snow cover further accentuates the contrast in freeze-thaw activity between coastal and interior locations.

In British Columbia and Yukon, sediment yields are not, in general, predictable from contemporary sediment production from bedrock. Rates of bedrock landform evolution are approximated only by the sediment yields of the smallest river systems (<100 km^2) which are of order 1–10 mm of lowering per 1,000 years (1 to 10 Bubnovs). There is no apparent connection between sediment yields of intermediate-scale systems (1,000–100,000 km^2), which are of order 100–500 Bubnovs, and rates of general landform evolution. Sediment yields in large-scale systems ($>100,000$ km^2), of order 10–50 Bubnovs, provide integrated measures of erosion and sediment redistribution rates throughout the mountain region (Slaymaker 1987b).

Intermediate-scale systems, with their high sediment yields, function as paraglacial systems, still responding to the high sediment–producing events of 9–12000 yr BP (Church and Ryder 1972).

Solute Fluxes

Conventional geomorphic wisdom suggests that mechanical weathering completely overshadows chemical weathering in siliceous high mountain regions. Because of thin alpine soil cover and the angularity of surface rock fragments, chemical weathering activity has been thought to be low in alpine and glacial regions. In addition, low average temperatures inhibiting chemical weathering reactions and low concentrations of carbon dioxide from biologic processes have been cited as confirming this observation (e.g. King 1986). By contrast, Carson and Kirkby (1972) noted large volumes of turbulent, well-aerated water constantly replenishing the acid potential of water. Their theoretical model shows chemical weathering of igneous rock most significant in areas of highest runoff, whether they are hot or cold. This model predicts broad-scale variation in chemical weathering consistent with the five-fold hydrologic regionalization proposed earlier.

Gallie and Slaymaker (1984; 1985), working in first- and second-order mountain watersheds in the BC Coast Mountains, report that most water there appears to bypass the soil matrix and moves either as overland flow or along discrete, saturated subsurface flow paths. These preferred routes limit primary soil and water reactions as well as ionic inputs from adjacent, relatively inactive zones. The hydrochemical balances of perhumid granitic watersheds there have been calculated on the basis of our understanding of preferred flow paths of water and solutes. In five cases examined to date, substantial silica removal of the order of 6–9 tonnes/km^2/yr has been found (Zeman and Slaymaker 1978). High altitudes and high relief thus appear to be associated not only with coldness but also with extensive chemical weathering caused primarily by substantial runoff.

MOUNTAIN ZONATION

There are two types of mountain zonation: vertical, resulting from changing altitude (e.g. Troll 1972; Caine 1984), and horizontal, caused by latitude, continentality, "mountain mass" effect, and aspect or pathways of gravity-driven denudation processes. Mountain zonation provides a framework for integrating morphological and dynamic aspects of coldness in mountain environments (Ives and Barry 1974; Slaymaker and McPherson 1977; Price 1981).

Altitudinal Zonation

"The transition from forest to tundra on a mountain slope is one of the most dramatic ecotones on earth ... For some reason, or combination of reasons, trees grow only

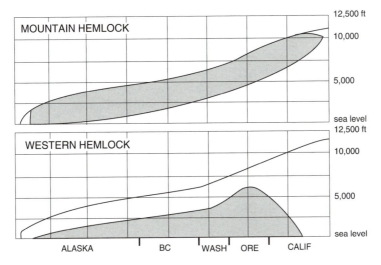

Figure 7.9
Elevational distribution of two species of trees: one timberline species and one lower-elevation (montane) conifer. *Source*: Arno and Hammerly (1984).

up to a certain elevation" (Price 1981: 271). The mean isotherm of 10°C in the warmest month corresponds fairly closely with timberline, though snow, wind, sunshine, and biotic factors cause local variations. In the Canadian Cordillera, vegetation zonation toward timberline varies according to location. In the perhumid "hypermaritime" locations of the west and south coasts, a mountain hemlock zone (mountain hemlock, amabilis fir, and yellow cedar) lies below timberline (Figure 7.9); in the more continental locations of the southern and central interior, an Engelmann spruce–subalpine fir zone lies adjacent to timberline, and in the north, a spruce-willow-birch zone adjoins alpine tundra. There is zonation by vegetation and also by geomorphic processes, associated with contrasts above and below the lower limits of permafrost and of periglacial activity.

Permafrost in the Cordillera is a continuous zone at higher elevation and latitudes. Below this zone lies a discontinuous one, down to the lower altitudinal limit of permafrost. In Yukon, permafrost occurs in valley bottoms; in southern British Columbia, distribution of mountain ("sporadic") permafrost is influenced by many factors, including relief, vegetation, hydrology, snow cover, fire, and soil and rock type.

Periglacial activity is sometimes thought to be coextensive with permafrost (e.g. Harris 1988). For the continental-climate mountains of the Eastern Cordillera, this approach may be useful; for the maritime-climate mountains of the Western Cordillera, however, it is confusing, because heavy snowfalls insulate the ground and inhibit permafrost development even where air temperature would normally indicate permafrost and where intense freeze-thaw activity has been monitored (e.g. Mackay and Mathews 1974). In Garibaldi Park therefore the lower limit of periglacial activity

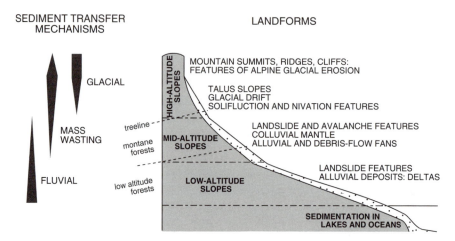

Figure 7.10
Zonation of major Holocene landforms and geomorphic processes, Coast Mountains.

is below timberline (about 1,650 m a.s.l.), whereas permafrost is probably restricted to the upper slopes of Mt Garibaldi (>2,300 m a.s.l.). There, snow accumulation early in autumn, before ground freezing, seems to be crucial. Permafrost is comparatively rare, and both timberline and the lower limit of periglacial activity are comparatively low. In the Canadian Rockies, however, Harris (1981) has shown closer association between the lower limit of permafrost and that of periglacial activity.

A conceptual model of such vertical zonation (Figure 7.10) demonstrates the huge contrasts among high-, mid-, and low-altitude slopes in the coastal mountains. High altitude is above timberline; mid- is from timberline to base of subalpine mountain hemlock zone, and low corresponds to the coastal western hemlock zone.

Harris (1988) defines the alpine periglacial region as low-latitude permafrost areas above 500 m a.s.l. He points out that most zonal (i.e. polar) permafrost landforms occur in alpine areas with a relatively moist climate but that in drier climates such landforms are commonly lacking. In the perhumid Coast Mountains permafrost landforms are absent because of insulating effects of snow.

One vital geomorphic process in the maritime mountains is needle ice growth (Mackay and Mathews 1974). This encourages substantial downslope movement (see chapter 6) (Figure 7.11). In the continental alpine areas, frost comminution is more common; the landforms most characteristic and best developed in the moister parts of continental alpine areas are hummocks, thufa, and rock streams.

Horizontal Zonation (Regional)

In general, timberline declines in elevation northward by 100 m per degree of latitude. Thus, at Haines Junction, Alaska (60°N), ten degrees north of Garibaldi, the tim-

Figure 7.11
Tree island movement in Garibaldi Provincial Park. *Photo*: J.R. Mackay.

berline is approximately 900 m a.s.l.; at 69°N in interior northern Yukon, timberline and the northern forest border merge (Figures 7.12 and 7.13). If summer warmth (specifically 10°C in the warmest month) controls timberline in a latitudinal sense, the mechanism whereby low temperatures stop tree growth is still unknown (Price 1981).

The lower limit of permafrost also declines northward, but more rapidly than the timberline, at 140 m per degree of latitude (Brown and Péwé 1973). In the Haines Junction area, the lower limit of permafrost is about 900 m a.s.l., almost coincident with local timberline.

Figure 7.12
Areas above timberline: (A) BC Coast Mountains, *photo*: British Columbia, BC 576–49; (B) Interior Cascades, *photo*: O. Slaymaker; and (C) Mackenzie Mountains, *photo*: O. Slaymaker.

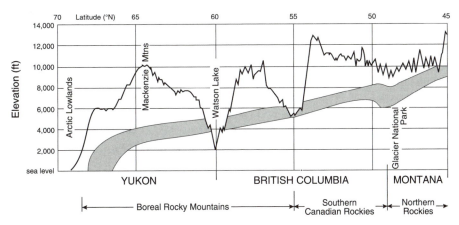

Figure 7.13
Timberline variations: transect of Canadian Cordillera. *Source*: Arno and Hammerly (1984).

At Garibaldi (50°N), the lowest visible indicators of permafrost are 450 m above timberline, but in the Rocky Mountains of Alberta, at approximately the same latitude, timberline is at 2,300 m a.s.l. and sporadic permafrost is recorded 1,000 m below timberline (Harris 1988). Thus timberlines are lower in maritime mountains and higher in continental mountains; the reverse is true for the lower limit of permafrost. Associated with the higher timberline is the higher glaciation level and equilibrium-line altitude associated with continentality. All these effects work together to generate an extensive zone of periglacial activity in the Rocky Mountains by comparison with the rather more limited periglacial zone in the Coast Mountains. On Vancouver Island and the Queen Charlotte Islands there is, effectively, no periglacial zone.

As a result of these latitude and continentality effects, mountain zonation varies as follows (Figure 7.8).

Insular Mountains.   On Vancouver Island and in the Queen Charlotte Ranges, periglacial activity, permafrost, and glaciers are effectively non-existent, although small islands of alpine tundra occur with precipitation greater than 3,000 mm/yr. High-intensity floods follow rain-on-snow events through the period October–February. These mountains are dominated by mass wasting and fluvial activity.

St Elias and Coast Mountains.   Because heavy snow in winter (>25 m) insulates the surface, permafrost is sporadic and occurs only at the highest elevations. Glacier ice is widespread, and surging glaciers are present in the northern part of this region. Chemical weathering associated with late-lying snow patches is relatively active, and large quantities of water may circulate through bedrock. Gelifluction is a function of frost in spring and fall plus large amounts of meltwater. Patterned ground is on a small scale, but on broad ridges, where wind keeps snow to minimal accumulations, thermal-contraction cracking probably takes place. The permafrost zone is substantially above timberline on the western slopes.

Subarctic Dry Interior Mountains.   Permafrost is widespread in areas close to timberline and continuous above timberline. Timberlines are substantially higher than those of the Coast Mountains – for example, in the St Elias Mountains 1,000 m a.s.l. on the interior side and 300 m a.s.l. at the coast. Annual and diurnal temperature ranges are great, and, despite low moisture, periglacial activity such as gelifluction is marked. However, few of these mountain ranges appear to have been glaciated, presumably because of low moisture supply throughout the Quaternary.

Temperate Interior Alpine Mountains. Timberline is higher than in either the Coast Mountains or the subarctic mountains; the lower limit of periglacial activity is lower than in the former but higher than in the latter. It is a transition zone in both west-to-east and south-to-north directions. Locally heavy snowfalls maintain glaciers and small ice caps in the Columbia Mountains.

Temperate Dry Interior Mountains. Harris (1988) describes the altitudinal distribution of periglacial activity in these mountains. Sporadic permafrost occurs over a wide elevation range, followed by a narrow zone of discontinuous permafrost and continuous permafrost above approximately 2,000 m a.s.l. Harris emphasizes the variability of permafrost conditions which result from cold-air drainage and snow phenology.

Horizontal Zonation (Local) and the Alpine Mosaic

The most obvious local-scale horizontal zonation of mountains is caused by gravity-driven processes of slope erosion, mass movement, and river action and by other influences such as aspect (Figure 7.14). These processes intersect the vertical zonation pattern of western Canada's mountains and subdivide the landscape into a series of slope facets bounded on the top and bottom by timberline, snowline, permafrost, or periglacial activity limit and bounded at the sides by a channel way or pathway of sediment, snow, ice, and water movement.

A more subtle effect, not easily documented in the Canadian Cordillera, is the so-called Massenerbehung (mountain mass) effect (Barry 1981), whereby timberline tends to reach higher altitudes on larger mountain massifs than on smaller ranges. On the former, winters are thought to be colder but summers warmer than on small mountains. Thus average temperature and effective growing climate may be more favourable at a given elevation there than in free air at the same altitude. However, this effect is difficult to demonstrate within the Canadian Cordillera because of the confounding effect of latitude, continentality, and glacial history.

Aspect controls timberline at the local scale such that in a typical southern Cordilleran mountain range it may be at 1,850 m a.s.l. on a northwest aspect; at 1,950 m a.s.l. on a northeast; at 2,100 m a.s.l. on a southwest; and at 2,250 m a.s.l. on a southeast aspect. Preferred cirque orientation illustrates longer-term effects of aspect.

Cold-climate mass-wasting processes such as talus development, debris slides, debris avalanches, and debris flows and torrents are widespread in the Canadian Cordillera (Eisbacher and Clague 1984). Debris torrents can be considered a special type of debris flow, characterized by steep channel slope, fast movement, and relatively little cohesive material in the moving water-debris mixture. They have become

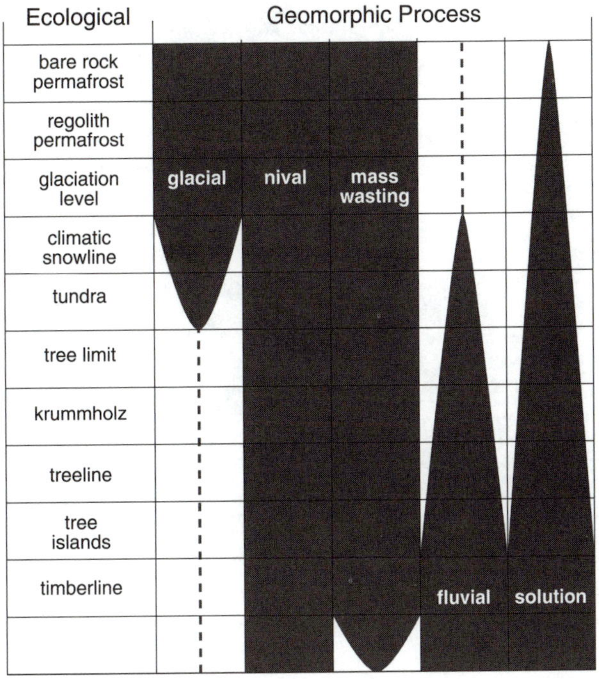

Figure 7.14
Vertical zonation, ecological and geomorphic, of Cordilleran alpine
regions.

of increasing concern over the past decade (see chapter 10 and Slaymaker 1988a).
Such processes emphasize the horizontal zonation of mountain landscapes.

The role of snow avalanche paths in Canada's western mountains is well known
in the context of natural hazard research (Schaerer 1977). The distinctive striped
pattern of high-mountain landscapes is commonly created by snow avalanche paths.
The amount of sediment that avalanches transport is less important than the distance
over which they carry sediment. Avalanche paths consist of three parts: the starting
zone, the track, and the runout deposition zone. The biggest problems occur where
the runout zone intersects major transportation routes. The Rogers Pass, in the Selkirk
Mountains, is one of the most intensively studied, but every mountain region experiences snow avalanche hazard.

Processes of alluvial fan construction appear to differ in various regions of the
Cordillera. There seems to be a relationship among climate and events and frequencies. For example, large-magnitude, high-frequency runoff produces alluvial fans
dominated by fluvial-sediment transport in the Coast Mountains and, to a lesser
degree, in the Rocky Mountains. Further north, however, alluvial fans on the eastern
slopes of the Richardson Mountains, in an environment of continuous permafrost
(Figure 7.15), suggest that debris flows dominate in the fanhead zone.

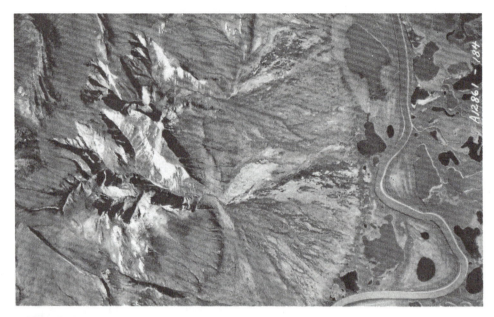

Figure 7.15
Alluvial fans on eastern slopes of Richardson Mountains. *Photo*: National Air Photo Library A12861–184.

The traditional model of mountain zones emphasizes their vertical divisions (Figure 7.10) – pioneers such as Carl Troll (1972) focused on regional scale comparisons, and many definitions of mountain landscapes incorporate altitudinal zonation. Five regionally characteristic mountain landscapes can be defined, four of which owe many distinctive properties to the influence of snow, ice, or cold temperatures. At the local scale, horizontal zonation is often more apparent and is recognized by the term *alpine mosaic*. Mesoscale alpine environments, which comprise this mosaic, can be identified. Sub-regional, local, and mesoscale sources of variability must be incorporated into any dynamic assessment of evolving mountain landscapes. A more realistic conceptual model of the zonation of mountain landforms and processes should look more like Figure 7.15. The effects of society, likely to be greater at the sub-regional scale, can then be assessed in their appropriate regional setting.

REFERENCES

Alford, D. 1985. "Mountain Hydrologic Systems." *Mountain Research and Development* 5: 349–63.
Arno, S.F., and Hammerly, R.P. 1984. *Timberline*. Seattle: The Mountaineers.

Atmospheric Environment Services (AES). 1982a. *Canadian Climate Normals. Temperature and Precipitation (1951–1980): The North*. Toronto: Environment Canada.

– 1982b. *Canadian Climate Normals. Temperature and Precipitation (1951–1980): B.C.* Toronto: Environment Canada.

– 1982c. *Canadian Climate Normals. Temperature and Precipitation (1951–80): The Prairie Provinces*. Toronto: Environment Canada.

Barry, R.G. 1981. *Mountain Weather and Climate*. London: Methuen.

Barsch, D., and Caine, N. 1984. "The Nature of Mountain Geomorphology." *Mountain Research and Development* 4: 287–98.

Brown, R.J.E., and Péwé, T.L. 1973. *Distribution of Permafrost in North America: Proceedings of the Second International Conference on Permafrost*, Washington, DC: National Academy of Sciences, 71–100.

Caine, C. 1984. "Elevational Contrasts in Contemporary Geomorphic Activity in the Colorado Front Range." *Studia Geomorphologica Carpatho-Balcanica* 18: 5–31.

Church, M. 1988. "Floods in Cold Climates." In V.R. Baker, R.C. Kochel, and P.C. Patton, eds., *Flood Geomorphology*, New York: J. Wiley and Sons Ltd., 205–29.

Church, M., and Ryder, J.M. 1972. "Paraglacial Sedimentation." *Geological Society of America Bulletin* 83: 3,059–72.

Claus, B.R., Russell, S.O., and Schaerer, P. 1984. "Variation of Ground Snow Loads with Elevation in Southern B.C." *Canadian Journal of Civil Engineering* 11: 480–93.

Eisbacher, G.H. 1979. "First Order Regionalisation of Landslide Characteristics in the Canadian Cordillera." *Geoscience Canada* 6: 69–79.

Eisbacher, G.H., and Clague, J.J. 1984. *Destructive Mass Movements in High Mountains*. Geological Survey of Canada Paper 84–16.

Fraser, J.K. 1959. "Freeze-Thaw Frequencies and Mechanical Weathering in Canada." *Arctic* 11: 40–53.

Gallie, T.M., and Slaymaker, O. 1984. "Variable Solute Sources and Hydrologic Pathways in a Coastal Subalpine Environment." In T.P. Burt and D.E. Walling, eds., *Catchment Experiments in Fluvial Geomorphology*, Norwich, UK: Geobooks, 347–57.

– 1985. "Hydrologic Controls in Alpine Stream Chemistry." *Proceedings of the 14th N.R.C. Hydrology Symposium*, Quebec City, 287–306.

Gardner, J.S. 1983. "Rockfall Frequency and Distribution in the Highwood Pass Area." *Zeitschrift fur Geomorphologie* 27: 311–24.

Harris, S.A. 1981. "Distribution of Active Glaciers and Rock Glaciers." *Canadian Journal of Earth Sciences* 18: 376–81.

– 1988. "The Alpine Periglacial Zone." In M.J. Clark, ed., *Advances in Periglacial Geomorphology*, New York: J. Wiley and Sons Ltd, 369–413.

Hewitt, K. 1972. "The Mountain Environment and Geomorphic Processes." In O. Slaymaker and H.J. McPherson, eds., *Mountain Geomorphology*, Vancouver: Tantalus Press, 17–34.

Ives, J.D., and Barry, R.G., eds. 1974. *Arctic and Alpine Environments*. London: Methuen.

Johnson, P.G. 1984. "Rock Glacier Formation by High Magnitude Low Frequency Slope Processes in Southwest Yukon." *Annals of the Association of American Geographers* 74: 408–19.

King, R.H. 1986. "Weathering in Holocene Volcanic Ashes." In S.M. Colman and D.P. Dethier, eds., *Rates of Chemical Weathering*, Orlando, Fla.: Academic Press, 239–64.

Mackay, J.R., and Mathews, W.H. 1974. "Movement of Sorted Stripes, Cinder Cone." *Arctic and Alpine Research* 6: 347–59.

Price, L.W. 1981. *Mountains and Man.* Berkeley: University of California.

Roberts, R.G., and Church, M. 1986. "The Sediment Budget in Severely Disturbed Watersheds, Queen Charlotte Ranges." *Canadian Journal of Forest Research* 16: 1,092–1,106.

Ryder, J.M. 1981. "Geomorphology of the Coast Mountains of British Columbia." In O. Slaymaker, ed., *High Mountains, Zeitschrift fur Geomorphologie* Supplement band 37: 120–47.

Schaerer, P.A. 1977. "Analysis of Snow Avalanche Terrain." *Canadian Geotechnical Journal* 14: 281–7.

Schreier, H., and Lavkulich, L.M. 1985. "Rendzina Type Soils in the Ogilvie Mountains." *Soil Science* 139: 2–12.

Slaymaker, O. 1974. "Alpine Hydrology." In Ives and Barry (1974) 133–58.

– 1987a. "Mountain Geomorphology." In R.D. Bedford, ed., *Geography: A Celebration*, Christchurch, NZ: University of Canterbury, 43–83.

– 1987b. "Sediment and Solute Yields in British Columbia and Yukon." In V. Gardiner, ed., *International Geomorphology 1986*, Chichester, UK: J. Wiley and Sons Ltd., 925–45.

– 1988a. "The Distinctive Attributes of Debris Torrents." *Journal of Hydrological Sciences* 33: 567–73.

– 1988b. "Slope Erosion and Mass Movement in Relation to Weathering." In A. Lerman and M. Meybeck, eds., *Physical and Chemical Weathering*, Dordrecht, Netherlands: Kluwer Academic, 83–111.

Slaymaker, O., and McPherson, H.J., eds. 1972. *Mountain Geomorphology.* Vancouver: Tantalus.

Slaymaker, O., and McPherson, H.J. 1977. "An Overview of Geomorphic Processes in the Canadian Cordillera." *Zeitschrift fur Geomorphologie* 21: 169–86.

Souch, C., and Slaymaker, O. 1986. "Temporal Variability of Sediment Yield." *Physical Geography* 7: 140–53.

Thompson, W.F. 1960. "The Shape of New England Mountains." *Appalachia* 145–59.

Troll, C. 1972. "Geoecology and the World-wide Differentiation of High Mountain Ecosystems." In *Geoecology of the High Mountain Regions of Eurasia*, 1–16.

Water Survey of Canada. 1989a. *Historical Streamflow Summary: British Columbia to 1988.* Ottawa: Inland Waters Directorate.

– 1989b. *Historical Streamflow Summary: Yukon and Northwest Territories to 1988.* Ottawa: Inland Waters Directorate.

Williams, L, 1964. "Regionalisation of Freeze-Thaw Activity." *Annals of the American Association of Geographers* 54: 597–611.

Woo, M.K. 1986. "Permafrost Hydrology in North America." *Atmosphere-Ocean* 24: 201–34.

Yarnal, B. 1984. "Synoptic Scale Circulation over B.C. in Relationship to Mass Balance of Sentinel Glacier." *Annals of the American Association of Geographers* 74: 375–92.

Young, G.J., and Ommanney, C.S.L. 1984. "Canadian Glacier Hydrology and Mass Balance Studies." *Geografiska Annaler* 66A: 169–82.

Zeman, L.J., and Slaymaker, O. 1978. "Mass Balance Model for Calculation of Ionic Input Loads in Atmospheric Fallout and Discharge from a Mountainous Basin." *Hydrological Sciences Bulletin* 23: 103–18.

CHAPTER EIGHT

Karst in Cold Environments

DEREK C. FORD

Canada's arctic and alpine environments display major temporal and spatial variations of climate and climatically controlled phenomena. During the past two million or more years, cycles of Quaternary glacier growth and decay wrought greater variety than exists today in the behaviour of the processes that shape landforms and establish or destroy ecological assemblages. From the perspective of a geomorphologist, glacial cycles proceed quite rapidly. As a result, many modern landforms are poly-genetic – they have been moulded by different processes at different times and may preserve (in modified form) features created long ago. The karst geomorphic and hydrologic system is well developed today in Canada's cold regions and preserves in its form and function the events of past glacial and periglacial cycles.

The nature of the karst system is illustrated in Figure 8.1. One process is dominant in it – aqueous dissolution of minerals. Karst is restricted therefore to the more soluble minerals and to rocks composed largely or entirely of them. Because the rocks are soluble, water can circulate underground via systems ("karst aquifers") of solution-enlarged fissures and conduits (caves) instead of discharging through river channels at the surface. Karst aquifers differ from other kinds of aquifers (e.g. "granular," such as spreads of glacial gravels and sands): they are modified erosionally by water flowing through them rather than functioning largely as inert water-storage containers. Although Figure 8.1 (being global in its comprehensiveness) indicates that karst aquifers may be constructed by expelled thermal and other deep waters, most Canadian examples have been created by meteoric waters circulating to depths of no more than a few hundred metres along flow paths ranging from hundreds of metres to several tens of kilometres in length. Thus, in hydrological terms, they are peculiarly behaved sub-basins of larger river basins.

Surface landforms of the karst system serve to funnel water to the underground plumbing system. Alternatively, they are created where water is discharged from it. In most karst areas, input landforms are the more important. Small-scale input features are patterns of solutional pits and grooves termed "karren" (Table 8.1). Individual features vary from one centimetre to a few metres in length or depth, but dense

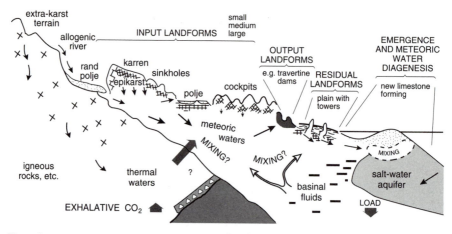

Figure 8.1
Components of comprehensive karst geomorphic and hydrologic system.

Table 8.1
Components of karst landform system

Input features	The aquifer	Output features	Residual features
Microkarren Karren	Diffuse flow components (pores, fissures, proto-caves)	Springs and springhead saps	Pinnacles
Dolines Poljes River valleys	Solution caves and rooms	Travertine and tufa constructions	Conical hills and towers

assemblages ("limestone pavements") can cover many square kilometres. At the intermediate scale, dolines (or sinkholes) are the principal features. They are cylinder-, funnel-, or bowl-shaped closed depressions ranging from one to about 1,000 metres in diameter. Poljes are larger, flat-floored closed depressions that may be seasonally inundated. Much karst water also reaches the underground via sink points in "dry" river valleys. Output landforms include small gorges created by springhead sapping and mounds of precipitated rock (travertine) accumulating where spring waters are depositing some of their dissolved load. Mature karst may also display alluviated corrosion plains below the springs, with residual hills scattered on them that are distinguished by their very steep sides.

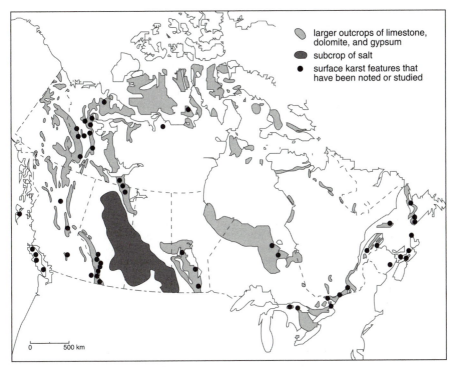

Figure 8.2
Larger outcrops of limestone, dolomite, and gypsum plus subcrop of Elk Point (prairie) salt in Canada.

The major karst rocks are salt, gypsum and anhydrite, limestone, and dolomite. Salt is so soluble that it is exposed as a rock at the ground surface only in the driest of places. Wherever water comes into contact with it underground (no matter how thick the overlying cover rocks may be) it will be dissolved en masse ("interstratal karst"). Gypsum and anhydrite have an equilibrium solubility between 2,000 and 3,000 mg/l in most waters – sufficient to ensure that karst features are rapidly established on them when they become exposed and that their interstratal dissolution is often significant. The equilibrium solubility of limestone and dolomite is lower (50 – 400 mg/l) so that thousands of years are required to establish karst groundwater circulation in them. However, they are areally predominant; the greatest extent and variety of karst landforms develop in them.

Figure 8.2 shows the distribution of karst rocks in Canada. For gypsum, limestone, and dolomite, only the "outcrop" is shown – i.e. where these rocks are not overlain by other consolidated rocks, although they may be mantled by, for example, veneers of glacial drift. These outcrops amount to approximately 1.2×10^6 km^2, or 10 per cent of Canada's landmass. Limestone and dolomite are abundant throughout the arctic islands. They are major mountain-forming rocks in the Rocky Mountains, Mackenzie Mountains, and Wernecke Range. There are lesser outcrops in the BC

Table 8.2
Effects of glacier action on karst geomorphic and hydrologic systems

---

Destructive, deranging
Erasure of karren, shallow (epikarst) aquifers and residuals
Dissection of integrated systems of caves by glacial cirques and valleys
Infilling of dolines and larger input features; aggradation of springs
Injection of clastic detritus into cave systems

Inhibitive
Shielding by carbonate- or sulphate-rich glacial drift protects bedrock surfaces from postglacial
    dissolution

Preservative
Sealing: cover by clay-rich deposits seals and confines aquifers

Stimulative
Focusing inputs and raising hydraulic head – with superimposed glacial streams or glacier aquifers
Lowering spring elevations by glacial entrenchment, to steepen hydraulic gradients
Deep injection: of glacial meltwaters/groundwater when bedrocks are being flexed during crustal
    rebound or depression?

Source: Ford (1987).

interior ranges, in Yukon and on Vancouver Island and other mountainous islands of the west coast. Sequences of limestone, dolomite, and gypsum outcrop cover vast lowland areas of the Mackenzie, Horton, and Anderson river basins.

Flowing glacier ice covered almost all of the lowland karst terrains during the Pleistocene glaciations. In the mountains, ice caps, cirques, and valley glaciers left only nunataks or mountain crest lines bare. Glacial processes entirely suppress the principal processes of most other landforming systems. For example, previous river channels will be erased or infilled by rock detritus. In the lowlands, entire bedrock valleys can be buried, and patterned ground and other periglacial forms are destroyed. The karst system is equally vulnerable, but fissures and caves of the karst aquifer normally survive (with minor truncation of extremities) because they lie deep beneath the glacier sole. Survival of the aquifer (the central component of karst) is the crucial difference between karst and other geomorphic systems active in the formerly glaciated terrains of Canada. However, glacial action can strongly modify the functioning of karst aquifers. Table 8.2 summarizes its effects, which can be stimulative in mountains when a melting glacier (itself an aquifer) is superimposed on a karst.

If we set aside temporarily the effects of past glaciers, we can model the impact of cold on karst. Deepening cold and permafrost progressively restrict the scope for the development of new aquifers. However, permafrost can thaw beneath glaciers. It is necessary therefore to investigate the interrelationship among karst, cold, and glaciation, over part or all of a glacial cycle, in considering Canada's karstic terrain. The presence of karst may affect economic development in cold regions, and human activity substantially affects karst landforms and water resources.

CLIMATIC CONTROLS OF RATES OF
KARST DISSOLUTION

Salt and gypsum dissolve by dissociation alone – the separation of cations and anions as they become exposed at the solid-liquid interface and their subsequent dispersion in the water. The equilibrium solubility of salt – the maximum amount that can be dissolved under given pressure and temperature conditions, such as 350,000 mg/l at 1 bar and 25°C – is so great that it is scarcely ever attained in nature. The equilibrium solubility of gypsum is much lower – 2,400 mg/l in pure water at 25°C, reducing to 2,100 mg/l at 0°C. This simple temperature effect is so small that it does not appear to create any noticeable difference in form and scale between landforms that develop in hot climates and those that develop in cold climates. The solubilities of salt and gypsum are high enough that the only major rate control appears to be the amount of runoff to which they are exposed each year (see chapters 5 and 7).

Dissolution of limestone and dolomite is more complex. Simple dissociation is ineffective. The most important mechanism involves carbon dioxide (a very soluble gas) in two steps:

$$\text{carbonation: } H_2O + CO_2 = H_2CO_3{}^\circ \tag{1}$$
$$\text{(carbonic acid)}$$

$$\text{dissolution: } CaCO_3 + H_2CO_3{}^\circ = Ca_{2+} + 2HCO_3{-} \tag{2}$$
$$\text{(calcite)} \qquad\qquad\qquad \text{(bicarbonate)}$$

The equilibrium solubility of these rocks is determined therefore, by the amount of CO_2 dissolved in the water.

The solubility of CO_2 is inversely proportional to temperature, doubling as water is cooled from 20°C to 0°C. The abundance of the gas in the ordinary atmosphere is approximately 0.03 per cent ($10^{-3.5}$) and does not vary significantly between the Equator and the poles. The equilibrium solubility of limestone and dolomite in waters that have encountered only the ordinary atmosphere is greatest in the coldest regions. This fact led French geomorphologist Jean Corbel (1959) to propose that limestone karst landforms develop most rapidly in arctic and alpine regions.

The solubility of a gas is also directly proportional to its abundance (partial pressure) in the atmospheric mixture of gases. The amount of CO_2 can be greatly increased in the pores of a soil because plant roots and bacteria discharge it there. Partial pressures of 3 per cent ($10^{-1.5}$, a 100-fold increase over the standard atmosphere), or higher, are frequent. Rainwater or meltwater passing through a soil will become enriched in this gas and capable of dissolving much more limestone than can water at the same temperature that has encountered only the ordinary atmosphere (Ford and Williams 1989). The amount of soil CO_2 production tends to increase with temperature. This factor works in opposition therefore to the inverse relationship between CO_2 solubility and temperature, and (as a global generalization) it overwhelms it.

From these considerations White (1984; 1988) developed a theoretical expression for the denudation rate of limestones:

$$D_{max} = \frac{100}{P\sqrt[3]{4}} \left[\frac{K_c K_1 K_{CO2}}{K_2} \right]^{1/3} P_{CO2}^{1/3} (P - E) \tag{3}$$

where D_{max} is the solution denudation rate in $mm/10^3$ yr for the system at equilibrium, P is rock density, K is equilibrium constants of reactions (see Ford and Williams, 1989: 56), P_{co2} is partial pressure of the gas, P is precipitation, and E is evapotranspiration. Temperature enters via the equilibrium constants.

Graphs of this theoretical equation for three different temperatures and three different partial pressures of CO_2 are given in Figure 8.3. Superimposed on the graph are empirical relationships obtained from analysis of more than one hundred published solution denudation studies in different regions of the world, including arctic Canada. It can be shown that, on average, real limestone dissolution rates are lowest in the cold region, where soil CO_2 production is least. For the mountainous regions of Canada we would expect denudation rates to be greatest on Vancouver Island, because its limestone mountains are exposed to the highest mean temperatures and the greatest runoff, and least in northern Ellesmere Island, with its low temperatures and runoff. This is true as a broad generalization (cf. chapter 7).

Detailed studies in the Rocky Mountains (Ford 1971; Drake and Ford 1974) and the High Arctic (Woo and Marsh 1977) reveal that local situations are more complicated: concentrations of dissolved limestone can be as high in the arctic tundra or alpine forest in high summer as anywhere in the world. These waters circulate through unconsolidated glacial sediments that contain large quantities of fresh rock fragments. As a result, rock, soil CO_2, and water can react together immediately. In this "coincidental system," as carbonic acid is removed by reacting with limestone, more can be created by dissolution of further CO_2 (Drake 1984). In extraglacial regions such as northern interior Yukon, soils are normally residual – they contain no fresh rock fragments in the zone of CO_2 gas production. Waters passing through these soils are first enriched in dissolved CO_2 and then encounter soluble rock at greater depth where no additional CO_2 is available to replenish the exhausted carbonic acid. The exhausted carbonic acid (= a "sequential system"). This complication is not allowed for in Figure 8.3. Local soil dissolution rates can be very high during the brief blooming season in the cold regions.

Development of karst landforms and aquifers in the glaciated regions of Canada, requires from tens to tens of thousands of years. This broader time scale overlaps the period required for growth and decay of the major glaciers and ice sheets in a glacial cycle. When glaciers waste rapidly, volumes of ice that might, say, represent one hundred years of net accumulation ($P - E$ in equation 3) may be discharged as

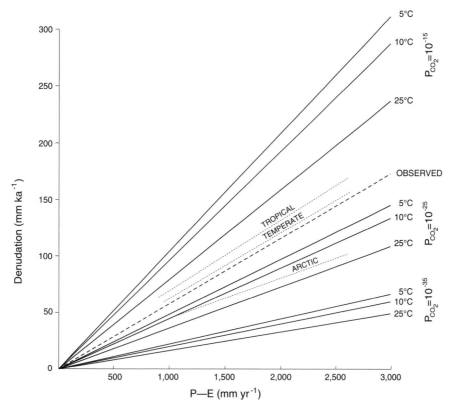

Figure 8.3
W.B. White's model of limestone denudation rates for three different temperatures and three different partial pressures of CO_2. See text re empirical relationships (dotted lines). *Source*: White (1984), with permission.

water within a few weeks. Large quantities of such glacier melt are often focused on a point or narrow band of soluble rock beneath a glacier or at its margins. Dissolution rates at such places are probably the highest in the world; thus the fastest development of individual karst features, such as a streamsink doline or a new section of cave passage, is usually associated with wasting glaciers (e.g. Lauritzen 1986).

KARST, MOUNTAINS, AND GLACIERS

In potential karst regions in any climate, karst processes directed underground into the heart of the rock compete with surficial erosion processes such as mass movement, river-channel cutting, and glacier scouring which destroy it from the outside. The karst system cannot function effectively in this situation until connected series of dissolutional fissures and caves have been created between sinkpoints and springs.

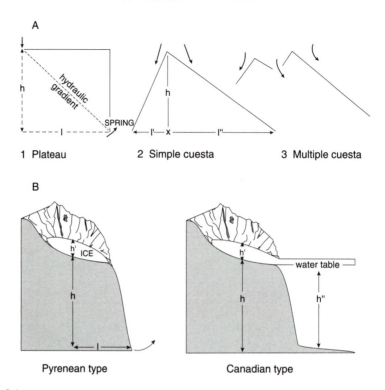

Figure 8.4
(A) Models depicting groundwater hydraulic gradients and surface runoff potentials in plateaus and cuestas; (B) Pyrenean and Canadian types of alpine karst-glacier relationships.

This process may take some tens of thousands of years in limestone and dolomite. Other things being equal (rock solubility and dissolution rate, as in equation 3), the potential for karst system development is greatest where the groundwater hydraulic gradient is steepest. This potential is best realized in mountainous regions.

Figure 8.4A defines the hydraulic gradient (h/l). A plateau setting (Figure 8.4A1) is optimal for karst development because there is a steep hydraulic gradient between potential sinkpoints and springs, while competing surface erosion processes are weakened by the low gradients on the plateau itself. There are many limestone and dolomite plateaus in the Main (central) Ranges of the Rocky Mountains and in the Mackenzie Mountains.

Cuesta-form mountains (figure 8.4A2) may offer comparably steep underground gradients, but here there will be powerful overground competition. River canyons or glacial cirques and valleys may dissect the cuesta before efficient caves can develop within it. Simple cuestas of this type are predominant in the Front (eastern) Ranges of the Canadian Rockies and on parts of Vancouver Island. Karst is comparatively limited in them. However, double or multiple cuestas (Figure 8.4A3) may be created where a few weaker rockbeds are interspersed with massive, strong limestones or where such limestones have been overthrust on themselves. That is common is the Front Ranges also.

In "temperate" or wet-based glaciers, the magnitude of glacier development affects the type and extent of karst system development in the mountains. There are two basic situations, Pyrenean and Canadian (see Figure 8.4B). In the Pyrenean, Quaternary glaciers never extended beyond the higher plateaus, cirques, or valleys; thus they occupied the input (recharge) zones of the system and might further stimulate it by adding additional head there (h' in Figure 8.4B). They did not extend down to major springs in trunk valleys or the piedmont zone and so could not obstruct them. This situation prevailed in most parts of the Pyrenees, the Alps, the Caucasus Mountains, the Colorado Rockies, and the Sierra Nevada of California.

The Canadian situation is more restrictive for karst development. Although there may be a Pyrenean-type superimposition of the glacier aquifer onto the karst recharge zone only when glaciers are first accumulating or are shrinking, for most of a glacial cycle the spring positions are also buried by ice. The result is major disruption of the aquifer, and both inputs and springs can be plugged with glacial detritus. During melt seasons, high water within the glacier (h" in Figure 8.4B) can reduce or even eliminate the groundwater hydraulic gradient. As the name suggests, the arctic and alpine ranges of Canada are the principal mountains where these more restrictive conditions have prevailed.

Canadian glaciers have receded from their late Pleistocene maxima, and there are today very few sites known where karst and glaciers are interacting. Nevertheless, their effects still determine much of the behaviour of modern aquifers. Relationships among karst, mountains, and glaciers are illustrated here with three contrasted examples. In each, the behaviour of the aquifer is investigated through an example of its flood hydrograph at the springs.

White Ridges overlooks the town of Gold River on northwestern Vancouver Island. It is a simple cuesta formed in limestone about 300 m thick (Figure 8.5). The summit altitude is comparatively low (1,200 m asl), so that at the height of glaciation it was buried by flowing ice. This truncated the cuesta to form a narrow plateau: thus the setting is mixed plateau-cuesta topography where, in addition, ice burial in each glaciation would have been brief because of the comparatively mild climate. As a consequence, Pyrenean-type conditions probably prevailed. The plateau is below the modern treeline, and its surface is densely riven with karren which drain into deep shafts. Mature fir forest grows on the karren, producing extra CO_2 at the roots. Precipitation is high, so that mean annual runoff probably exceeds 2,000 mm.

The cave system (and the aquifer that it constitutes) is of the most simple kind. Systems of vertical shafts up to 150 m in depth drain the karren surface directly to the base of the limestone. There they are halted by underlying, insoluble, and impervious volcanic rocks. Cave passages carry their flow along the limestone-volcanic contact for distances of 1,000–2,000 m to a major spring, Quatsino Spring. The passages are blocked by fallen rock or water seals at two places but otherwise are simple, open pipes that may drain dry during a drought. The maximum value of h in the system is 600 m and of l, 2,500 m.

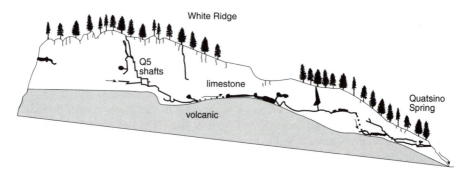

Figure 8.5
Schematic section through White Ridge karst, Vancouver Island.

Figure 8.6A shows the hydrograph at Quatsino Spring for a flood in October 1983 (Ecoch 1984). Before the onset of precipitation, spring discharge was 0.02 m^3/s (20 l/s) and diminishing slowly. In a few hours on 20 October approximately 70 mm of rain fell. It was a typical heavy storm for that time of year at White Ridge. Discharge began to rise almost immediately as shafts in the forest just above the spring fed early rains to it. Discharge peaked at 0.52 m^3/s some ten hours after onset of the rain and approximately three hours after it had ceased. At that time the maximum flow from the largest recharge shafts (Q_5 – see Figure 8.5) was arriving at the spring. Averaged velocities of flow between groundwater sink points and the spring were 300–500 m/hr. Twelve hours later, discharge had fallen to 0.1 m^3/s, and thereafter it declined exponentially ("groundwater baseflow decline") to return to the pre-storm value in about five days.

In summary, the flood produced by four to five hours of heavy frontal rainfall was cleared through the Quatsino karst system within forty-eight hours. This is typical for a small to medium-sized alpine karst where the hydraulic gradient is steep (about 1:4 or 0.25 in this instance), the configuration of passages in the aquifer is simple, and there is little retention of excess water is soils or detrital deposits in the recharge zone. It is a very rapid response; few aquifers in warmer, extraglacial karst regions can clear storm waters so quickly.

Crowsnest Pass is a major east-west breach through successive cuestas of the Front Ranges in the southern Rocky Mountains of Alberta and British Columbia. The centre of the pass is an overdeepened glacier scour filled with water to form Crowsnest Lake (1,300 m a.s.l.). Limestone mountains north and south of it are intensely dissected by glacial cirques and trough valleys. The latter entrenched through ancient aquifers, leaving remnants exposed as dry, fragmentary caves at all altitudes up to the crestline at 2,400–2,700 m a.s.l. There are no modern glaciers of consequence and few long-lasting snowbanks. Many small modern karst aquifers are contained within particular valleys, and two major regional aquifers drain from north and south to two big springs at the lake or close to it. These springs have approximately equal

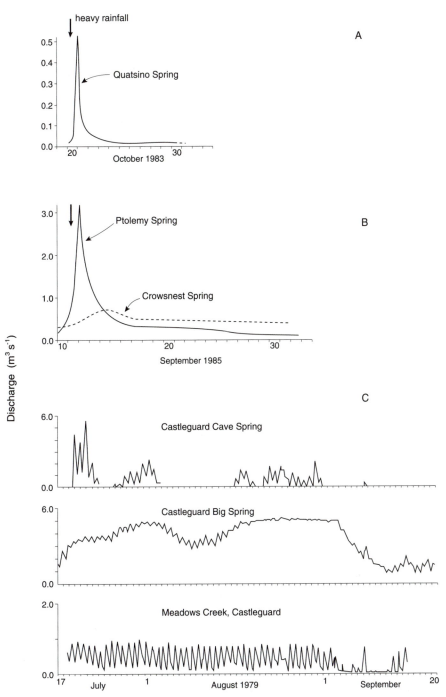

Figure 8.6
Flood hydrographs for (A) Ouatsino Spring (Vancouver Island), (B) Crowsnest and Ptolemy springs (Alberta Rockies), and (C) Castleguard springs (Alberta Rockies). Data are simplified for illustrative purposes.

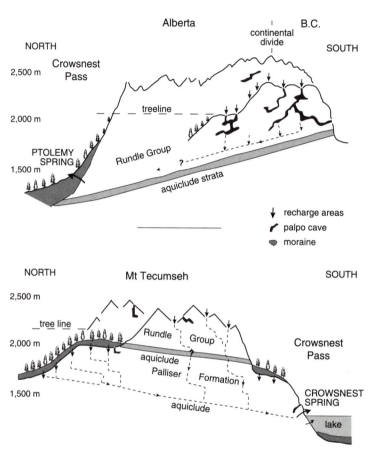

Figure 8.7
Karst aquifers: schematic sections through (A) Ptolemy Spring and (B) Crowsnest Spring, both at Crowsnest Pass, Alberta and BC.

mean annual discharges of approximately 0.7 m³/s – about five times more than the mean discharge at Quatsino Spring, which has less than one-tenth of their catchment.

The north and south systems (Crowsnest Spring and Ptolemy Spring), respectively, are illustrated in Figure 8.7 and their sample hydrographs in Figure 8.6B (from Worthington 1990). A rainstorm on 10 September 1985 affected both systems equally; their greatly contrasting hydrographic responses reflect the complex geological, geomorphological, and vegetational factors affecting alpine karst hydrology. Ptolemy Spring is confined to an upper aquifer in the Rundle Group of limestones (Mississipian age). These are about 400 m thick and rest on a thinner sequence of cherty limestones, shales (clay rocks), and thin coals that constitute an impermeable base, or "aquiclude" (Figure 8.7A). Recharge for the aquifer is limited largely to high plateaus and cirque floors that are above the treeline. Recharge surfaces are mostly bare rock, there being little till and sparse-to–non-existent tundra vegetation. The principal recharge area, Andy Good Plateau, lies west of the continental divide in British Columbia, but the spring is east of it, in Alberta. This is but one instance where groundwater drainage does not accord with surface topography – a common phenomenon in mountain karsts.

The hydrograph for Ptolemy Spring is similar in form to that of Quatsino, representing a simple, quickly responding system. However, the rise to peak discharge ($3.3 \, m^3/s$) takes approximately twenty-four hours because flow paths from the main recharge areas are comparatively long (at least 8–14 km). Recession to pre-storm discharge is largely exponential in form and takes about twenty-one days; water is being stored temporarily in paleocaves.

Crowsnest Spring (Figure 8.7B) discharges from a prominent flooded cave at the edge of the lake. Its catchment is probably a little smaller than that of Ptolemy Spring. However, extensive spreads of glacial till serve as gravel aquifers that overlie the karst aquifer, and at least half the recharge area is forested. Most important, it is a double aquifer, collecting water from 200 m of Palliser limestone (Upper Devonian age) as well as the Rundle limestones. The intervening aquiclude is breached underground (Ford 1983a), possibly by some major fractures.

Storm runoff is thus retained by forest, glacial gravels, paleocaves, and temporary impoundment above bottlenecks where flow must pass through the aquiclude. These factors combine to produce the slow and moderate response displayed by the hydrograph. Peak discharge ($0.70 \, m^3/s$) is less than one-quarter of the Ptolemy peak and occurs some seventy-two hours later. Correspondingly, the recession is slow; it will take at least forty days to return to the pre-storm discharge. Crowsnest Spring is a good example of a large alpine aquifer drawing much of its recharge from forested detrital deposits.

The Castleguard Mountain karst is located in the Main Ranges at the north end of Banff National Park. It extends beneath a substantial part of the Columbia Icefield, the largest ice cap remaining in the Rocky Mountains. The karst is now in a Pyrenean state, with glaciers occupying much of the recharge zone but not obstructing the springs (Figure 8.8). However, the latter were buried and suppressed beneath at least 800 m of flowing ice during the more recent glaciations. This is the most significant known alpine karst because it retains part of its former ice cover and as a consequence has been studied intensively (Ford 1983b; Smart and Ford 1986).

The karst aquifer is developed in 800 m of massively bedded limestones and dolomites of Middle Cambrian age. It has at least three major components: Castleguard, Castleguard II, and Castleguard III. Castleguard is currently the longest explored cave in Canada. There are 20 km of mapped passages. From the explorers' entrance at its downstream end, the cave passes underneath Castleguard Mountain and expands into a warren of small inlet passages beneath the icefield. Many of these are blocked with injections of glacier ice or of till. The groundwater streams that created the cave abandoned it more than 700,000 years ago. However, during the melt season, invasion waters cascade in from shafts in the roof and disappear into

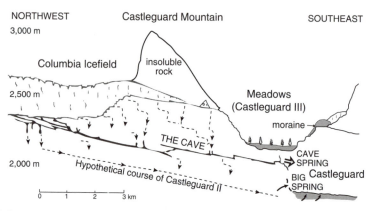

Figure 8.8
Castleguard alpine karst aquifer, Banff National Park, Alberta: schematic section.

slots in the floor. One thousand metres of galleries at the downstream end can also flood very quickly; water then pours out of the cave mouth as Cave Spring (Figure 8.6C). Castleguard II is the modern drain of the central icefield and small glaciers and snowfields surrounding Castleguard Mountain. It is a system of inaccessible galleries approximately 200–300 m beneath the cave. Castleguard III drains extensive meadows east of the mountain (Figure 8.8) and also part of the meltwater that sinks into moulins on Saskatchewan Glacier, the largest of the valley glaciers that descend from the icefield. The meadows are dominated by grasses and krummholz growing in moraine ridges and alluvial fans.

Castleguard II and III discharge their waters through approximately 100 springs scattered along 4.5 km of the Castleguard river valley. The highest spring (The Cave) is more than 340 m above the lowest. Their maximum aggregate discharge is about 20 m³/s. Many of the springs bubble up through gravel in the valley floor, but the most prominent, Big Spring, bursts from a narrow fissure in a rock wall 40 m higher.

Hydrographic records for the aquifer over the period 17 July–20 September 1979 (Smart 1983) are shown in Figure 8.6C. Meadows Creek is a surficial melt stream fed by snowfields and a small glacier and is typical of the streams that recharge the aquifers. It can be duly demonstrated that, in contrast to White Ridge and Crowsnest Pass, the daily cycle of snow and ice melt completely dominates the hydrograph's behaviour. When rainfall occurs (with its associated lack of sunshine) flow in the creek actually diminishes.

In response to these rather stable daily cycles of meltwater recharge, discharge at Big Spring is seen to increase in a step-like manner between 17 July and 3 August. There is a dampled diurnal cycle superimposed on the rise. Maximum daily discharge at Meadows Creek occurs between 1400 and 1600 hours, which is typical of glacier melt. At Big Springs this pulse is registered between 0300 and 0500 hours the next day. Thus much of the aquifer is in a vadose condition – i.e. streams flow along the floors of air-filled passages. If the latter were filled with water (phreatic conditions) the pulse would be transmitted nearly instantaneously. After 3 August discharge fell a little at Big Spring (in response to a spell of cloudy, rainy weather) and then climbed to a maximum of 4.8 m^3/s on 19 August. This maximum was maintained for fifteen days; it is a plateau that indicates that the cave passage supplying the spring was filled to its maximum capacity.

Between 18 and 28 August the Cave Spring was also active. It was functioning as an overflow, and, although it is no less than 280 m above Big Spring, the two springs then were hydrodynamically linked. The aquifer was largely filled up with water. However, the biggest discharges of the record at Cave Spring took place between 19 and 22 July, when Big Spring was well below its peak. At that time the two springs were entirely out of phase with each other. The July floods at Cave Spring are to be attributed to a seasonal peak of snowmelt between 2,000 and 2,400 m a.s.l. on the alpine meadows and Saskatchewan Glacier. This inundated the Castleguard III aquifer. By the end of the month those snows were largely consumed. The melt peak moved higher, onto the icefield, where its waters filled the Castleguard II aquifer and overspilled into downstream parts of Castleguard I and III.

Complexity on this scale (particularly with so great a range of elevation of seasonal springs) is almost unknown in extraglacial regions. It is a product of the impact of great meltwater pulses released in mountainous terrain and the effects of large valley glaciers that (cyclically during the Quaternary) have infilled parts of karst aquifers and obstructed their springs, compelling diversion of groundwaters to other outlets. At Crowsnest Pass the behaviour of the modern aquifers is simpler because their recharge zones lack glaciers and are smaller in area, having been reduced by glacial dissection. At White Ridge the behaviour is the most simple because the aquifer is very small and has suffered the least disturbance from glaciation.

KARST AND COLD

The effects of cold temperatures on the operation of modern karst systems can be considered without reference ot any complicating features inherited from past glaciers. A general model for this simple situation is given in Figure 8.9. It can be applied to outcrops of the karst rocks in lowland subarctic Canada (Figure 8.2) and throughout the arctic islands. In the Western Cordillera of Northwest Territories and Yukon it may be replaced by a mountain and permafrost model (Figure 8.10).

Under certain circumstances large volumes of ground ice may accumulate in caves far south of the limits of conventional permafrost (see Figure 8.9). These ice masses (termed "glacières"; Balch 1900) are mixtures of firn, regelation ice, and/or frozen pondings. They form where a large cave chamber with a narrow entry descends steeply from the surface and is blocked at its base, preventing flow of water or air further downward. Such chambers are cold traps; the coldest (most dense) winter

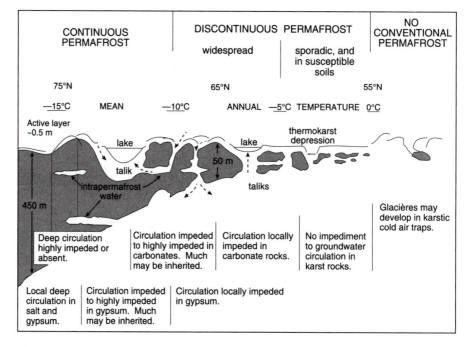

Figure 8.9
Model for karst-permafrost relationships in lowland Canada.

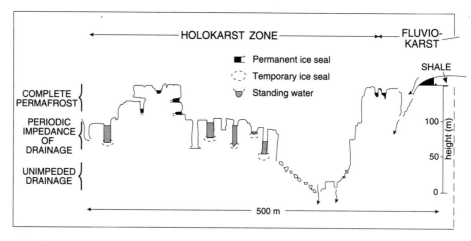

Figure 8.10
Model for karst-permafrost relationships in mountainous subarctic/arctic Canada, based on findings in Nahanni North Karst.

air comes to fill them and neither direct sunlight nor warm air can penetrate them in the summer. Runoff or snow that enters is frozen in. Glacières are static ice masses; there are no instances recorded where they exhibit the plastic flow of normal glaciers.

Small glacières are known in Canada's southern cordillera, including a few examples below modern timberline. Marshall and Brown (1974) describe Coulthard Cave, at timberline in Crowsnest Pass, where the ice probably accumulated at the end of the Hypsithermal (postglacial warm peak about 4000 yr BP) and is now evaporating slowly in a perennial $-2°C$ atmosphere. The most celebrated glacière survives in the forests of southern Slovakia, where mean annual external temperature is as high as $+6°C$.

There is little or no impediment to karst development in the zone of sporadic permafrost. As the latter becomes more widespread and deeper, however, karst activity apparently becomes progressively restricted in limestone and dolomite, the least soluble karst rocks. This is especially true in the lowlands, where the vigour of groundwater circulation is also limited by low hydraulic gradients. In the continuous permafrost zone, development of limestone and dolomite karst appears to be confined to the active layer or to scarp edges (where local hydraulic gradients are high), unless there is a glacial inheritance as described in the next section.

Karst groundwater circulation is more readily established and maintained in gypsum and salt because of their greater solubility and because the addition of large quantities of solutes to fresh water depresses its freezing point. There are large areas of gypsum beneath thin covers of dolomite in the Colville Hills, around the west end of Great Bear Lake, and in the Horton and Anderson plains north of it. The discontinuous-continuous permafrost boundary bisects this region (Figure 8.2). Van Everdingen (1981) studied the southern half of it, mapping 1,400 dolines, 27 larger karst depressions such as poljes, and at least 63 substantial karst springs signified by icings. It appears that karst groundwater circulation is vigorous and perennial and that this interstratal karst is extending, despite low hydraulic gradients and annual runoff of less than 200 mm.

Large dolines and some springs associated with gypsum occur at least 200 km inside the continuous permafrost zone on the Horton Plain. Study of aerial photographs suggests that they may be becoming immobilized by ingrowth of permafrost – i.e. the extent of karst circulation here may be shrinking. Salt is exposed in diapirs on Axel Heiberg Island in the High Arctic. Superficial solution channels are well developed, and analysis of aerial photographs suggests that local aquifers several hundred metres in length can exist.

Figure 8.10 is a model for karst development in mountainous regions with widespread permafrost. It is based on conditions in the Nahanni karst, a belt of limestone and dolomite terrain extending for 80 km north of South Nahanni National Park in the Mackenzie Mountains, NWT. This extremely rugged terrain contains dolines as much as 150 m deep and groundwater flow paths that have been traced for 15 km (Brook and Ford 1980). The country has not been glaciated for at least 300,000 years; for most of that time it will have experienced greater cold than the present mean annual temperatures of -5 to $-7°C$.

The karst displays three distinct levels. In the highest, features such as sinkholes and caves are inert; caves are filled entirely with ground ice (glacières) or frozen silt. The intermediate level experiences periodic impedance of the sinking waters; dolines there have limited catchments (a few hectares at most), and so runoff received in a given year is small. It tends to become ponded above a basal ice blockage until the waters of several or many successive years accumulate sufficient head to rupture it; drainage is then very rapid. In any year some Nahanni sinkholes in this zone will contain 5–10 m of standing water while neighbours a few tens of metres distance and at the same elevation are dry. In the lowest level of unimpeded karst drainage, surface runoff from the high karst and surrounding, non-karstic terrain is focused into large sinkholes and is able to keep them open every year.

Further north in the Mackenzie Mountains, Tsi-It-Toh-Choh is a similar limestone karst in a region that has never been glaciated (Cinq-Mars and Lauriol 1985). Mean annual air temperature is -10 to $-11°C$. In comparison to Nahanni, the completely frozen zone appears more extensive, and the zone of periodic impedance is reduced or absent. However, where the runoff of large areas can be focused into individual dolines on valley floors there is unimpeded groundwater circulation to regional springs. Thus the combination of steep hydraulic gradients and focusing of runoff in mountains maintains karst groundwater circulation in the zone of continuous permafrost.

Figure 8.11 illustrates the range of mountain karst phenomena in Canada's western mountains.

KARST, COLD, AND GLACIERS

The thermal diffusivity of glacier ice is low. Where ice is thick, this factor effectively insulates underlying land from atmospheric low temperatures; the land then is warmed by the geothermal heat flux of approximately 2 cal.cm^2/yr. Where ice cover is maintained for a long period, any earlier permafrost may thaw completely.

One can speculate that the Laurentide ice cover of subarctic Canada and some of the southern, low-lying islands of arctic Canada was sufficiently thick and long-lasting during recent glaciations to permit complete or partial thawing of permafrost. Probably the general ice cover over the northern islands was too thin, and the permafrost too cold, for similar large-scale degradation there. There was complete permafrost thawing probably only in glacial trough valleys where the ice was deeper and the rates of glacier flow were higher.

Subglacial thawing of permafrost in karstic terrains raises two questions. To what extent, if any, is modern (postglacial) karst activity in the permafrost regions inherited from the more favourable thermal conditions that may have prevailed in the sub-glacial or deglacial environments? In periglacial environments, are there karst landforms that can develop only beneath warm-based glaciers, subsequently becoming inert as permafrost aggrades into the karst system?

A   B

D

Figure 8.11

Mountain karst development with variable permafrost influences. (A) Karst spring at Moraine Polje, Keele River, Mackenzie Mountains, at 65°N in continuous permafrost zone; (B) Sinkhole Col, Nahanni North Karst, Mackenzie Mountains, at 61°30′N in zone of widespread permafrost; (C) beneath 500 m of rock and glacier ice, gallery in centre of Castleguard Cave, Banff National Park – an alpine glaciated aquifer; (D) Medicine Lake, Jasper National Park, at close of winter season – lake basin, approximately 8 km^2 and averaging 25 m deep, is believed to occupy karst polje infilled by glacial outwash but now partly re-excavated through sinkholes.

The latter question is more readily answered, and the response is affirmative. When glaciers spread over a region, whether they are warm or cold at the base, operation of other geomorphic process systems is usually largely or entirely suppressed, and the landforms they have created will be modified or destroyed. It was shown above that the Pyrenean type of glaciation can actually accelerate karst development, but that is a case of only partial ice cover where the karst hydrological system remains dominant over the glacial hydrological system. Beneath extensive ice caps or continental sheets such as the Laurentide glacier, any karst groundwater circulation must be subordinate; but in both mountain and lowland settings in arctic Canada it has apparently been able to create karst features that became frozen and inert following the Holocene deglaciation.

The northern mountain example is found in the Nanisivik zinc/lead mine on the Borden Peninsula, Northwest Baffin Island. This is a rugged terrain, with deep glacial valleys and fjords between narrow plateaus or ridges. Mean annual ground temperature in the mine is $-11°C$. Its south wall approaches and partly underlies an adjoining glacial valley. There, for a length of 200 m, dolomite and ore are a jumble of giant blocks in a state of arrested collapse. The arresting agent is a firm cement of ground ice. Evidently waters were injected (to a depth of at least 60 m) into the bedrock beneath a valley glacier and were able to dissolve the dolomite, so inducing the collapse. This process halted when the water froze following recession of the ice.

There is a perennial lake upstream, and a large creek flows along the glacial valley each summer. The potential hydraulic gradient through the collapse is 0.30, a high value. Yet these favourable circumstances could not maintain karstic groundwater circulation upon deglaciation. This situation may be contrasted with that at never-glaciated Tsi-It-Toh-Choh, where circulation may have been maintained throughout the Quaternary on similar gradients in a mean annual temperature that is only 2–3C° higher today. Its maintenance is probably attributable to the existence of a more mature aquifer (bigger and more efficiently linked cave passages) before the advent of the permafrost. Tsi-It-Toh-Choh is thus a preglacial inheritance, whereas the Nanisivik collapse is wholly subglacial.

The lowland karst example comprises patterns of intersecting corridors in limestone and dolomite on King William, Prince of Wales, and Somerset islands and in the Adelaide Peninsula (Fraser and Henoch 1959; Bird 1967). They are straightwalled features, 5 to 30 m wide and as much as 1,000 m long. They are now blanketed by perennially frozen glacial deposits and entirely inactive. They are interpreted here as giant grikelands (a karren landform assemblage) created perhaps by subglacial meltwater. They certainly antedate the postglacial period.

These grikelands should be compared to the gypsum sinkhole karst of the Colville Hills and Horton Plain which maintains active groundwater circulation and, in its southern regions, is expanding. Groundwater hydraulic gradients and mean annual temperatures are scarcely more favourable than they are on the Adelaide Peninsula. The larger karstic depressions are all partly filled with glacial deposits and thus must have existed before the last deglaciation of the region was completed. Much of this karst may have been initiated either subglacially or at the margins of melting ice sheets. Circulation is being maintained because of the greater solubility of gypsum and because summer runoff is notably warmer than on the Adelaide Peninsula. The karst is not expanding in its northern areas and may actually be shrinking there as permafrost slowly re-establishes itself following the last subglacial thaw.

The general conclusion must be that a number of factors control modern operation of the karst system in the cold, formerly glaciated mountain and northern environments of Canada. These include the passive variables of rock solubility and groundwater hydraulic gradient and the active variables of rock and water temperature and magnitude of available runoff. Only in a few instances can their interaction be understood without reference to the historic variables of glacial initiation or glacial suppression of karstic circulation.

HUMAN USE OF COLD KARST

In the mountainous limestone regions of the Old World the greatest visual impact has been caused by deforestation. It permitted loss of much soil into underlying karren pits and troughs. Overgrazing with sheep and goats has inhibited the lengthy rebuilding of an organic base sufficient to support large trees. This explains the typically rocky, gleaming white hillsides of classical Greece, southern Italy, and western Croatia – human-made landscapes appealing to the eye but impoverished. In northern Vancouver Island the Mediterranean experience is now being repeated. Steeply dipping limestones with high densities of karren pits and troughs are common and support fine stands of timber. The standard policy of clearcutting and then burning the slash has ensured that much nutrient matter has been washed underground, where it fouls caves and springs. Many of the bare rock slopes that are left (Figure 8.12A) will not be productive until the next glaciation deposits a veneer of till. Sample clearcut sites in the Benson River valley show that even locations unusually favourable for restoring growth have in seventy years regained only 15 per cent of the volume of timber removed.

A less commonly recognized, more site-specific implication of human use of the mountains concerns caves. Caves appeal to many people because of their stark rock scenery, their mysterious recesses, the dynamics of the running water, and the beauty of their speleothem (stalactite and stalagmite) deposits. Hundreds of them have been developed for tourists in other mountain regions. Several hundred are now known in the Canadian mountains, chiefly on Vancouver Island and in the Rocky Mountains. A handful were developed as tourist show places – chief among them Nakimu Caves in Glacier National Park, which closed in 1928 along with the CPR resort hotel at Rogers Pass. Reopening has been delayed because their "remoteness" (700 m above the Trans-Canada Highway) supposedly makes them too "hard" for modern visitors. The mountain caves thus are seen only by small numbers of explorers. Fortunately, most exploration has taken place during the past twenty-five years, at the same time

Figure 8.12
(A) Recently deforested and burned slopes on Quatsino Limestone, Benson River valley, Vancouver Island; (B) polygonal ice crystals in a Rocky Mountain alpine grotto, emphasizing extreme fragility of perennial ice deposits in caves.

as a strong conservation ethic has developed. Canada's caves are little vandalized in comparison with many others. Being cold, the mountains contain some caves where ice crystals have grown slowly over many centuries; ice crystal caves are among the most beautiful natural features on the planet. They also pose the ultimate problem in conservation – they are melted by the viewer's body heat (Figure 8.12B).

The chief problems that karst poses in the mountains are those of exceptional environmental hazard and of difficulties in construction (see chapter 10). The exceptional hazard is the catastrophically rapid bedrock landslide. This may happen because dissolution along bedding planes and fractures weakens the bonding of the adjoining rock masses. In fold mountains many bedding planes dip steeply and function readily as sliding surfaces. This is a hazard throughout the cuesta-form mountain regions of Canada. Cruden (1985) has analysed the one hundred largest landslides that have occurred in the Rocky Mountains since their general deglaciation some ten thousand years ago. All of them are slides of limestone and dolomite, often with shale interlayers that abetted slipping; a majority resulted from bonding failures on bedding planes, including the most destructive historic slide, which destroyed the village of Frank at Crowsnest Pass one night in 1903. The greatest recent slide in Canada, at Trench Lake, North Nahanni River, Mackenzie Mountains, in 1986 was also a limestone bedding plane failure.

Major difficulties may be encountered during construction of railways, highways, and (pre-eminently) hydro-electric power dams in karstic mountains. Common problems include leakage via caves and fissures, collapse of footings and retaining structures, and induced landslides. Many areas of mountain karst can show a number of dam and reservoir projects that never held water. In Jasper Park, Alta, Medicine Lake, on the Maligne River, is a karst polje 6 km long that fills with water in most summers but drains nearly dry each winter. This played havoc with efforts to stock it with sports fish. During the 1930s many truck loads of fill, including old mattresses, were dumped into the sink ponds in vain attempts to seal them. The Maligne karst is one of the greatest underground river systems in the world, and modern park managers wisely leave it alone.

REFERENCES

Balch, E.S. 1900. *Glacières or Freezing Caverns*. Philadelphia: Lane and Scott.

Bird, J.B. 1967. *The Physiography of Arctic Canada, with Special Reference to the Area South of Parry Channel*. Baltimore: Johns Hopkins University Press.

Brook, G.A., and Ford, D.C. 1980. "Hydrology of the Nahanni Karst, Northern Canada, and the Importance of Extreme Summer Storms." *Journal of Hydrology* 46: 103–21.

Cinq-Mars, J., and Lauriol, B. 1985. "Le karst de Tsi-It-Toh-Choh: notes préliminaires sur quelques phénomènes karstiques du Yukon septentrional, Canada." *Comptes rendus du Colloque international de Karstologie appliquée, Université de Liège*, 185–96.

Corbel, J. 1959. "Érosion en terrain calcaire." *Annales géographiques* 68: 97–120.

Cruden, D.M. 1985. "Rock Slope Movements in the Canadian Cordillera." *Canadian Geotechnical Journal* 22: 528–40.

Drake, J.J. 1984. "Theory and Model for Global Carbonate Solution." In R.G. LaFleur, ed., *Groundwater as a Geomorphic Agent*, London: Allen and Unwin, 210–26.

Drake, J.J., and Ford, D.C. 1974. "Hydrochemistry of the Athabasca and North Saskatchewan Rivers in the Rocky Mountains of Canada." *Water Resources Research* 10 no. 6: 1192–8.

Ecock, K.R. 1984. "The Hydrology of an Alpine Karst: White Ridge, Vancouver Island." MSc thesis, McMaster University, Hamilton, Ontario.

Ford, D.C. 1971. "Characteristics of Limestone Solution in the Southern Rocky Mountains and Selkirk Mountains, Alberta and British Columbia." *Canadian Journal of Earth Sciences* 8 no. 6: 585–609.

– 1983a. "Alpine Karst Systems at Crowsnest Pass, Alberta–British Columbia, Canada." *Journal of Hydrology* 61: 187–92.

– ed. 1983b. "Castleguard Cave and Karst, Columbia Icefields Area, Rocky Mountains of Canada: A Symposium." *Arctic and Alpine Research* 15 no. 4: 425–554.

– 1987. "Effects of Glaciations and Permafrost upon the Development of Karst in Canada." *Earth Surface Processes and Landforms* 12 no. 5: 507–21.

Ford, D.C., and Williams, P.W. 1989. *Karst Geomorphology and Hydrology*. London: Unwin Hyman.

Fraser, J.K., and Henoch, W.S. 1959. *Notes on the Glaciation of King William Island and Adelaide Peninsula, NWT*. Department of Mines and Technical Surveys, Geographical Branch Paper 22.

Lauritzen, S.-E. 1984. "A Symposium: Arctic and Alpine Karst." *Norsk Geografisk Tidsskrift* 38: 139–214.

Marshall, P., and Brown, M.C. 1974. "Ice in Coulthard Cave, Alberta." *Canadian Journal of Earth Sciences* 11 no. 4: 510–18.

Smart, C.C. 1983. "The Hydrology of the Castleguard Karst, Columbia Icefields, Alberta, Canada." *Arctic and Alpine Research* 15 no. 4: 471–86.

Smart, C.C., and Ford, D.C. 1986. "Structure and Function of a Conduit Aquifer." *Canadian Journal of Earth Sciences* 23: 919–29.

Van Everdingen, R.O. 1981. *Morphology, Hydrology and Hydrochemistry of Karst in Permafrost near Great Bear Lake*. National Hydrological Research Institute of Canada, Calgary, Paper 11.

White, W.B. 1984. "Rate Processes: Chemical Kinetics and Karst Landform Development." In R.G. LaFleur, ed., *Groundwater as a Geomorphic Agent*, London: Allen and Unwin, 227–48.

– 1988. *Geomorphology and Hydrology of Karst Terrains*. Oxford: Oxford University Press.

Woo, M.-K., and Marsh, P.E. 1977. "Effects of Vegetation on Limestone Solution in a Small High Arctic Basin." *Canadian Journal of Earth Sciences* 14 no. 4: 571–81.

Worthington, S.R.M. 1991. "Karst Hydropology of the Canadian Rocky Mountains." PhD thesis, McMaster University, Hamilton, Ontario.

CHAPTER NINE

Mountain Paleoenvironments of Western Canada

JOHN J. CLAGUE

Some of the distinctive physical attributes of Canada's western mountains are attributed to processes and events that took place during the last two million years, and earlier. Equally, some mountain hazards are partially the result of the Quaternary legacy, such as glacially oversteepened mountain walls and glacial, lacustrine, alluvial, and other loose sediments. This chapter focuses on three groups of paleoenvironmental information: landforms, sediments, and fossilized remains of plants and animals.

MECHANISMS AND TEMPO OF LANDSCAPE CHANGE

Diastrophism and Volcanism

The major endogenous processes that effect landscape change are diastrophism and volcanism, linked to plate interactions in the eastern North Pacific Ocean. The region is located at the edge of the America lithospheric plate and consists of the deformed western margin of the North American craton and a collage of crustal fragments, or terranes, that were accreted to the craton and subsequently fragmented and displaced northward along major strike-slip faults (Jones et al. 1982). These processes have given the Cordillera a strong northwest–southeast structural grain and are largely responsible for the present physiography and complex distribution of rocks, faults, and other structures in the region. Many mountain ranges have been uplifted hundreds, perhaps even thousands, of metres during the Quaternary (Ford et al. 1981; Parrish 1983), while much of the continental shelf has subsided (Yorath and Hyndman 1983). Volcanism has occurred sradically throughout the Quaternary and produced pyroclastic cones, stratovolcanoes, and intra-valley flows in four relatively narrow belts in British Columbia and Yukon (Souther 1977).

Subaerial Erosion and Mass Movement

Subaerial erosion and mass movement, accompanying and following late Cenozoic uplift, did much to develop the major valleys of the Cordillera (Clague 1989). Most existing Cordilleran valleys were initiated by streams prior to the Quaternary. Valleys were widened and deepened by glaciers during Pleistocene cold periods and further modified by streams during interglaciations. In unglaciated northern Yukon, streams alone controlled erosion. However, even there periodic glacial activity to the east and south influenced stream behaviour by adding to the sediment load and encouraging aggradation of streams such as Yukon River. Mass movement processes have helped denude mountains and widen valleys during the Quaternary by transferring material from oversteepened mountainsides to sites where debris could be carried away by streams. Weathering and soil formation appear to have been less important in this respect.

Rates of fluvial erosion and mass movement are controlled, in part, by diastrophism and climate. Valley incision proceeded relatively rapidly during extended periods of rapid uplift and more slowly during periods of tectonic quiescence. Periglacial processes – such as solifluction, cryoplanation, and thermokarst development, limited today to high latitudes and high altitudes (see chapter 6) – were more widespread during cooler nonglacial stages of the Pleistocene and early parts of glaciations.

Glaciation

Isotopic and magnetic studies of deep-sea sediments have shown eight major climatic cycles in the last 800 thousand years (ka), each about 100 ka in duration and marked by sharp fluctuations in climate on shorter time scales; these cycles were preceded back to the Tertiary by shorter, more frequent cycles (Shackleton and Opdyke 1976). The colder parts of many of these cycles were accompanied by widespread glaciation in the Canadian Cordillera (Fulton 1971; Clague 1981).

Glaciation has profoundly altered the Cordilleran landscape. During major Pleistocene cold periods, glaciers advanced from high mountains to bury most of British Columbia and parts of Yukon, District of Mackenzie, Alberta, and the northwestern United States (Figures 9.1, 9.2). Occasionally, ice thickened so much that one or more domes, with surface flow radially away from their centres, became established over the BC interior. In the process, the drainage network of western Canada was repeatedly disrupted and rearranged, and prodigious quantities of sediment were deposited in valleys and on plateaus, coastal lowlands, and the sea floor. Never-glaciated parts of Yukon have a somewhat muted topography that contrasts sharply with the rugged mountains and fresh drift-covered lowlands of formerly glaciated areas.

Glaciations terminated with sharp climatic amelioration and rapid melting of glaciers. The Cordilleran Ice Sheet decayed by complex frontal retreat and downwasting (Fulton 1967; Clague 1981). Frontal retreat dominated at the periphery of the ice

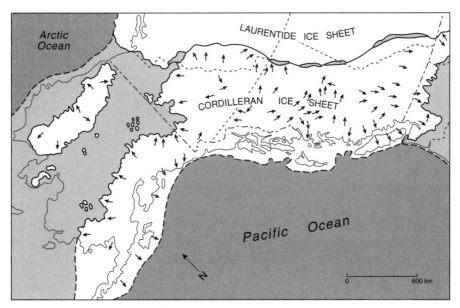

Figure 9.1

Maximum extent of Pleistocene glaciation in Canadian Cordillera and adjacent areas, and ice-flow pattern during Late Wisconsinan; *source*: Flint (1971: Figure 18.1). Glacier complex includes Cordilleran Ice Sheet and independent and nearly independent glacier systems in some peripheral mountain ranges. Nunataks were small in ice-sheet areas at climax of glaciation. In contrast, there were large ice-free areas in some peripheral glaciated mountain ranges.

sheet, as for example in westernmost British Columbia where glaciers grounded on the continental shelf calved back into fjords and mountain valleys. In areas of low and moderate relief in the interior, deglaciation progressed by stagnation, with uplands appearing through the ice cover first and dividing the ice sheet into a series of tongues that retreated in response to local conditions. Active ice eventually became restricted to major mountain ranges.

Not all glaciations advanced to the continental ice-sheet stage of development. Some, possibly most, glaciations climaxed with the ice cover consisting of a series of separate glacier complexes confined to, or spreading some distance beyond, mountain ranges; at such times, significant plateau areas, some coastal lowlands, and parts of the continental shelf remained ice-free.

Quaternary nonglacial periods (interglaciations and interstades) varied in length and, like glacials, were marked by fluctuations in climate. Glaciers in the Cordillera were restricted to mountain ranges during interglaciations, and lowland and plateau areas were continuously ice-free (Clague 1981). Physiography and sedimentary environments during the last nonglacial period (Middle Wisconsinan) were similar to

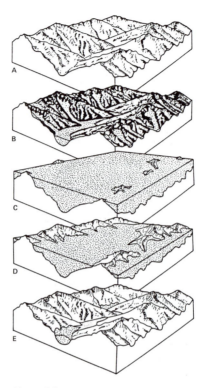

Figure 9.2
Growth and decay of Cordilleran Ice Sheet: (A) mountain area at beginning of glaciation; (B) development of network of valley glaciers;
(C) coalescence of valley and piedmont lobes to form ice sheet;
(D) decay of ice sheet by downwasting, with upland areas deglaciated
before adjacent valleys; and (E) residual dead ice masses confined to
valleys.

those of the present (Fulton 1971), and the same probably is true for earlier nonglacials
as well.

The impact of glaciation on the Cordilleran landscape has varied through time.
Glacial erosion and deposition during Pleistocene interglaciations, as at present, were
restricted to localized alpine settings. More widespread and perhaps rapid erosion
characterized cool interstades and the early parts of glaciations when snow cover and
ice cover were more extensive than today. These periods and deglacial intervals were
times of enhanced sedimentation in glaciofluvial and glaciolacustrine settings. At
glacial maxima, mountains were extensively eroded by temperate glaciers. The products of this erosion were deposited as till and stratified drift in lowlands and on
plateaus.

LANDFORMS

Pleistocene Landforms

Classic alpine features, including cirques, overdeepened valley heads, horns, and comb ridges (Figure 9.3) are common in all mountain systems, except those in northern Yukon, and are products of several Pleistocene glaciations. Cirques associated with existing glaciers have been modified during the Holocene, but the degree of modification, in most instances, is slight. U-shaped valleys and fjords record the passage of Pleistocene ice streams from the gathering grounds of the Cordilleran Ice Sheet onto intermontane plateaus and the continental shelf. Smaller-scale glacial landforms, present in most mountain ranges and on adjacent plateaus and lowlands, include drumlins, drumlinoids, flutings, and other streamlined features produced by subglacial erosion and/or deposition. Morainal ridges are rare, except in front of existing glaciers. Glaciofluvial landforms, including abandoned meltwater channels, eskers, kames, kame terraces, and valley trains, are found locally in most mountain ranges and record the decay of the Cordilleran Ice Sheet and satellite glaciers at the close of the Pleistocene.

Holocene landforms

Holocene river terraces and relict alluvial and colluvial fans are common in valleys throughout the Canadian Cordillera. Geomorphic and sedimentological studies of southern BC terraces and fans have provided information on early and middle Holocene paleohydrology and sediment yields (Ryder 1971; Ryder and Church 1986). These studies also have been instrumental in development of the concept of "paraglacial sedimentation" (Church and Ryer 1972), according to which sediment yields at the end of the last glaciation were much higher than at present. During and shortly after deglaciation, valleys became choked with glaciofluvial, fluvial, and colluvial sediments eroded from poorly vegetated, unstable drift deposits mantling upland slopes and valley walls. As slopes stabilized and vegetation became better established, the supply of sediment decreased, causing streams to incise their late-glacial and early-postglacial floodplains. As a result, many streams today flow in valleys that are cut deeply into Quaternary fills (Figure 9.4). Contemporary channel patterns and sediment transport are strongly influenced by these earlier Holocene events. Much of the present sediment load of many Cordilleran streams, for example, has been eroded from incised valley fills (Slaymaker and McPherson 1977; Church, Kellerhals, and Day 1989). On a more general level, it can be concluded that present day fluvial sediment yield is a consequence of late Pleistocene glaciation and that many tens of thousands of years may pass before this phenomenon dissipates (Church and Slaymaker 1989).

A variety of mass movement processes, including rockfalls, debris and rock slides and slumps, debris flows, and solifluction, have modified the late Pleistocene land-

Figure 9.3
Mackenzie Mountains, NWT. *Photo*: National Air Photo Library T22L–81.

Figure 9.4
Fraser River valley, west of Clinton, BC. Fraser River trenched valley fill during early Holocene, shortly after end of last glaciation. *Photo*: Province of British Columbia BC 1087–46.

scape. Many landslides in mountain areas date to the end of the Pleistocene when slopes lost their supportive cover of ice and were infiltrated by meltwater. Most landslides, however, are younger and have resulted from gradual loss of strength of rocks and surficial sediments underlying marginally stable slopes (see chapter 10).

Lateral and end moraines situated short distances beyond the margins of existing glaciers provide information on postglacial climatic change. These moraines were produced by relatively minor advances of cirque and valley glaciers during Holocene and, possibly, latest Pleistocene time. The distribution, morphology, stratigraphy, and ages of the moraines have been studied in several areas in the Cordillera – see

Figure 9.5
Neoglacial moraines bordering Tiedemann Glacier, southern Coast Mountains, BC. Arrowed moraines date to eighteenth or nineteenth century. See Ryder and Thomson (1986) for details.

Luckman and Osborn (1979) and Osborn and Luckman (1988) for summaries of work done in Alberta and British Columbia and Denton and Stuiver (1966) and Rampton (1970) for Yukon. Moraines of more than one age are present at many sites, indicating multiple advances during the Holocene (Figure 9.5).

In general, older moraines are more subdued and have better developed soils than younger moraines. Additional evidence for multiple advances has come from studies of superposed drift units exposed in lateral moraines (Figure 9.6). Dendrochronology, lichenometry, radiocarbon dating, and tephrachronology have been used to date Holocene advances. Collectively, these studies indicate that glaciers expanded during middle and late Holocene time after a warm period that spanned much or all of the early Holocene. In western Canada, the best documented Neoglacial advances are those between 3.3 and 1.9 ka BP and during the last millennium. There may also

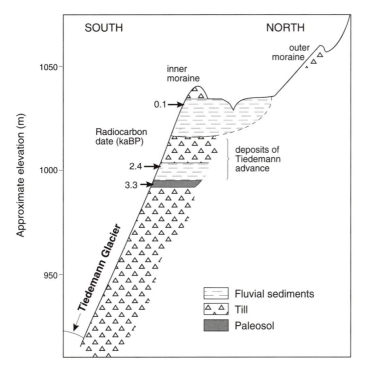

Figure 9.6

Tiedemann Glacier: schematic cross-section through a lateral moraine; *source*: Ryder and Thomson (1986: Figure 7). During middle-Neoglacial Tiedemann advance, ice-marginal fluvial sediments were deposited on lowest till; wood interpreted to be part of overridden forest has been dated at 2355 ± 60 yr BP (S-1471). Till immediately above fluvial sediments also was deposited during Tiedemann advance, and higher sediments during a more recent Neoglacial advance.

have been an expansion of ice between 6 and 5 ka BP. In most areas, glaciers achieved their maximum Holocene extent during the last several centuries (Little Ice Age), destroying or burying evidence of earlier advances.

The Neoglacial period was also a time of intensified physical weathering and mass movement in alpine areas. This resulted in increased sediment production and sediment delivery to streams and lakes. Sedimentation rates in some lakes in the Rocky Mountains, for example, varied during the Holocene as a function of upvalley glacial activity; periods during which rates were high correspond to times when glaciers were relatively advanced (Leonard 1986a; 1986b). Streams draining glaciated mountain ranges also may have undergone complex changes in channel and floodplain morphology in response to variations in sediment supply (Church 1983; Gottesfeld and Gottesfeld 1990).

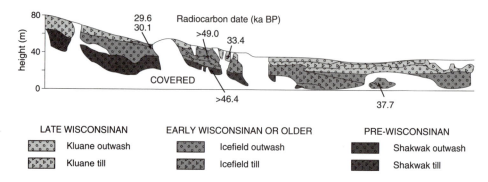

Figure 9.7

Deposits of three glaciations: exposure along Silver Creek at eastern front of St Elias Mountains, Yukon; *source*: Denton and Stuiver (1967: Plate 4A). Such sections provide information on glaciation history in mountain areas of western Canada.

STRATIGRAPHIC RECORDS

Distribution and Character

Thick sequences of Quaternary glacial and nonglacial sediments are present in many valleys and lowlands in the Canadian Cordillera (Figure 9.7) (Clague 1989). In most areas, the upper parts of these sequences are Late Wisconsinan and Holocene and the lower parts Middle Wisconsinan and older. Sediments that antedate the last glaciation are common where ice flow was transverse to valley axes, where bedrock projections sheltered downstream areas from scour, and where ice flow was sluggish, as for example in some lowlands near the margin of the Cordilleran Ice Sheet and in some valleys near its centre. At the northern limit of ice-sheet glaciation in Yukon, pre–Late Wisconsinan drift occurs at the surface beyond the Late Wisconsinan glacial limit.

The stratigraphic relationships and internal characteristics of Quaternary sediment sequences are complex. Valleys and lowlands experienced several cycles of infilling and partial or complete re-excavation; thus younger sediments commonly are inset into older deposits or are draped over an irregular older landscape. Younger sediments typically have been reworked from older Quaternary deposits; thus units of different age within a given area may be physically similar. However, individual units may display abrupt facies changes because of local variability of depositional environments within rugged terrain.

The terrestrial Quaternary stratigraphic record is a product of brief sedimentation events separated by long periods of nondeposition and erosion (Clague 1986). Most Quaternary deposits are of glacial origin and contain stratified sediments deposited in proglacial and ice-contact fluvial, marine, and lacustrine environments as well as till deposited subglacially and supraglacially.

Glacial sequences are separated by nonglacial sediments or unconformities. Nonglacial sediments generally are thin and patchy because sedimentation was more restricted and slower during nonglacial periods than during glaciations. Unconformities, produced by glacial and fluvial erosion, define former land surfaces similar in morphology and relief to the present.

All major Quaternary stratigraphic units in the Canadian Cordillera are time-transgressive, largely as a result of diachronous glacier growth and decay during Pleistocene glaciations. Sedimentation and erosion during periods of glacier growth propagated outward from mountain ranges into plateau and lowland areas. In contrast, deglacial sedimentation began earlier in peripheral glaciated areas than in the core area of the Cordilleran Ice Sheet.

Quaternary Events and Environments

Stratigraphic and sedimentological studies have provided the following information on late Quaternary events and environments in the Canadian Cordillera (Clague 1989). The present interglaciation was preceded by a period during which an ice sheet covered almost all of British Columbia and southern and central Yukon. Climatic deterioration at the onset of this glaciation about 25–30 ka BP led to increased sediment production in the mountains of the Cordillera. As glaciers grew, many valleys became aggraded with outwash, and thick deltaic and marine sediments accumulated locally in offshore areas. Lakes were impounded by advancing glaciers, and thick deposits of sand, silt, and clay collected in them. Those within the limits of glaciation eventually were overridden and obliterated, and their fills either eroded or covered with drift. In contrast, large lakes in northern Yukon dammed by the Laurentide Ice Sheet were not subsequently overridden by Late Wisconsinan glaciers, and thus their sedimentary fills are better preserved. Advancing glaciers also forced streams from their nonglacial courses. Swollen with meltwater and laden with sediment, these streams followed complex paths along constantly shifting glacier margins. This situation contrasts with that in unglaciated Yukon, where streams generally maintained their courses during the last glaciation, although they carried large amounts of meltwater and additional sediment.

At the climax of the last glaciation, mountains, fjords, and valleys parallel to the direction of ice flow were subject to intense scour by glaciers, and the unconsolidated fills of many valleys were partly or completely removed. Some eroded materials were redeposited as till, whereas others were carried englacially to the periphery of the ice sheet and laid down in ice-contact, proglacial, and extraglacial environments.

This period ended about 9–13 ka BP with rapid deglaciation, valley aggradation, and drainage change. Large glacial lakes formed and evolved as the Cordilleran Ice Sheet downwasted and separated into valley tongues. In coastal areas, thick glaciomarine sediments accumulated on isostatically depressed lowlands vacated by retreating glaciers and covered by the sea. Similar sediments also were deposited on the adjacent continental shelf.

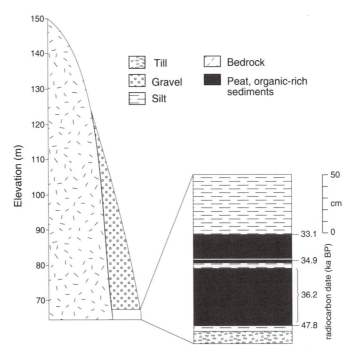

Figure 9.8
Exposure of organic-rich, Middle Wisconsinan sediments in Lynn Canyon on southern flank of Coast Mountains; *source*: Armstrong, Clague, and Hebda (1983: Figure 6). This exposure provides evidence for continuous nonglacial conditions adjacent to a major mountain range from before 48 ka BP until after 33 ka BP. See Figure 9.9 for pollen record from this site.

In many places, Late Wisconsinan drift overlies Middle Wisconsinan nonglacial sediments deposited in terrestrial, deltaic, and marine settings. The presence of these sediments within and at the margins of mountain ranges suggests that glaciers did not extend into plateau and lowland areas during the Middle Wisconsinan (Figure 9.8). Some of the sediments contain abundant fossil and animal remains which have provided a wealth of paleoenvironmental and paleoclimatic information (see below).

Middle Wisconsinan sediments are underlain by drift deposited during an Early Wisconsinan or older glaciation. These glacial deposits are similar to drift of Late Wisconsinan age and may record comparable glaciation. In a few areas in British Columbia, Early Wisconsinan or older drift is underlain by interglacial (Sangamonian?) sediments.

Pre-Sangamonian glacial and nonglacial sediments are present in the Canadian Cordillera but are fragmentary and difficult to correlate from area to area. The evidence suggests, however, that the climatic oscillations of the Wisconsinan stage, manifested in the growth and decay of the Cordilleran Ice Sheet and its satellite glaciers, occurred earlier during the Pleistocene as well.

PALEOECOLOGICAL RECORDS

Studies of fossil pollen, plant macrofossils, paleosols, speleothems, and vertebrate and invertebrate faunal remains have supplied data on Quaternary environments and climates in the Canadian Cordillera (Clague 1989). Although most of these studies have been conducted in lowlands and plateaus outside mountain ranges, the findings are pertinent to mountain paleoenvironments. The paleoecological record of the mountains may also be compared to contemporary northern and cold assemblages (see chapter 4).

Pre–Late Wisconsinan Records

Information for early and middle Quaternary time is sparse, although recent paleo-ecological studies in northern Yukon have shed light on the character and evolution of circumpolar floras and faunas during this period. Fossil pollen and plant macro-fossils from the Porcupine River area, for example, indicate that a flora dominated by *Picea* (spruce), *Pinus* (pine), and *Betula* (birch), but also containing *Larix* (tamarack) and *Corylus* (hazel), was present in northern Yukon during the Pliocene and perhaps the early Pleistocene (Lichti-Federovich 1974; Pearce, Westgate, and Robertson 1982; Ritchie 1984). The climate thus was much warmer and probably moister than at present.

In contrast, there is abundant paleoenvironmental information for the late Quaternary. The climate of the warmest part of the Sangamonian interglaciation (marine oxygen isotope stage 5e) was warmer than the present. Environmental conditions during the long Middle Wisconsinan nonglacial interval were variable in both space and time; the climate at times was colder and at times similar to that of the present. Some Middle Wisconsinan strata from low-elevation sites in south-coastal British Columbia, for example, are dominated by trees that currently grow near sea level, whereas others contain assemblages indicative of subalpine forests or grass-herb meadows (Figure 9.9) (Clague 1978; Alley 1979). In comparison, an isotopic study of dated cave deposits on southern Vancouver Island suggests that the climate of southwestern British Columbia gradually cooled from 64 to 28 ka BP (Gascoyne, Schwartz, and Ford 1980; Gascoyne, Ford, and Schwartz 1981). Mean temperatures in the caves at the beginning and end of this period may have been 4°C and 8°C, respectively, below the present mean temperature.

Late Wisconsinan Records

During the Late Wisconsinan, very cold and probably arid conditions prevailed at the eastern and northern margins of the Cordilleran Ice Sheet. Yukon, for example,

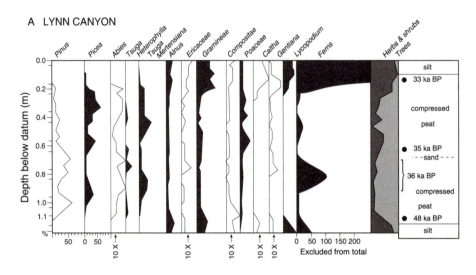

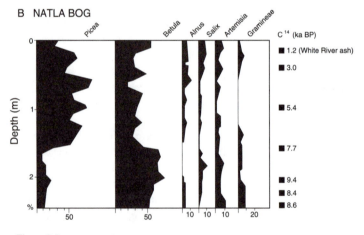

Figure 9.9

Pollen records: paleoenvironmental information for mountain areas of the Canadian Cordillera. (A) Middle Wisconsinan organic-rich sequence from Lynn Canyon, BC; *source*: Armstrong, Clague, and Hebda (1983). Pollen assemblages at base and top are those of subalpine or alpine meadows and record cold climate; forested conditions and variable, but generally cold climate prevailed when middle part accumulated. (B) Holocene peat from Natla Bog, Mackenzie Mountains, NWT; *source*: MacDonald (1983). Spruce forest replaced birch-shrub tundra about 7.7 ka BP. High percentages of spruce pollen in middle of sequence suggest climate warmer than at present from 7.7 ka BP until at least 5 ka BP.

was occupied by tundra, characterized in the pollen and plant macrofossil record by Cyperaceae (sedges), Graminaeae (grasses), *Artemisia* (sage), and various herbs. This vegetation was widely established in Yukon by 30 ka BP (Cwynar 1982). It has been suggested that the vegetation from 30 to 14 ka BP resembled a sparse, fell-field tundra typical of the modern vegetation of the arctic islands (Cwynar and Ritchie

1980; Cwynar 1982; Ritchie 1984). This interpretation runs counter to the view that a highly productive steppe tundra existed there during this period (Matthews 1982). Although there is continuing dispute over this issue, it generally is agreed that a reasonably large mammal population was supported by available plant cover and that the climate was very cold and arid. Mountain ranges in the northern and eastern Cordillera had more extensive ice cover than at present; unglaciated alpine sites probably were extremely cold and virtually lifeless.

In contrast, the climate along the Pacific coast was more moderate, and diverse floras may have persisted in localized refugial areas. Most of the Coast, St Elias, and Insular Mountains, however, were covered by snow and glacier ice throughout the Late Wisconsinan. Palynological investigations of sediments deposited during the early part of the last glaciation in south-coastal British Columbia indicate gradual deterioration of climate after about 29 ka BP (Alley 1979; Mathewes 1979). By 21 ka BP, subalpine parkland and possibly alpine plant communities were dominant on the lowlands of eastern Vancouver Island, and large vertebrates, including *Mammuthus imperator* (mammoth) and *Symbos cavifrons* (Muskox), were relatively common there (Harington 1975; Hicock, Hobson, and Armstrong 1982). Some lowland areas were covered by glaciers flowing from the Coast Mountains. In the BC interior, major climatic deterioration accompanying the onset of the last glaciation began about 25 ka BP, but a full glacial climate was not attained until 19 ka BP, when large areas of the Interior Plateau were covered by ice derived from the Coast, Selkirk, and Cariboo Mountains (Clague 1981).

Diverse plant communities existed in some BC coastal areas near the climax of the last glaciation. About 18 ka BP, *Abies lasiocarpa – Picea cf. engelmannii* (subalpine fir – Engelmann spruce) forest and parkland grew under cold humid continental conditions at sea level near Vancouver (Hicock, Hebda, and Armstrong 1982). Mean annual temperature was depressed by about 8°C, and timberline was 1,200–1,500 m lower than today. The nearby Coast Mountains were probably heavily covered by snow and ice. About 15 ka BP, the lowlands of the Queen Charlotte Islands supported a varied nonarboreal terrestrial and aquatic flora characteristic of alpine herbaceous and shrub meadows (Warner, Mathewes, and Clague 1982).

Records of Late Wisconsinan–Holocene Transition

Major floral changes in the Canadian Cordillera during and immediately following terminal Pleistocene deglaciation are attributable to climatic change and to different rates of plant migration and plant succession on freshly deglaciated terrain. Nonarboreal plant communities adapted to cold and probably dry conditions were the first to appear. These were replaced by forests as climate improved.

In northern Yukon, herb and willow tundra was replaced by birch tundra about 12 ka BP (Ritchie 1982; Ritchie, Cinq-Mars, and Cwynar 1982). *Populus* (poplar), *Myrica* (bog myrtle), and *Typha* (cat-tail) grew north of their present limits in this region 10–11 ka BP (Cwynar 1982; Rampton 1982; Ritchie, Cwynar, and Spear 1983), indicating a warmer climate than the present.

The earliest postglacial pollen assemblages in coastal British Columbia, dating from 13–13.5 ka BP, are dominated by herbs and shrubs adapted to a cold, relatively dry climate. After 13 ka BP, forest became established and began to change in composition as climate improved and plants migrated into the region from extraglacial areas.

A sparse herbaceous tundra existed in the foothills of the Rocky Mountains early during deglaciation (MacDonald 1982; Mott and Jackson 1982). This vegetation was replaced by shrub tundra before the end of the Pleistocene. Coniferous forests dominated by *Picea* and *Pinus* invaded lower valleys in the southern Rocky Mountains between about 12 and 10 ka BP, replacing nonarboreal vegetation and, in places, pioneering stands of *Populus* (poplar-aspen). *Pinus contorta* (lodgepole pine) was growing near present timberline in this area by 9.7 ka BP (Kearney and Luckman 1983b).

Holocene Records

The general warming trend that accompanied deglaciation continued well into the Holocene, though perhaps interrupted by brief cool intervals. Shrub birch–tundra prevailed near the crest of the Mackenzie Mountains until 7.7 ka BP, when *Picea* became established (Figure 9.9) (MacDonald 1983). Timberline was higher than at present, and the climate thus warmer, from 7.7 ka BP until at least 5 ka BP. Farther north, in the Richardson Mountains, birch shrub–tundra existed from 11 to 9 ka BP (Ritchie 1982). At about 9 ka BP, *Picea* expanded into this area, and within 1 ka it was well established. *Alnus* appeared at about 7.5 ka BP and achieved its present abundance by 6.5 ka BP. *Betula neoalaskansis* (Alaskan paper birch) apparently attained its modern extent and abundance in the Richardson Mountains about 6 ka BP. While these changes may be the result of migrational lags, pollen, plant macrofossil, and invertebrate fossil data from sites on the Mackenzie River delta and Yukon Coastal Lowland indicate that the climate in areas bordering the Richardson Mountains was significantly warmer than at present from about 11 ka BP until 5–6 ka BP (Ritchie, Cwynar, and Spear 1983).

Pseudotsuga menziesii (Douglas fir) and *Alnus* (alder) dominated south-coastal BC forests during the early Holocene (Mathewes 1985), indicting a climate similar to, or warmer than, the present. Maximum warmth and dryness occurred between 10 and 7 ka BP (Mathewes and Heusser 1981). There also was a xerothermic interval in the BC interior during the Holocene, although the proposed timing varies considerably from locality to locality (Clague 1981; Mathewes 1985). For south-central British Columbia, Hansen (1955), for example, identified a warm, dry period between 7.5 and 3.5 ka BP, whereas Alley (1976) placed it from about 8.4 to 6.6 ka BP. Miller and Anderson (1974) proposed that the period of maximum Holocene warmth at Atlin lasted from about 8 ka until 2.5 ka BP.

Fossil tree remains found above present timberline in the southern BC and Alberta mountains provide some additional evidence for a Holocene xerothermic interval

A

B

Figure 9.10
(A) Castle Peak cirque, southern Coast Mountains, BC; (B) large log on cirque floor, well above timberline (dot in A shows location of site). This log yielded radiocarbon date of 8580 ± 90 yr BP (RIDDL 885), which, with other evidence, indicates that climate was warmer than today.

(Figure 9.10). Most of the radiocarbon-dated remains are early to middle Holocene in age (Figure 9.11) (Clague 1980; Clague and Mathewes 1989). In the St Elias Mountains, stumps above present timberline have been dated at 5.35, 5.25, 3.0–3.6, and 1.23–2.1 ka BP (Rampton 1970; Denton and Karlen 1977); at these times climate probably was warmer than it is today.

Studies of pollen and paleosols in the southern Rocky Mountains generally support these conclusions (Reeves 1975; Kearney and Luckman 1983a; 1983b; Beaudoin 1986; Luckman and Kearney 1986). They indicate that timberline was higher than at present during most of the early and middle Holocene, and climate thus warmer. This is also in agreement with data from the southern Rocky Mountains and Interior Plains which show that the regional molluscan fauna contained a higher complement of south-ranging species from approximately 9 to 6 ka BP than after 6 ka BP (Harris and Pip 1973; MacDonald 1982).

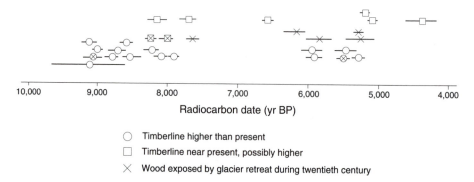

Figure 9.11

Radiocarbon dates on fossil wood and charcoal recovered near or above present timberline, southern Canadian Cordillera; *source*: Clague and Mathewes (1989: Figure 4). Horizontal lines through dates are laboratory error terms; data points are arranged vertically to allow discrimination. Distribution of dates suggests warmer climate during first half of Holocene.

The xerothermic interval was followed by wetter and cooler conditions. On the BC south coast, modern forests dominated by *Tsuga heterophylla* and *Thuja plicata* (western red cedar) were established by 4–5 ka BP (Mathewes 1985). Since then, there has been very little change in vegetation. In the BC interior, the xerothermic interval was followed by a generally wetter, cooler period marked by sharp, but relatively minor fluctuations in climate. Essentially modern climatic conditions were attained in the southern interior about 4–5 ka BP, after a period of cooling and increased precipitation that began sometime between 8 and 6 ka BP (Hebda 1982). A cool climate, similar to the present, was achieved after 5 ka BP in the southern Rocky Mountains. There were complex, but minor timberline fluctuations in this area during the late Holocene, associated with Neoglacial climatic oscillations (Kearney and Luckman 1983a; Beaudoin 1986).

SUMMARY

The landscape of the Canadian Cordillera has been shaped, and continues to be modified, by tectonic forces, volcanism, fluvial denudation, mass movement, and glaciation. These processes, operating over millions of years on a geologically complex suite of accreted and autochthonous rocks, have given the Cordillera its unique physiographic character.

Climate is arguably the most important factor to have affected mountain environments in western Canada during the Quaternary. Mountains were blanketed and eroded by glaciers during Pleistocene cold periods, and many of the distinctive features of the landscape developed at these times. There also is a close link between climate and sediment production in mountain areas. In general, denudation caused by solifluction and other forms of mass movement was greater during cool interstades and the early and late stages of glaciations than it was during interglaciations. Braided streams and gravelly and sandy floodplains dominated mountain valleys at these times, much as they do in the St Elias Mountains today.

Quaternary climatic fluctuations also produced changes in the distribution and character of vegetation in the Cordillera. Altitudinal treeline was much lower during glaciations and cool interstades than at present. Tundra, currently restricted to high elevations except in northern areas, was widespread in all mountain ranges at these times. The species composition of forests also varied during the Quaternary. Contemporary forests at low elevations through much of southern British Columbia, for example, are dominated by Douglas fir, but this species probably was not present in western Canada during Pleistocene glacial periods.

The effectiveness of the geological processes discussed in this chapter in changing the landscape has varied through time. Most Quaternary sediments and many landforms in the Cordillera are products of relatively brief, but intense episodes of erosion and sedimentation associated with glaciation. Quaternary valley fills, for example, are dominated by glaciofluvial, glaciolacustrine, and glaciomarine sediments deposited rapidly during periods of growth and decay of the Cordilleran Ice Sheet and by paraglacial sediments deposited during and shortly after deglaciation. In contrast, interstades and interglaciations, times of relatively restricted ice cover and sedimentation, are poorly represented in most stratigraphies in the Cordillera.

Montane environments in western Canada changed rapidly during periods of climatic deterioration and amelioration. Forested mountains in the southern Cordillera were transformed into unbroken expanses of snow, ice, and tundra between about 30 and 20 ka BP. In less than 4 ka at the end of the Pleistocene, mountain ranges were deglaciated, experienced a major paraglacial transfer of sediment from slopes to valley floors, and acquired something approaching their present vegetation cover. In addition, the Cordilleran drainage system was re-established during this short period. The last 8–9 ka of the present interglaciation have seen changes in mountain landscapes and vegetation that are minor when compared with those of the preceding glaciation.

REFERENCES

Alley, N.F. 1976. "The Palynology and Palaeoclimatic Significance of a Dated Core of Holocene Peat, Okanagan Valley, Southern British Columbia." *Canadian Journal of Earth Sciences* 13: 1,131–44.

Alley, N.F. 1979. "Middle Wisconsin Stratigraphy and Climatic Reconstruction, Southern Vancouver Island, British Columbia." *Quaternary Research* 11: 213–37.

Armstrong, J.E., Clague, J.J., and Hebda, R.J. 1983. "Late Quaternary Geology of the Fraser Lowland, Southwestern British Columbia." In D.J. Templeman-Kluit, ed., *Field Guides to Geology and Mineral Deposits in the Southern Cordillera*, Geological Society of America, Cordilleran Section, Field Trip Guidebook, 15–1–25.

Beaudoin, A.B. 1986. "Using Picea/Pinus Ratios from the Wilcox Pass Core, Jasper National Park, Alberta, to Investigate Holocene Timberline Fluctuations." *Geographie physique et quaternaire* 40: 145–52.

Church, M. 1983. "Pattern of Instability in a Wandering Gravel Bed Channel." In J.D. Collinson and J. Lewin, eds., *Modern and Ancient Fluvial Systems*, International Association of Sedimentologists, Special Publication 6, 169–80.

Church, M., Kellerhals, R., and Day, T.J. 1989. "Regional Clastic Sediment Yield in British Columbia." *Canadian Journal of Earth Sciences* 26: 31–45.

Church, M., and Ryder, J.M. 1972. "Paraglacial Sedimentation: A Consideration of Fluvial Processes Conditioned by Glaciation." *Geological Society of America Bulletin* 83: 3,059–72.

Church, M., and Slaymaker, O. 1989. "Disequilibrium of Holocene Sediment Yield in Glaciated British Columbia." *Nature* 337: 452–4.

Clague, J.J. 1978. "Mid-Wisconsinan Climates of the Pacific Northwest." In Current Research, Part B, Geological Survey of Canada, Paper 78–1B, 95–100.

– 1980. *Late Quaternary Geology and Geochronology of British Columbia. Part 1: Radiocarbon Dates*. Geological Survey of Canada, Paper 80–13.

– 1981. *Late Quaternary Geology and Geochronology of British Columbia. Part 2: Summary and Discussion of Radiocarbon-Dated Quaternary History*. Geological Survey of Canada, Paper 80–35.

– 1986. "The Quaternary Stratigraphic Record of British Columbia – Evidence for Episodic Sedimentation and Erosion Controlled by Glaciation." *Canadian Journal of Earth Sciences* 22: 256–65.

Clague, J.J., comp. 1989. "Quaternary Geology of the Canadian Cordillera." In R.J. Fulton, ed., *Quaternary Geology of Canada and Greenland*, Geological Survey of Canada, Geology of Canada, no. 1, 16–94.

Clague, J.J., and Mathewes, R.W. 1989. "Early Holocene Thermal Maximum in Western North America: New Evidence from Castle Peak, British Columbia." *Geology* 17: 227–80.

Cwynar, L.C. 1982. "A Late-Quaternary Vegetation History from Hanging Lake, Northern Yukon." Ecological Monographs 52, 1–24.

Cwynar, L.C., and Ritchie, J.C. 1980. "Arctic Steppe-Tundra: A Yukon Perspective." *Science* 208: 1,375–7.

Denton, G.H., and Karlen, W. 1977. "Holocene Glacial and Tree-line Variations in the White River Valley and Skolai Pass, Alaska and Yukon Territory, Canada." Geological Society of America, *Bulletin* 78: 485–510.

Denton, G.H., and Stuiver, M. 1966. "Holocene Climatic Variations – Their Pattern and Possible Cause." *Quaternary Research* 3: 155–205.

– 1967. "Late Pleistocene Glacial Stratigraphy and Chronology, Northeastern St. Elias Mountains, Yukon Territory, Canada." *Geological Society of America Bulletin* 78: 485–510.

Flint, R.F. 1971. *Glacial and Quaternary Geology*. New York: John Wiley & Sons.

Ford, D.C., Schwarcz, H.P., Drake, J.J., Gascoyne, M., Harmon, R.S., and Latham, A.G. 1981. "Estimates of the Age of the Existing Relief within the Southern Rocky Mountains of Canada." *Arctic Alpine Research* 13: 1–10.

Fulton, R.J. 1967. *Deglaciation Studies in Kamloops Region, an Area of Moderate Relief, British Columbia*. Geological Survey of Canada, Bulletin 154.

– 1971. *Radiocarbon Geochronology of Southern British Columbia*. Geological Survey of Canada, Paper 71–37.

Gascoyne, M., Ford, D.C., and Schwarcz, H.P. 1981. "Late Pleistocene Chronology and Paleoclimate of Vancouver Island Determined from Cave Deposits." *Canadian Journal of Earth Sciences* 18: 1,643–52.

Gascoyne, M., Schwarcz, H.P., and Ford, D.C. 1980. "A Paleotemperature Record for the Mid-Wisconsin in Vancouver Island." *Nature* 285: 474–6.

Gottesfeld, A.S., and Gottesfeld, L.M.J. 1990. "Floodplain Dynamics of a Wandering River: Dendrochronology of the Monice River, British Columbia." *Geomorphology* 3: 159–79.

Hansen, H.P. 1955. "Postglacial Forests in South Central and Central British Columbia." *American Journal of Science* 253: 640–58.

Harington, C.R. 1975. "Pleistocene Muskoxen (Symbos) from Alberta and British Columbia." *Canadian Journal of Earth Sciences* 12: 903–19.

Harris, S.A., and Pip, E. 1973. "Molluscs as Indicators of Late- and Post-glacial Climatic History in Alberta." *Canadian Journal of Zoology* 51: 209–15.

Hebda, R.J. 1982. "Postglacial History of Grasslands of Southern British Columbia and Adjacent Regions." In A.C. Nicholson, A. McLean, and T.E. Baker, eds., *Grassland Ecology and Classification Symposium Proceedings*, Victoria: BC Ministry of Forests, 157–91.

Hicock, S.R., Hebda, R.J., and Armstrong, J.E. 1982. "Lag of Fraser Glacial Maximum in the Pacific Northwest: Pollen and Macrofossil Evidence from Western Fraser Lowland, British Columbia." *Canadian Journal of Earth Sciences* 19: 2,288–96.

Hicock, S.R., Hobson, K., and Armstrong, J.E. 1982. "Late Pleistocene Proboscideans and Early Fraser Glacial Sedimentation in Eastern Fraser Lowland, British Columbia." *Canadian Journal of Earth Sciences* 19: 899–906.

Jones, D.L., Cos, A., Coney, P., and Beck, M. 1982. "The Growth of Western North America." *Scientific American* 247 no. 5: 70–84.

Kearney, M.S., and Luckman, B.H. 1983a. "Holocene Timberline Fluctuations in Jasper National Park, Alberta." *Science* 221: 261–3.

Kearney, M.S., and Luckman, B.H. 1983b. "Postglacial Vegetational History of Tonquin Pass, British Columbia." *Canadian Journal of Earth Sciences* 20: 776–86.

Leonard, E.M. 1986a. "Varve Studies at Hector Lake, Alberta, Canada and the Relationship between Glacial Activity and Sedimentation." *Quaternary Research* 25: 199–214.

– 1986b. "Use of Lacustrine Sedimentary Sequences as Indicators of Holocene Glacial Activity, Banff National Park, Alberta, Canada." *Quaternary Research* 26: 218–31.

Lichti-Federovich, S. 1974. *Palynology of Two Sections of Late Quaternary Sediments from the Porcupine River, Yukon Territory*. Geological Survey of Canada, Paper 74–23.

Luckman, B.H., and Kearney, M.S. 1986. "Reconstruction of Holocene Changes in Alpine Vegetation and Climate in the Maligne Range, Jasper National Park, Alberta." *Quaternary Research* 26: 244–61.

Luckman, B.H., and Osborn, C.D. 1979. "Holocene Glacier Fluctuations in the Middle Canadian Rocky Mountains." *Quaternary Research* 11: 52–77.

MacDonald, G.M. 1982. "Late Quaternary Paleoenvironments of the Morley Flats and Kananaskis Valley of Southwestern Alberta." *Canadian Journal of Earth Sciences* 19: 23–5.

– 1983. "Holocene Vegetation History of the Upper Natla River Area, Northwest Territories, Canada." *Arctic and Alpine Research* 15: 169–80.

Mathewes, R.W. 1979. "A Paleoecological Analysis of Quadra Sand at Point Grey, British Columbia, Based on Indicator Pollen." *Canadian Journal of Earth Sciences* 16: 847–58.

– 1985. "Paleobotanical Evidence for Climatic Change in Southern British Columbia during Late-Glacial and Holocene Time." In C.R. Harington, ed., *Climatic Change in Canada 5: Critical Periods in the Quaternary Climatic History of Northern North America*, National Museums of Canada, National Museum of Natural Sciences, Syllogeus Series 55, Ottawa, 397–422.

Mathewes, R.W., and Heusser, L.E. 1981. "A 12,000 Year Palynological Record of Temperature and Precipitation Trends in Southwestern British Columbia." *Canadian Journal of Botany* 59: 707–10.

Matthews, J.V., Jr. 1982. "East Beringia during Late Wisconsin Time: A Review of the Biotic Evidence." In D.M. Hopkins, J.V. Matthews Jr., C.E. Schweger, and S.B. Young, eds., *Paleoecology of Beringia*, New York: Academic Press, 127–50.

Miller, M.M., and Anderson, J.H. 1974. "Out-of-phase Holocene Climatic Trends in the Maritime and Continental Sectors of the Alaska-Canada Boundary Range." In W.C. Mahaney, ed., *Quaternary Environments: Proceedings of a Symposium*, York University, Geographical Monograph 5, Toronto, 33–58.

Mott, R.J., and Jackson, L.E., Jr. 1982. "An 18,000 Year Palynological Record from the Southern Alberta Segment of the Classical Wisconsinan 'Ice-Free Corridor'." *Canadian Journal of Earth Sciences* 19: 504–13.

Osborn, G., and Luckman, B.H. 1988. "Holocene Glacier Fluctuations in the Canadian Cordillera (Alberta and British Columbia)." *Quaternary Science Review* 7: 115–28.

Parrish, R.R. 1983. "Cenozoic Thermal Evolution and Tectonics of the Coast Mountains of British Columbia. 1. Fission Track Dating, Apparent Uplift Rates, and Patterns of Uplift." *Tectonics* 2: 601–31.

Pearce, G.W., Westgate, J.A., and Robertson, S. 1982. "Magnetic Reversal History of Pleistocene Sediments at Old Crow, Northwestern Yukon Territory." *Canadian Journal of Earth Sciences* 19: 919–29.

Rampton, V.N. 1970. "Neoglacial Fluctuations of the Natazhat and Klutlan Glaciers, Yukon Territory." *Canadian Journal of Earth Sciences* 7: 1,236–63.

– 1982. *Quaternary Geology of the Yukon Coastal Plain*. Geological Survey of Canada, Bulletin 317.

Reeves, B.O.K. 1975. "Early Holocene (ca. 8000 to 5500 B.C.) Prehistoric Land/Resource Utilization Patterns in Waterton Lakes National Park, Alberta." *Arctic and Alpine Research* 7: 237–48.

Ritchie, J.C. 1982. "The Modern and Late Quaternary Vegetation of the Doll Creek Area, North Yukon, Canada." *New Phytologist* 90: 563–603.

– 1984. *Past and Present Vegetation of the Far Northwest of Canada*. Toronto: University of Toronto Press.

Ritchie, J.C., Cinq-Mars, J., and Cwynar, L.C. 1982. "L'environnement tardiglaciaire du Yukon septentrional, Canada." *Géographie physique et quaternaire* 36: 241–50.

Ritchie, J.C., Cwynar, L.C., and Spear, R.W. 1983. "Evidence from North-west Canada for an Early Holocene Milankovitch Thermal Maximum." *Nature* 305: 126–8.

Ryder, J.M. 1971. "The Stratigraphy and Morphology of Para-glacial Alluvial Fans in South-Central British Columbia." *Canadian Journal of Earth Sciences* 8: 279–98.

Ryder, J.M., and Church, M. 1986. "The Lillooet Terraces of Fraser River: A Palaeoenvironmental Enquiry." *Canadian Journal of Earth Sciences* 23: 869–84.

Ryder, J.M., and Thomson, B. 1986. "Neoglaciation in the Southern Coast Mountains of British Columbia: Chronology Prior to the Late Neoglacial Maximum." *Canadian Journal of Earth Sciences* 23: 273–87.

Shackleton, N.J., and Opdyke, N.E. 1976. "Oxygen-Isotope and Paleomagnetic Stratigraphy of Pacific Core V28–239, Late Pliocene to Latest Pleistocene." In R.M. Cline and J.D. Hayes, eds., *Investigation of Late Quaternary Paleoceanography and Paleoclimatology*, Geological Society of America, Memoir 145, 449–64.

Slaymaker, O., and McPherson, J.H. 1977. "An Overview of Geomorphic Processes in the Canadian Cordillera." *Zeitschrift fur Geomorphologie* 21: 169–86.

Souther, J.G. 1977. "Volcanism and Tectonic Environments in the Canadian Cordillera – A Second Look." In W.R.A. Baragar, L.C. Coleman, and J.M. Hall, eds., *Volcanic Regimes in Canada*, Geological Association of Canada, Special Paper 16, 3–24.

Warner, B.G., Mathewes, R.W., and Clague, J.J. 1982. "Ice-Free Conditions on the Queen Charlotte Islands, British Columbia, at the Height of Late Wisconsin Glaciation." *Science* 218: 675–7.

Yorath, C.J., and Hyndman, R.D. 1983. "Subsidence and Thermal History of Queen Charlotte Basin." *Canadian Journal of Earth Sciences* 20: 135–59.

CHAPTER TEN

Mountain Hazards

JAMES S. GARDNER

WHAT ARE NATURAL HAZARDS?

Natural hazards are understood as those interactions between the natural environment and society that have net costs to society (White 1974). Usually the natural environment is seen as the primary actor, and people are seen as respondents or victims. Thus when a flash flood occurs as a result of heavy rain in the Coast Mountains of British Columbia and a residential subdivision is damaged, we think of the flood as being a natural hazard, the event as a disaster, and the people and community as victims. However, on exploring the event in more detail we may find that the flood was itself caused, at least in part, by extensive clearcutting of forest in the watershed. The subtle interaction of society and environment often precludes the possibility of assigning blame or primary cause.

Natural hazards are the antithesis of natural resources. Whereas resources are useful, hazards impose costs. Society defines what is hazardous. Thus to understand natural hazards we must call on knowledge, from both natural and social sciences. A common characteristic of natural hazards is the uncertainty associated with their magnitude and time of occurrence.

Natural hazards exist in all environments. Here, the first objective is to define those attributes of the mountain environment that give rise to hazardous processes (Table 10.1). The second is to illustrate how mountain environments have attracted activities and land uses that are particularly vulnerable to hazards. The third is to assess the extent to which human use of mountain land has altered natural conditions, thus changing the frequency and extent of potentially hazardous processes.

The impact of a mountain range on precipitation may result in distinctive precipitation gradients leading to hazard potential. For example, the BC Coast Mountains induce orographic precipitation by intercepting warm, moist Pacific air masses travelling eastward (chapter 7). This contributes to a debris flow hazard. Farther east, in the Selkirk Mountains, a similar though less pronounced orographic precipitation effect is the basis of a severe snow avalanche hazard. Still farther east, the topography

Table 10.1
Attributes of mountain environments contributing to natural hazards

Attribute	Contribution	Hazard type
Geology	Tectonism: Correspondence of mountain areas to crustal plate boundaries leads to seismic activity. Volcanism: Volcanic activity in some mountain areas leads to volcanic explosions and lava flows.	Earthquakes Volcanic eruptions Lava flows Lahars Ashfalls
Topography	Great local relief and steep slopes create potential for gravity-induced movements of rock, soil, snow, ice, and water (mass wasting).	Rockfalls Rock avalanches Slumps debris flows and torrents Snow and ice avalanches Flash floods
Climate	Temperature: Great variability over short distances and through time creates thaw and freeze conditions; cold seasonally and permanently (at high elevation) permits presence of snow, glacier ice, and ground ice. Precipitation: Orographic and convective effects in mountains lead to abundant and/or intense rainfall or snowfall.	Snow and ice avalanches Floods and flashfloods Debris flows and torrents Rockfall slumps
Hydrography	Presence of water in all states of aggregation (gas, liquid, solid) and several forms (streams, wetlands, glaciers, snow, ground ice, surface ice, river and lake ice) creates potential for large variety of hazards.	Floods Jökulhlaups Snow and ice avalanches Debris flows and torrents
Surface sediments (include soils)	Glacial, alluvial, lacustrine, marine, and other loose sediments as well as soils provide material for erosional events when subject to wetting.	Debris flows and torrents Landslides Slumps
Vegetation	Topoclimatic variability leads to great variability in vegetation from valley locations through high-elevation areas. Where present as forest cover, vegetation generally acts as a stabilizing factor. Where forest or other continuous vegetation cover is not present, by virtue of natural climatic conditions or cutting and fire, stability is lost, thus increasing potential for certain hazardous processes.	Snow avalanches Debris flows and torrents Floods and flash floods Forest fires

of the Rocky Mountains induces localized, intense convective rainstorms which may lead to flash floods and debris flows.

In addition to having the steep slopes and high elevation that define mountains (chapter 7), these regions often present a potential for earthquake and volcanic activity. This is especially relevant to southwestern British Columbia, which has a large concentration of population. The hazard potential is for a megathrust earthquake related to movement of the Pacific, American, and Juan de Fuca plates. Although no earthquake of magnitude 8 and greater on the Richter scale has been observed in the past 200 years in the region, an earthquake of magnitude 7.2 occurred on Vancouver Island in 1946 (Mathews 1979). Other earthquake areas include the St Elias Mountains and parts of eastern Yukon and westernmost District of Mackenzie. One recent example was the 1985 Nahanni earthquakes (Horner et al. 1987). On 5 October 1985, a magnitude-6.6 earthquake occurred west of Fort Simpson. A magnitude-6.9 event followed on 23 December 1985. Fortunately, the sparse population in the epicentre region and the preponderance of one or two-storey wood-frame or log structures in the mountains limited serious damage. Subsequent surveys revealed large landslides and rockfalls in the interior mountains: in one locality, a rock avalanche had moved an estimated 5–7 million m^3 of rock to form an avalanche scar over 70 m high.

The widespread occurrence of Quaternary volcanic rocks and ash indicates past volcanism in the Canadian Cordillera. As part of this potential, the area is also glacierized. For example, the 1980 eruption of Mt St Helens in Washington's Cascade Mountains showed the ability of volcanic activity to melt large quantities of snow and ice which, in combination with sediments, produce volcanic mudflows or lahars, the effects of which may be translated great distances away from the point of eruption.

This chapter focuses on hazards that develop as a result of the terrain, weather and climate, and hydrological conditions of mountain environments. Events usually happen suddenly, but conditions leading to them may have developed over days, weeks, months, or years, depending on the hazard type. The process or its consequences may be translated downslope or far downstream into quite different environments. This translation is the direct result of the geopotential energy produced by the steep slopes and great local relief that characterize Canada's mountain environments.

MOUNTAIN HAZARD TYPES AND CASE STUDIES

Several types of hazards are relatively distinctive to mountain areas and commonly associated with mountains. The classification shown in Figure 10.1 uses a number of criteria: a distinction between geophysical and biological natural hazards; the sphere or domain of primary operation and impact (e.g. atmosphere, biosphere); the degree of interaction between elements; and the type of material, objects, or processes involved. While many natural hazards listed in Figure 10.1 occur in mountain environments they are not limited to such environments. Examples include drought, infestations and diseases, heatwaves, coldwaves, predatory animals, and forest fires.

The mountain hazards specifically treated here are snow avalanches, floods, debris flows, and rock avalanches. Snow avalanches and floods can be grouped in terms of a recurrence interval ranging from less than 1 year to 10 years at a given location. The frequency scale for debris flows is 10–100 years, and that for rock avalanches

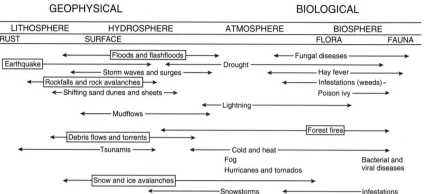

Figure 10.1
Types of natural hazards, by sphere of occurrence, medium, and materials.

is 1,000–10,000 years at a given location. The frequency of hazard occurrence is a useful criterion in differentiating society's perception of, and response to, natural hazards.

Though sparsely inhabited and having a short recorded history of occupance, the Canadian Cordillera provides examples of all four hazard types.

SNOW AVALANCHES

The majority of snow avalanches, which involve rapid downslope sliding or flowing of large masses of snow, take place on open slopes and in gullies at high elevations, away from most facilities and concentrations of people. Avalanches become hazardous when they begin at high elevations, sweep down lower, often forested, slopes, and intrude on facialities and settlements at lower elevations (Figure 10.2). Snow avalanches have been the most widespread and persistent of natural hazards in mountain environments in Canada, a generalization true also about other mid-latitude mountain areas such as the Alps of Europe (Fraser 1978).

Whenever enough snow falls to produce a snowpack or snow cover on the ground and wherever slopes are steep, snow avalanches can result. The behaviour of the avalanche and the avalanche type depends on the characteristics of the snow (e.g. dry, wet, powdery, compacted, or loose) and the timing (i.e. directly after or during a snowstorm = direct action; delayed some time from several days or weeks to months = delayed action).

Figure 10.2
Avalanche paths Front Ranges, Canadian Rockies, Alberta. *Photo*:
J. Gardner.

Three sets of factors contribute to snow avalanches: terrain conditions, meteorological conditions, and snowpack conditions. Terrain provides the platform on which
sufficient snow can accumulate and the gradient down which snow is transported.
Most snow avalanche paths can be divided into three areas: the starting zone (snow
accumulation and avalanche initiation); the track (avalanche transport); and the runout
or deposition zone (snow deposition and, usually, main area of damage). Figure 10.2
illustrates all three zones and shows that the process cuts through several altitudinally
arranged ecological zones.

Being passive, such terrain conditions set the stage for repeated avalanches at the
same locations. Starting-zone terrain must ensure or enhance snow accumulation
while providing sufficient gradient to permit snowpack failure and movement.
Starting-zone gradients are generally in the range of 20–50°; steeper gradients will
not permit accumulation, while gentler ones will not support failure and movement.
Track gradients must either maintain or enhance motion. Depending on snow conditions, gradients of 15° will do this. Often, tracks have vertical cliffs and steep steps
which enhance motion and may initiate airborne transport. Runout zones are generally
less than 15° in gradient.

The typical avalanche path is often seen on mountain valley-side slopes below
timberline. Less obvious are avalanche paths on open or unconfined slopes at high

Table 10.2
Avalanches' effects on human life, by type of land use (Canada)

Category	Deaths	Injuries	Burials	Total	Percentage
Transportation	206	70	108	384	43
Trails, roads, highways	114	61	81	256	29
Railways	92	9	27	128	14
Recreation	122	49	74	245	28
Mountaineering	53	20	27	100	11
Back-country skiing	43	14	29	86	10
Dowhill resort skiing	17	11	14	42	5
Showmobiling	7	4	1	12	1
Tobogganing	2	–	3	5	1
Resource development	70	47	116	233	26
Mining	60	47	115	222	25
Logging	9	–	1	10	1
Trapping	1	–	–	1	<1
Town/service buildings	14	6	5	25	3
Total	412	172	303	887	100
Percentage	47	19	34	100	100

Source: McFarlane (1985:40).

elevation and above treeline. Apart from having terrain characteristics that permit snow accumulation, failure, and transport, there is little to distinguish such avalanche paths from surrounding terrain. They have become lethal hazards in the recreation sector.

Meteorological conditions include variables such as snowfall, wind, and temperature changes. Rain may be a triggering and loading factor in some cases. High-intensity and copious snowfall raises the probability of avalanche release. Strong and sustained wind enhances the probability even more. Sharp, upward temperature changes to above freezing act to trigger avalanche release, while sustained temperatures above freezing break down internal snowpack bonding, thereby reducing resistance to failure and enhancing avalanche occurrence.

The snowpack itself is not a totally passive agent. Equi-temperature metamorphism of snow grains, besides causing visible "settling" of the snow, "strengthens" the snowpack and reduces probability of avalanche. Temperature-gradient metamorphism causes redistribution of mass within the snowpack, growth of hoar-like grains, development of weak layers in the snowpack, and increased probability of avalanche.

Detailed inventories of damaging snow avalanches have been produced by Stethem and Schaerer (1979; 1980). The importance of snow avalanche hazard is illustrated by the impact on human life as summarized in Table 10.2 (McFarlane 1985). The

Table 10.3
Persons per year affected by avalanches, by type of land use

Category	Prior to 1930	1930–74	1975–84
Transportation	6.2	1.2	1.2
Trails, roads, highways	3.7	0.9	0.7
Railways	2.5	0.3	0.5
Recreation	0.1	2.6	13.1
Mountaineering	0.1	0.9	5.2
Back-country skiing	0	1.2	4.8
Downhill resort skiing	0	0.4	1.8
Snowmobiling	0	0.1	0.8
Tobogganing	0	0.1	0.5
Resource development	2.3	1.6	0.2
Mining	2.3	1.5	0.1
Logging	0	0.1	0.1
Trapping	<.1	0	0
Town/service buildings	0	0.6	0
Total average	8.6	6.0	14.5
As percentage	30	20	50

Source: McFarlane (1985:63).

data convey land uses and activities most affected by the snow avalanche hazard in the Canadian context. The vulnerability of transportation facilities and land uses is clear, a point that can made for debris flows as well. Although data measuring damage to facilities are difficult to obtain, McFarlane (1985) estimates direct damage from snow avalanches at about $10 million (1984 dollars). While less than that attributed to other hazards, the figure does not include indirect costs of disrupted services. An example of the latter is the stoppage of traffic on Highway 1 at Rogers Pass during high avalanche hazard and avalanches. Public safety (Table 10.2) and indirect costs justify the huge expenditures on mitigation of snow avalanche hazard in Canada's mountain areas.

The land uses most affected in terms of effect on human life have changed through time (Table 10.3). Prior to 1930 the primary areas of concern were in transportation and resource development, principally mining. Since the early 1970s, the principal sector affected has been recreation, primarily mountaineering and back-country skiing, including helicopter skiing.

The character of the avalanche hazard can be illustrated with a few examples. The Rogers Pass area in the Selkirk Mountains of British Columbia is the location of the

Figure 10.3
Illecillewaet valley near Rogers Pass, Selkirk Mountains, BC. Photograph
shows confined nature of valley bottom, with convergence of (Highway
1, CPR), stream channel, and snow avalanche paths. *Photo*: J. Gardner.

most serious avalanche hazard to transportation (Figure 10.3). At risk are Highway 1
(Trans-Canada Highway) and the Canadian Pacific Railway line. The hazard results
from a combination of high snowfall rates and amounts and steep, long slopes leading
to confined valley floors. Susceptibility was established early: during the first winter
of operation, in 1885–86, the railway line was closed by snow avalanches. Then,
in March 1910, a single avalanche killed sixty-two men. This disaster, plus many
long periods of closure, led the CPR to construct the Connaught Tunnel under the
most dangerous avalanche terrain in Rogers Pass. In 1989 a second tunnel was
completed to increase rail traffic flow. In summary, between 1885 and 1916, when
the original tunnel was completed, over 200 people lost their lives to avalanches at
Rogers Pass. Snow avalanches continue to plague railway operations in the area
despite construction of protection sheds and other structures.

From the point of view of avalanche hazard, Rogers Pass was one of the least
likely candidates for the routing of the Trans-Canada Highway in the late 1950s. In
spite of this, the highway was completed in 1962, with careful forethought, research,
and planning for the avalanche hazard (Schaerer 1962). The hazard has been mitigated
through careful route selection, snowsheds, a sophisticated weather monitoring and
avalanche forecasting system, a mobile artillery defence for pre-release, road closures,

dams, dykes, mounds, and heavy snow-clearance machinery. Other avalanche hazard locations are on Highway 16 between Terrace and Prince Rupert, on Highway 3 over Kootenay Pass between Salmo and Creston, and, most recently, along the newly constructed Coquihalla highway between Meritt and Hope, all in British Columbia.

Avalanches have significantly affected mining activities in Canada's mountain areas (Peck 1970; Gardner 1986). This was demonstrated dramatically during the early "silver rush" days in the BC interior. This period was characterized by numerous small operations, often located well above treeline. In Carpenter Creek valley in the Slocan district of south-central British Columbia, in a relatively small area, over twenty people were killed between 1895 and 1925 and the entire mining operation was constrained by, and adjusted to coping with, snow and snow avalanches (Gardner 1986). This situation was echoed in many locations during the height of the pre-Depression mining area.

Snow avalanches have continued to affect the mining industry. In February 1965, an avalanche on the access road to Granduc Mine near Stewart, BC, killed twenty-six men and injured twenty-three others. Subsequent operation of the mine has relied on development of a sophisticated hazard-forecasting and -management program for the road.

Growth of the winter outdoor recreation industry in the 1970s and 1980s in western Canada caused escalation of the avalanche hazard (Table 10.3). Expansion of major ski resorts has been partly responsible. While some form of avalanche-hazard management occurs at most ski resorts in the Cordillera, major resorts such as those at Sunshine and Lake Louise in Banff National Park and the Whistler-Blackcomb area, north of Vancouver, require careful and constant attention to the avalanche hazard. Designated ski areas are usually designed to avoid avalanche hazard. Most recent accidents at ski resorts have occurred outside the designated area when people have sought more challenging out-of-bounds skiing.

The "out-of-bounds" phenomenon in winter snow-based recreation now extends well beyond the immediate vicinity of ski resorts. This movement to back-country (cross-country) ski touring and high-country (mountaineering) explains the trend toward increased hazard shown in Table 10.3.

Approaches to mitigating the snow avalanche hazard vary greatly. The first approach is identification of potentially hazardous locations and sites. Mapping of hazardous locations is integral to hazard mitigation with respect to most transportation, resource extraction, and recreation land uses and activity. This process includes production and use of detailed hazard maps for ski areas and highway routes and less formalized hazard evaluation and route selection during mountaineering and ski expeditions in remote areas.

The second approach involves short-term knowledge of weather and snow conditions. Ongoing (daily) collection of meteorological and snowpack data at representative sites in the mountains is the basis of avalanche forecasting. A forecast is then used to justify road closures, closing of parts of ski areas and certain slopes in helicopter-skiing operations, mine closures, and evacuation. Closure – behavioural adjustment applied to the user – also provides time whereby natural stabilization in the snowpack or active stabilization through use of explosives may reduce the hazard.

The third approach uses protective devices for facilities such as roads where environmental and/or locational considerations provide no alternatives in route selection. Protective devices that have been employed include snowsheds over roads and railroads, dykes or other deflecting devices, and earth mounds in the upper runout zone.

A final set of approaches addresses the aftermath, where people are buried in a snow avalanche, and is usually referred to as emergency measures: rescue procedures that entail a trained crew or a set of trained supervisors and or use of search and rescue dogs maintained by the RCMP in the Canadian Cordillera.

FLOODS

Floods in Canadian mountain areas take three principal forms: spring freshets (snowmelt and rainfall), rainfall-generated flooding, and jökulhlaups or sudden outbursts of water from glaciers (chapter 7). Many in small drainage basins ($<10^2$ km^2) are described as "flash floods": sudden responses to rapid production of water and its removal from the land surface to a stream channel, a feature encouraged by mountain drainage basins. Other aspects of mountain environments that enhance "flash floods" are presence of a deep snowpack, propensity for radical upward temperature changes producing rapid snowmelt, intense and/or prolonged rain produced by mountain convective and orographic effects, and, occasionally, frozen ground.

In contrast, some of the larger rivers in the Cordillera are subject to large-scale flooding, usually as a result of the combination of rainfall and rapid melting of a deep snowpack over large areas of the watershed. In the eastern and interior Cordillera, ice jams enhance this type of flooding. The most serious flood hazard in Canada occurs on the lower reaches of the Fraser River between Hope and Vancouver (Sewell 1965). Other locations of severe floods include the lower reaches of the Columbia River in south-central British Columbia; several places on the upper Fraser, Peace, and Liard rivers in the northwest of the province; the Yukon River; and a few locations along the Athabasca, North Saskatchewan, and Bow rivers in Alberta.

Large riverine floods illustrate the "hinterland" effect of mountain areas. It has long been recognized that the physical characteristics of many mountain areas permit them to serve as huge "catchments" and storage places for water in various forms. As such, mountains become source areas for many rivers that flow into semi-arid and arid plains where they are major water resources. The same hinterland effect occurs with the flood hazard. It is often the cumulative effect over an entire watershed rather than extreme runoff in the small, individual tributary watersheds that produces a flood in the riverine context. The flood may be most pronounced in the lower reaches of a river such as the Fraser because of the cumulative hydrological effect and the presence of an extensive floodplain, or it may be translated entirely out of the mountain area, as on the Bow River which drains the east slope of the Rockies. Such a situation demonstrates that the different and unique environments of the high mountains can generate processes and events that are hazardous in environments and locations far removed from where the processes originated.

Three examples reveal the widespread potential of the flood hazard in the Canadian Cordillera: the Lower Fraser valley (large riverine flooding); the Bow River valley in the Rockies (mountain river flooding and flash floods in small valley-side tributaries); and the Fyles Glacier–Ape Lake jökulhlaup floods on the Noeick River in the Coast Mountains.

The extreme flood hazard in the Lower Fraser River valley is produced by three sets of factors: a large (230,000 km^2) mountainous watershed of which 75 per cent supports a seasonal snowpack and much of which is subject to rapid warming and/or rainfall during spring melt; a low-lying, extensive floodplain between Hope and Richmond (Vancouver) which provides some of the best land for agricultural, industrial, and residential purposes in the Cordillera; and extensive population growth and land-use intensification in the lower mainland region of which the Lower Fraser valley forms the largest part.

The most damaging flood on the Fraser occurred in May and June 1948. The flood was caused by a delayed and then sudden melting of a deep snowpack throughout the mountainous part of the watershed. On the basis of the snowpack, its delayed melting and the potential for a sudden rise in temperature, a flood warning was issued on 1 May, 1948. By 25 May the river had reached the critical level, above which extensive flooding could be expected. By the end of May the major dykes from Mission downstream had been breached and both (CPR and CNR) rail lines had been inundated. Rail service to and from Vancouver was not resumed until 30 June, over a month after the dykes were first breached. Total damage and losses of property and income are difficult to estimate but were in excess of $20 million in 1948 dollars. This disaster of national proportions raised the level of concern for the flood hazard across Canada.

Mitigation of the Fraser River flood hazard has been piecemeal, and many of the forces that escalated the risk prior to 1948 still remain in effect (Sewell 1965). Compensation by various levels of government has covered flood losses, rebuilding and further development of dyking systems, some structural flood-proofing and building-grade raising, and continuation and additional development of the flood warning system. The presence of many municipal jurisdictions has inhibited widespread and effective land-use regulation in flood-prone areas, and intensification of land use in flood-prone areas continues to the present.

The Bow River is a watershed smaller than the Fraser and experiences a drier climate. As a result, it has not witnessed a disaster of the magnitude of the 1948 flood on the Fraser. In addition, downstream from the town of Banff, flow of the Bow River is regulated by several hydroelectric installations. Flooding has been of two types: major valley-bottom flooding on the Bow and its major tributaries, and

Figure 10.4
Floodplain settlement: Exshaw, Alberta, in Front Ranges of Canadian
Rockies, and residential area Lac des Arcs are located on side valley
alluvial fans on Bon River's floodplain. *Photo*: J. Gardner.

flash flooding on small tributary streams. As in most mountain terrain, land for
building purposes is at a premium. The flat land on the floodplain and the gently
sloping alluvial fan surfaces that characterize the small tributary streams are attractive
building sites (Figure 10.4).

Evidence of flooding prior to 1880 is sketchy. However, forest fires in the Bow
valley associated with construction of the CPR line (1880–85) are thought to have
increased runoff and exacerbated the flood problem (Nelson and Byrne 1966). There
is evidence of flooding in the Banff-Canmore area at least sixteen times between
1878 and 1974, the most significant being in 1904, 1916, 1923, and 1974. The 1904
flood, on a tributary of the Bow, destroyed much of Anthracite, a small coal-mining
community near Banff. The 1923 flood, the largest on record at Banff (Figure 10.5),
demonstrated the hydrological importance of the large area of wetlands and lakes in
the Bow valley immediately west of the Banff townsite. The brunt of the 1923 flood
was "absorbed" by this wetland area and emphasized the wisdom of officials of
Dominion Parks (now National Parks) during the previous two decades; they consistently turned down proposals to drain and develop the wetland area for building
and agricultural purposes.

Figure 10.5
Flood: Bow River, Banff, Alberta, 1923. *Photo*: courtesy Whyte Foundation, Archives of the Canadian Rockies, Banff.

The 1974 flood caused serious damage in Canmore and resulted from a large and sudden June runoff. It escalated public concern for the flood hazard and led to hydrologic studies, channelization, dyking, landscaping, and formation of the Canmore Flood Control Advisory Committee (Alberta 1977). The flood took place at a time of unprecedented growth in Banff, Canmore, and several other small townsites. In some cases, flood-prone land was being developed for residential and tourism purposes. The flood gave pause for thought and some systematic consideration of the increasing risks and vulnerability caused by townsite growth. While there has been no serious flooding since 1974, the risk of flood damage continues. In the Canmore area, one response to the Bow River flooding has been to develop land on valley-side terraces and alluvial fans, raising the spectre of flash floods.

All valley-side tributaries draining from the east to the Bow River between the Front Range and Banff have experienced flash floods in the past century. Most of these tributaries, in their present or earlier form, have produced alluvial fans which have become building sites and routes for road and rail transport. The historical record indicates recurring damage to highway and railroad bridges and lines adjacent to such streams as Cougar and Carrot creeks. Today, a large residential subdivision has developed on the fan of Cougar Creek. The flood hazard has been dealt with through minor channel adjustments (channelization and some dyking). The town of Exshaw, on the fan of Exshaw Creek, provides another example (Figure 10.4); by 1910 a well-developed community had grown there around mining operations (Portland cement). The stream course across the fan was dyked with concrete walls from the beginning. Nonetheless, the community has been damaged by flooding on at least three occasions since 1906.

Numerous jökulhlaups have been documented for the Cordillera. Jökulhlaups are sudden outbursts of large quantities of water from behind or beneath dams of glacier ice. Some release enough water to produce major floods on large rivers far from the glacial source.

The Fyles Glacier and adjacent Ape Lake in the BC Coast Mountains produced two major floods of this type in 1984 and 1986. The first was described in detail by Jones et al. (1985). When filled and under normal drainage conditions, Ape Lake drains to the east into the Talchako River. The two events produced catastrophic drainage under the Fyles Glacier to the west and into the Noeick River. During high-water levels in Ape Lake, water seeps beneath Fyles Glacier. This seepage accelerates to an outburst flood as the subglacial channel enlarges. Both events released about 46×10^6 m^3 of water. The resulting floods on the Noeick River were unprecedented in 300 years, as indicated by the age of forest vegetation destroyed in the two events. The river's floodplain, channel geometry, and valley morphology were radically altered. Since the valley is essentially uninhabited, structural damage was minimal. However, this example clearly underlines the need for careful examination of all aspects of a watershed that could contribute to flooding prior to development of downstream areas. It shows also that place-specific hydrological events in mountains can be translated very quickly, and with little warning, over great distances and into totally different environments.

DEBRIS FLOWS

Debris flows are rapid downslope movements of a mixture of water and debris ranging from fine-grained inorganic sediments to rock blocks and organic debris. They can happen in saturated materials with continuously deforming movement (fluid-like motion) and/or in well-defined channels where water flow mobilizes large quantities of debris from the sides and bed of the channel. The key ingredients in Canada's mountain areas include availability of inorganic and organic debris, presence of water, and a steep slope or channel (Evans and Lister 1984; Van Dine 1985) – in other words, enough debris must collect where it can be mixed with, or entrained by, water. This collection process describes the "loading" essential for debris flow. The flushing of a channel of this load usually decreases the probability of subsequent debris flows until other erosional processes "load" the channel again. Loading may take several years or decades.

The other ingredient – large amounts of water – is delivered by several mechanisms which are mostly related to climate (Eisbacher 1979). Prolonged and intense rainstorms with or without melting snow is an important mechanism in the BC coastal and interior mountains. Intense, convective rainstorms in summer and sudden, copious snowmelt in spring are major sources of water throughout the Rocky Mountains. More generally, in glaciated areas throughout the mountain regions of Canada, sudden outbursts of large volumes of water either dammed behind moraines and glaciers or held within glacier basins may cause debris flows and flash floods downstream (Jackson 1979; Clarke 1982; Blown and Church 1985; Clague, Evans, and Blown 1985).

The most complete inventory of the debris-flow hazard in the southern Cordillera is that of Van Dine (1985). Debris flows are by no means restricted to the southern Cordillera, but their effect as a hazard and our knowledge of them as such is greater for this area. Indeed, the summer of 1988 in northern British Columbia and Yukon demonstrated the vulnerability of the Alaska Highway to this type of process. Traffic was held up for periods of up to a week at several locations because of debris flow. Van Dine's (1985) inventory includes about forty debris flows. He attributes seventeen deaths directly or indirectly to these events and estimates damage in excess of $100 million since 1962. Most often, damage is sustained to roads, railways and bridges. The main occurrences are along Highway 99 on Howe Sound north of Vancouver; on Highway 1 (Trans-Canada) between the BC-Alberta border and Revelstoke, BC; and along the Coquihalla Highway.

Highway 99 on Howe Sound has presented the most serious debris-flow hazard in the Canadian mountains. Most of the seventeen deaths and property damage have occurred on a stretch of road between Horseshoe Bay and M Creek, a distance of about 30 km. The principal contributing drainage basins are small (less than 10 km^2), and the gulley or creek profiles are steep (20–50° in the initiating area to 8–20° in the depositional area). Virtually all the creeks have small, steep depositional fans where they debouch into the Sound. These fans provide relatively flat surfaces for residential developments in a region (West Vancouver) where residential land is at a premium.

Virtually all the Howe Sound drainage basins have obvious sources of debris. The forested slopes contribute organic material in the form of logs which may also dam the stream to produce small dam-burst floods. Exposure of these drainage basins to airflow off the ocean, coupled with the orographic lift by the topography, ensures rainfall that is both copious and intense in conditions of extreme atmospheric instability. Damaging events occur most frequently from October through February, confirming that snow in the higher-elevation parts of the basins also contributes to water generation.

In recent times, three storms appear responsible for damaging debris flows and the resulting deaths of twelve people along Highway 99. These events took place on 28 October 1981, 4 December 1981, and 11 February 1983. These disasters – coupled with the prior history of debris flow and the recognition that, given time (i.e. debris loading), many of these basins would be repeat performers – led to a major effort to control debris flow along Highway 99 (Hungr, Morgan, and Kellerhals

Figure 10.6
Debris-flow protective structures, Lions Bay, BC: single-span bridge and oversized concrete channel with basins on Highway 99. *Photo*: J. Gardner.

1984). At the same time, the economic importance of Highway 99 has increased: it is the only access route to the Whistler area, a major resort focused primarily on skiing.

The most obvious approach to mitigation and control of debris flows is through rational land-use planning and avoidance of hazardous areas. However, at Highway 99 there is already an established facility, and there are few alternatives for relocation. In such cases, active remedial measures must be used. In addition to warning systems, attempts are made to prevent debris flow through regulation of forest cutting and reforestation. Other methods include containment of the flow in check dams and receiving basins; dyked, deepened, and straightened channels; and construction of deflection dykes, oversized culverts, and single-span or free-span bridges (Figure 10.6).

Linear facilities such as roads and railways are especially vulnerable to debris flows. This is demonstrated along the Trans-Canada Highway between the Alberta-BC border and Revelstoke. The principal sites include Cathedral Mountain and Mt Stephen near Kicking Horse Pass (Figure 10.7) and along the Illecillewaet valley between Rogers Pass and Revelstoke (Figure 10.3) (Van Dine 1985). Debris flows have blocked and damaged rail and highway transport, derailed trains, and damaged bridges and other structures. In contrast to the Highway 99 cases, these debris flows all occurred during the summer, often as a result of intense rainfall.

The Cathedral Mountain site in addition includes catastrophic glacial meltwater (jökulhlaup) release as a generating mechanism. There, the meltwater from a small, high-elevation glacier is stored in an interior basin during most summers. Normal

Figure 10.7
Debris-flow site, Mt Stephen, Yoho National Park, BC. Debris-flow deposits cross CPR main line and approach Highway 1 on left. In background is Cathedral Mountain (Cathedral Crags), site of one of the Cordillera's most hazardous debris-flow locations. *Photo*: J. Gardner.

drainage is via a harmless northeasterly meltwater stream, but sudden releases on to the north flank cause meltwater to mix with snow and old, unconsolidated sediments to produce a debris slurry. On several occasions, debris flow has engulfed the three sets of double track on the CPR line and Highway 1 (Jackson 1979).

There were especially damaging debris flows at the Cathedral Mountain site in 1925, 1946, 1962, 1978, 1982, and 1984. Rail and road traffic has been delayed for from one day to as much as a week. The large event on 6 September 1978 was made up of two surges, the first of which collided with a westbound train. Both CPR and Parks Canada have made substantial efforts to mitigate the hazard. Today, a warning system is in place for the railway, and the condition of the glacial meltwater reservoir is monitored during summer. When water in the latter reaches a critical level, supervised drainage by pumping occurs. So far this has proved effective in reducing the hazard.

ROCK AVALANCHES

Rock avalanches involve detachment of rock from a slope and its subsequent rapid downslope transport. Once detached, the rock breaks up into finer pieces. When a

Figure 10.8
Rock-avalanche deposit, Front Ranges, Canadian Rockies, Alberta. Deposit is "splashed" up opposing valley side (lower right) and forms dam behind which small lake has developed (lower left). *Photo*: J. Gardner.

mixture of rock, dust, and air, moving at velocities as great as 250 km/hr, is achieved, the moving mass takes on a flow-like motion such that it can travel great distances (several kilometres), down gentle slopes, over flat surfaces, and up opposing slopes before coming to rest (Figure 10.8). The resulting deposit is usually a huge, hummocky mass of jumbled rocks of various sizes mixed with fine dust.

A variety of factors may cause the rock detachment that sets the rock avalanche in motion. Usually, the rock mass breaks away from the mountain slope along a line of geological weakness – a bedding plane, a joint, a fracture, or a plane of exfoliation. Buildup for detachment may occur slowly over a long period because of slow release of pressure within the rock mass. Detachment may be aided by physical and chemical weathering. The moment of release may depend on a trigger such as loading of the rock mass by water, ice, or snow, reduction of resistance along the geological boundary by water or ice, a disturbance such as that caused by formation of ice, and/or an earthquake (e.g. Keefer 1984; Evans and Gardner 1989).

There are several reviews of rock avalanches in the Cordillera (Eisbacher and Clague 1984; Evans 1984; Cruden 1985). Table 10.4 provides a summary of known

Table 10.4
Major rock avalanches in the Cordillera

Location/name	Date	Estimated volume $(10^6\ m^3)$	Previous occurrence?
Rubble Creek, Garibaldi, BC	1855–56	25	Yes
Frank, Alta	1903	31	No
Kaouk River, Vancouver, BC	1928	8	No
Devastation Glacier, BC	1931	~20	Yes
Brazeau Lake, Alta	1933	4.5	No
Devastation Glacier, BC	1947	~4	Yes
Dusty Creek, BC	1963	5	Yes
Hope, BC	1965	47	Yes
Devastation Glacier, BC	1975	12	Yes
North Nahanni, NWT	1985	5	Yes
Mt Meager, BC	1986	1	No

Source: Evans and Gardner (1989).

rock avalanches in the past century. At least 140 deaths have resulted from rock avalanches in Canadian mountain areas since mid-nineteenth century (Evans and Gardner 1989). The most devastating occurred at Frank, Alta, in 1903 (seventy-six deaths); Jane Camp, BC, in 1915 (fifty-six deaths); Hope, BC, in 1965 (four deaths); and Devastation Glacier, BC, in 1975 (four deaths).

Rock avalanches at a particular location can indicate that later avalanches may follow (see Table 10.4). Recognizing such potential is one of the few measures that can reduce the risk from such high-magnitude processes. Thus, at Turtle Mountain, the source of the Frank rock avalanche in 1903, efforts are made to monitor movement along some of the joints and fractures that could mark future large-scale rock detachments and avalanches.

The site of the Rubble Creek rock avalanche (1855–56), 80 km north of Vancouver, retains a potential for rock avalanches and landslides (Moore and Mathews 1978). Construction in the early 1960s of a power dam and a residential area at the toe of the rock avalanche deposit showed little concern for such potential. In 1972, when application was made for the second stage of development, it was rejected on the basis of the hazard. This rejection was upheld by the BC Supreme Court in March 1973 – one of the few examples in the Cordillera of prohibitive land-use zoning to reduce the risk of such a geological hazard. Nonetheless, the rock-avalanche deposit area is traversed by a power line and a major provincial road (Highway 99), and the power dam and reservoir remain. Figure 10.9 shows a sign on Highway 99 that explains the hazard and the reason for the warning to motorists stopping along the highway in the Rubble Creek hazard area.

Figure 10.9
Warning sign at Rubble Creek rock-avalanche area, Highway 99, Coast
Mountains, BC. *Photo*: J. Gardner.

REFERENCES

Alberta, Government of 1977. *The Canmore Flood Protection Proposal*. Edmonton: Alberta
    Environment and Canmore Flood Control Advisory Committee.

Blown, I., and Church, M. 1985. "Catastrophic Lake Drainage within the Homathko River
    Basin, British Columbia." *Canadian Geotechnical Journal* 22: 551–63.

Clague, J.J., Evans, S.G., and Blown, I. 1985. "A Debris Flow Triggered by Breaching of
    a Moraine Dammed Lake, Klattasine Creek, BC." *Canadian Journal of Earth Sciences* 22:
    1,492–1,502.

Clarke, G.C.K. 1982. "Glacier Outburst Floods from 'Hazard Lake' Yukon Territory and the
    Problem of Flood Magnitude Prediction." *Journal of Glaciology* 28: 3–21.

Cruden, D.M. 1985. "Rock Slope Movements in the Canadian Cordillera." *Canadian Geo-
    technical Journal* 22: 528–40.

Eisbacher, G.H. 1979. "First Order Regionalization of Landslide Characteristics in the Ca-
    nadian Cordillera." *Geoscience Canada* 6: 69.

Eisbacher, G.H., and Clague, J.J. 1984. *Destructive Mass Movements in High Mountains:
    Hazard and Management*. Geological Survey of Canada, Paper 84–16.

Evans, S.G. 1984. "The Landslide Response of Tectonic Assemblages in the Southern Ca-
    nadian Cordillera." *Proceedings, IV International Symposium on Landslides, Toronto*, VI,
    495–502.

Evans, S.G., and Gardner, J.S. 1989. "Geological Hazards in the Canadian Cordillera." In
    L.E. Jackson, ed., "The Influence of the Quaternary Geology of Canada on Man's Envi-
    ronment," in R.J. Fulton, ed. *Quaternary Geology of Canada and Greenland*, Geology of
    Canada No. 1, 702–13.

Evans, S.G., and Lister, D.R. 1984. "The Geomorphic Effect of the July 1983 Rainstorms
    in the Southern Cordillera and Their Impact on Transportation Facilities." In Current Re-
    search, Part B, Geological Survey of Canada, Paper 84–1B, 223–35.

Fraser, C. 1978. *Avalanches and Snow Safety*. London: Murray.

Gardner, J.S., 1986. "Snow as a Resource and Hazard in Early 20th Century Mining, Selkirk Mountains, BC." *Canadian Geographer* 30: 217–28.

Hungr, O., Morgan, C.C., and Kellerhals, R. 1984. "Quantitative Analysis of Debris Torrent Hazards for Design of Remedial Measures." *Canadian Geotechnical Journal* 21: 663–77.

Jackson, L.E. 1979. "A Catastrophic Outburst Flood (Jokulhlaup) Mechanism for Debris Flow Generation at the Spiral Tunnels, Kicking Horse River Basin, B.C." *Canadian Geotechnical Journal* 16: 806–13.

Jones, D.R., Ricker, K.E., Desloges, J.R., and Maxwell, M. 1985. *Glacier Outburst Flood on the Noeick River: The Draining of Ape Lake, B.C.* Geological Survey of Canada Open File Report 1139.

Keefer, D.K. 1984. "Rock Avalanches Caused by Earthquakes: Source Characteristics." *Science* 223: 1288–90.

McFarlane, R.C. 1985. "Snow Avalanche Impacts and Management in Canada." PhD thesis, University of Waterloo, Waterloo, Ont.

Mathews, W.H. 1979. "Landslides of Central Vancouver Island and the 1946 Earthquake." Seismological Society of America, *Bulletin* 69: 445–60.

Moore, D.P., and Mathews, W.H. 1978. "The Rubble Creek Landslide, Southwestern British Columbia." *Canadian Journal of Earth Sciences* 15: 1039–52.

Nelson, J.G., and Byrne, A.R. 1966. "Fires, Floods and National Parks in the Bow Valley, Alberta." *Geographical Review* 56: 226–38.

Peck, J.W. 1970. "Mining vs. Avalanches, British Columbia: Ice Engineering and Avalanche Forecasting and Control." National Research Council of Canada, Technical Memo No. 98, Ottawa, 79–83.

Schaerer, P.A. 1962. *The Avalanche Hazard Evaluation and Prediction at Rogers Pass*. National Research Council, Division of Building Research, Paper No. 142, Ottawa.

Sewell, W.R.D. 1965. *Water Management and Floods in the Fraser River Basin*. University of Chicago, Department of Geography, Research Paper No. 100, Chicago.

Stethem, C.J., and Schaerer, P.A. 1979. *Avalanche Accidents in Canada; A Selection of Case Histories of Accidents, 1955–1976*. National Research Council of Canada, Division of Building Research, Paper No. 834, Ottawa.

– 1980. *Avalanche Accidents in Canada II: A Selection of Histories of Accidents, 1943 to 1978*. National Research Council of Canada, Division of Building Research, Paper No. 926, Ottawa.

Van Dine, D.F. 1985. "Debris Flows and Debris Torrents in the Southern Canadian Cordillera." *Canadian Geotechnical Journal* 22: 44–68.

White, G. 1974. *Natural Hazards*. Oxford: Oxford University Press.

PART FOUR

The Changing Cold Environments

CHAPTER ELEVEN

Climate Variability, Change, and Sensitivity

ELLSWORTH F. LEDREW

The well-publicized trend in background concentrations of carbon dioxide in the atmosphere and model extrapolations into the future (Figure 11.1) have focused public attention on the possibility of a changing climate and its implications. Within Canada's cold environments, transportation sectors, the livelihood of native peoples, natural gene pools, and the development and use of park reserves will probably be affected. The replacement of highly reflective sea ice with open ocean will amplify the warming trend. To illustrate the consequences of such feedback links, Figure 11.2 compares observed temperature departures from the northern hemisphere mean with similar data for the Arctic since 1880. The hemispheric trend of temperature is magnified three times in the Arctic (Kelly et al. 1982). The rate of change in the central Arctic since 1976 has been in excess of $0.75C°$ per decade, while mid-latitude figures are typically less than $0.50C°$ per decade (Jones 1988).

This amplification also has global significance (Goody 1980). The role of the Arctic as an energy sink ties it strongly to the general circulation, which may communicate any perturbations in the arctic climate to other regions. An apparent link, for example, has been demonstrated between an anomalous cold regime over Hudson Bay and drier-than-normal conditions over South America (Hartman 1984).

The fact that Canada's cold environments will be affected by greater climatic change than experienced elsewhere seems probable on the basis of the most reliable simulations on global climate models. This chapter examines the nature of that climate change and explores its variability and sensitivity in Canada's cold environments. It seeks in particular to identify the uncertainties surrounding any attempt to plan for future sustainable economic and social development in those areas.

CONCEPTS OF CHANGE, VARIABILITY, AND SENSITIVITY

A number of aspects of change should be recognized. First, there is geophysical change. The climate has changed in the past because of interaction between com-

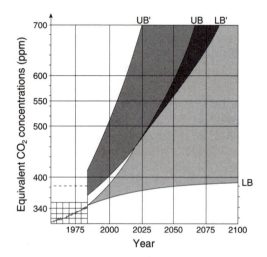

Figure 11.1
Projected trends in greenhouse gases. UB–LB: range of estimates for
CO_2; UB'–LB' range when other greenhouse gases are included. *Source*:
Canadian Climate Centre.

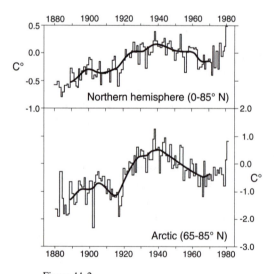

Figure 11.2
Annual temperature departures from 1946–60 mean (°C), averaged over
northern hemisphere (0–85°N) and Arctic (65–85°N). *Source*: Jones and
Kelley (1983).

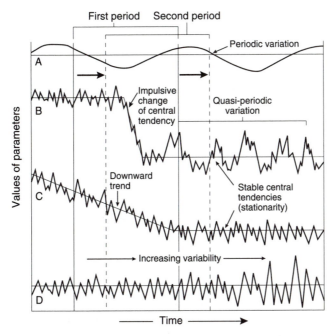

Figure 11.3
Time series of hypothetical climate element that is continuous in time, such as temperature. Vertical bars represent averaging interval updated each decade. Individual curves (A–D) are discussed in text. After Hare (1979).

ponents of the natural system, and it will continue to do so. To anticipate future change from "natural" sources, we shall need to understand how these components interact and how the system has produced the historical record of change.

Second, there is human-induced change. We now recognize that human beings can influence climate significantly. Warming as a result of the so-called greenhouse effect would be inadvertent modification of climate on a global scale. We have created huge industrial power plants that have waste-energy densities approaching those of natural net radiation fluxes at Earth's surface.

Third, there is society's perception of change. One image is that change is a force over which society has little control. Change is to be accepted passively and fatalistically. As an illustration, the people of Bali perceived the eruption of Mt Agung during the purification ceremony of Eka Dasa Rudra in 1963 as divine retribution. Alternatively, a proactive view anticipates future regimes of nature – either planning to adapt to them or attempting to modify them.

Any discussion of change must take into account the concept of variability. The time series in Figure 11.3 may represent the temporal course of a climatic parameter such as temperature. The periodic variation in curve A about a mean or central

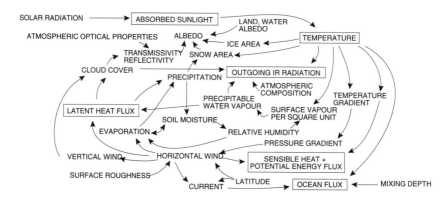

Figure 11.4
Possible process links within climate model illustrating interdependent feedback paths. *Source*: Schneider (1977).

tendency may be a response to processes with diurnal or annual rhythms. Typically this pattern is complicated by other processes with different time scales of response. Curve B illustrates a more typical climate signal in which there may be a high-frequency variation about the central tendency. This variability may appear to be either random or quasi-periodic. Also indicated is an abrupt change in the central tendency to a new state. We now know from deep-sea cores that climate may change very abruptly in a geological frame of reference – perhaps within a hundred years. This climate change is distinct from the variability superimposed on it. Curve C illustrates a gradual trend in climate. Over a short observational period, the natural variability of the system may obscure this trend.

To understand the distinction between change and variability, we may consider the climate system as a delicate clock work–type of mechanism. A number of process loops join climate parameters which may be linked to and nested within other loops in a flow diagram (Figure 11.4). The processes differ in time scales, thresholds, and rates of response to change. As a result of interaction among all the process loops, the system may be continually changing. If forcing functions for the system change, however, the internal system must adjust to a new state of quasi-equilibrium. Variability will continue, but the central tendency will shift and the climate will change. Forcing functions may include amount of extraterrestrial solar radiation, configuration of the continents, volcanic dust loading in the atmosphere, or concentration of CO_2. Because of the difference in temporal scales between changes in forcing functions and internal processes of the climate system, variability will probably mask any change in the central tendency over a period of a few decades and will be evident only over the longer term. Consequently, it is difficult to identify conclusive evidence for the greenhouse effect in the climate data that would correspond to the increase in CO_2 observed since 1958.

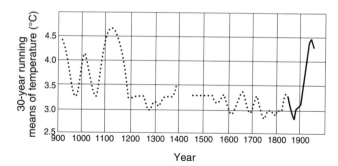

Figure 11.5

Running thirty-year means of temperature in Iceland. Reconstructed data for 1591–1846 (dotted line) are based on relationship with incidence of drift ice developed from observed data since 1846 (solid line). Relationship between ice incidence and number of severe ice years provides reconstruction between 930 and 1591 (dotted line). *Source*: Bergthorsson (1969).

The temperature trends in Figure 11.2 represent the most complete compilation of observational data currently available. Warming up to the 1940s and subsequent cooling are evident for both arctic and hemispheric averages – either in response to changes in the forcing functions or as a natural expression of variability (Schneider 1977) within the climate system. The years 1980, 1981, 1983, and 1987 were the warmest since 1880 (Hansen and Lebedeff 1988).

For the longer term, we must turn to proxy data. The record in Figure 11.5 is derived from data on ice incidence collected from AD 930 to the present in Iceland (Bergthorsson 1969). The temperature scale is based on correlation with observed temperature during the period of recent instrumental record. Of interest is the last thirty years, which is designated a "normal" climate for statistical purposes. This evidence suggests that we are in a rather benign climate interval which is far from the "normal" of the past millennium.

For understanding climate change on the time scale of glacial periods, data from ice cores – in which the deuterium composition can be related to temperature, and encapsulated aerosol and CO_2 concentrations can be analysed – are useful. The record from the 2,200-m-long core drilled at the Vostok station in east Antarctica clearly shows the present Holocene interglacial (at A in Figure 11.6) (Barnola et al. 1987), the past interglacial (at G), and three temperature minima separated by warmer episodes in the intervening glacial epoch. Spectral analysis of this time series reveals periodicity of 100, 40, and 20 thousand years. This finding lends credence to the role of astronomical forcing as a result of the Milankovitch variations to explain the Pleistocene glacial-interglacial intervals. Concentration of CO_2 is strongly correlated with this record, the first conclusive statistical evidence of the connection between CO_2 and temperature.

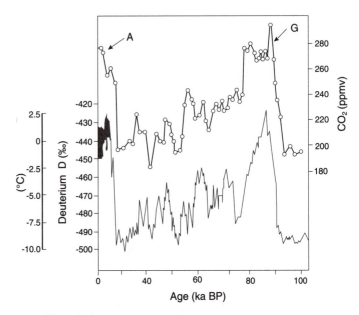

Figure 11.6

Carbon dioxide concentrations ("best estimates") and atmospheric temperature changes derived from isotopic profile for Vostok record. See text for explanation. *Source*: Barnola et al. (1987).

These illustrations of variability and change prompt a critical question: can we define the governing forces of climate sufficiently to develop the useful long-range forecasts for variability and change necessary for planning in Canada's cold environments? The answer at the moment is *no*, primarily because of the sensitivity of processes to change. In our most elaborate numerical models we can deal only with a subset of the suite of global processes, and then only as equilibrium solutions or snapshots in time. We cannot yet produce the transient solutions that may replicate the more realistic gradual changes of the atmosphere (Schneider 1984) in response to external forcing, such as carbon dioxide enrichment.

Some progress, however, is being made. A simple model for the albedo/sea-ice feedback is given in Figure 11.7A. The symbolic conventions used in feedback theory are described in the caption. According to this model, a regional temperature rise resulting from warm air advection from the south would be accompanied by a rise in the heat flux from atmosphere to surface. This would prompt melting of the highly reflective ice cover which is replaced by the water surface of lower albedo or reflectivity. Increased absorption of solar irradiance forces greater temperature increase. A strong surface inversion, typical of the Arctic, inhibits vertical mixing of this warmed air and augments the near-surface heating rate. Models incorporating this positive feedback concept with an initial warming of 5C° simulate complete disappearance of the arctic pack ice in August and September and its reappearance the following winter (Parkinson and Kellogg 1979). This displays the sensitivity of arctic processes to small perturbations through amplification as a consequence of such feedbacks.

Recent investigations have indicated that there may be negative feedback loops. One proposed interaction can dampen the albedo feedback effect at the margins of the pack ice of the central Polar Basin. Enhanced evaporation over open water, as well as steering of cyclones along the sea-ice boundary, can lead to increased cloudiness (Barry, Henderson-Sellers, and Shine 1984). If the cloud is optically thick, the increase in the planetary albedo associated with increased cloud cover would result in net negative feedback (Figure 11.7B), while the reverse may be the case in the more rare instance when cloud is optically thin. A second but related linkage may include infrared cooling from open water with a net negative feedback (Figure 11.7B) or positive feedback from infrared warming if we consider the increased cloud cover that may be formed from evaporation.

If we extend the albedo/sea-ice feedback loop to include changes in biomass, which will be effective on a much longer time scale, we may have another negative feedback loop (Figure 11.7C). Initial temperature amelioration would be conducive to growth of forests (tempered by slow retreat of permafrost). In the absence of fossil-fuel sources in the future, this sink may be sufficient to reduce the density of atmospheric carbon dioxide, with a corresponding effect on atmospheric temperature. The net result of this model may be stability, an ice-free Arctic Ocean, and northward shift of the treeline (Kellogg 1983).

There is no agreement on cause and effect in such cryosphere-atmosphere-biosphere process linkages (Walsh 1983). Nevertheless, because of the sensitivity illustrated above, the cryosphere may be the most significant focus of research into climate variability and change in cold environments. For example, in a recent simulation by Hansen et al. (1984), doubling of the CO_2 concentration warms the global surface by a mean of $4.2C°$; snow/ice-albedo feedback accounts for $1.3C°$ of that total.

The results of two different global simulations are plotted in Figures 11.8 and Figure 11.9 for Canada for January temperatures. These state-of-the-art simulations, from the Geophysical Fluid Dynamics Laboratory (GFDL) and the Goddard Institute of Space Studies (GISS) (both in the United States), yield markedly different patterns at this scale. Disagreement of up to $5C°$ over the western provinces and of smaller but still significant amounts over Northwest Territories highlights the difficulty of using global simulations at the regional scale. Although these simulations provide good qualitative agreement for many zonal and globally averaged parameters, there is considerable quantitative disagreement over geographical distributions of the data (Schlesinger 1984).

A measure of the discrepancies between simulations is illustrated in Figure 11.10. The zonal mean surface air temperature simulated by six GCMs is complicated as one includes additional processes to imitate more realistically the situation for the atmosphere (Figure 11.7D). The net sign of the feedback becomes more difficult to

A

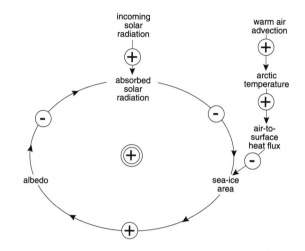

B

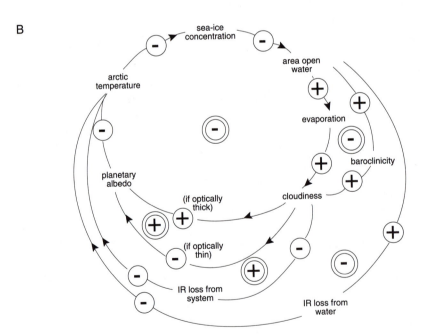

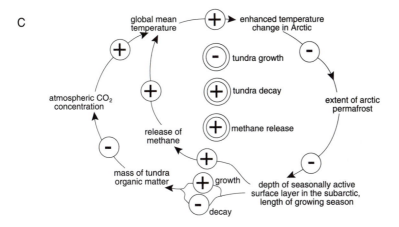

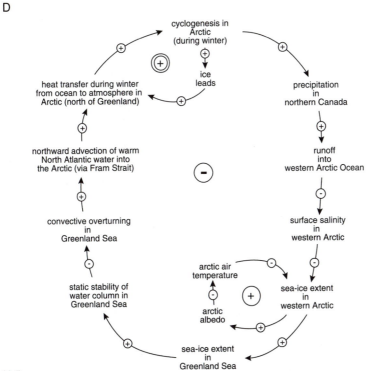

Figure 11.7
Feedback loops of increasing complexity: (A) albedo–sea ice feedback loop; (B) sea ice–baroclinicity, cloud cover feedback loop; (C) carbon dioxide–sea ice–biomass feedback loop; and (D) a more complex and realistic feedback loop. From Kellogg (1983). A plus in the diagram indicates that a change in parameter causes change of similar direction in effect; entire loop is negative with odd number of negative linkages.

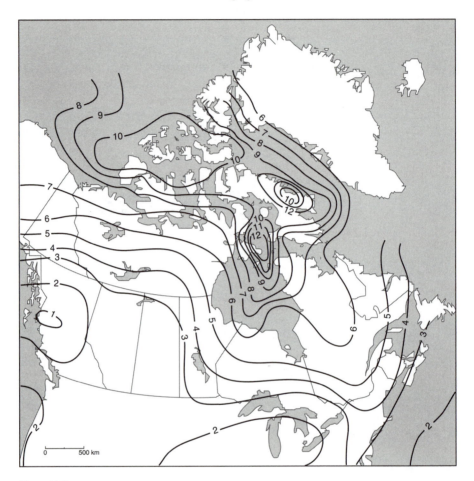

Figure 11.8

Two-times CO_2 warming for Canada, from GFDL experiment: simulation results of consequent January surface temperature change. *Source*: Data courtesy Canadian Climate Centre, Atmospheric Environment Service, Toronto.

determine. Nevertheless, because of the sensitivity illustrated above, the cryosphere may be the most significant focus of research into climate variability and change in cold environments.

This discussion of sensitivity and the cryosphere emphasizes the consequences of small perturbations, whether natural or human-induced. The points of human intervention are particularly worthy of study. In effect, we are performing experiments on the "real world," from deliberate acts such as diversion of rivers flowing into the Arctic Basin, with unknown consequences (Barry 1983), to inadvertent processes such as the "greenhouse" effect of carbon dioxide enrichment.

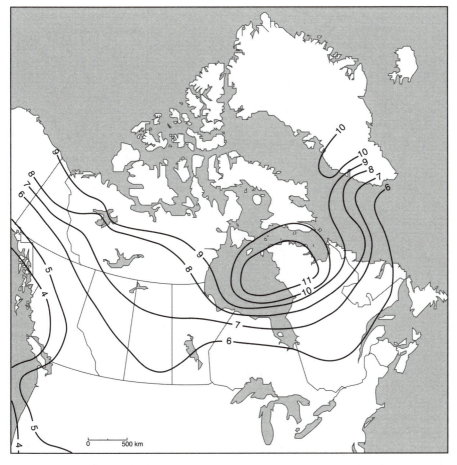

Figure 11.9
Two times CO_2 warming for Canada, from GISS experiment: simulation results of consequent January surface temperature change. *Source*: Data courtesy Canadian Climate Centre, Atmospheric Environment Service, Toronto.

GCMs

One approach to determining the result of carbon dioxide enrichment on the atmosphere is numerical simulation using a global general circulation model, or GCM. The model's resolution, the nature of the parameterization, and the method of handling the temporal scales of atmospheric and oceanic processes may all affect the results. In all cases, the sensitivity of the Arctic mean temperatures are plotted for six GCM models for an atmosphere with doubled CO_2. The greatest ranges are in the Arctic and Antarctic. In the Arctic, the range may be attributed to the different modelling techniques for dealing with sea-ice growth, decay, and transport and the consequent effect on temperature through the ice-albedo feedback.

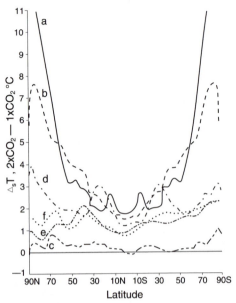

Figure 11.10
Changes in zonal mean surface air temperature simulated by six GCMs
for double CO_2 atmosphere; from Schlesinger (1984) with permission.
Sources: a and b: Manabe and Wetherald (1975 and 1980, respectively);
c: Gates, Cook, and Schlesinger (1981); d: Schlesinger (1986); and e
and f: Washington and Meehl (1983), with predicted clouds and pre-
scribed clouds, respectively.

   Future planning based on these models must take into account the role of the time-
lag in thermal response of the atmosphere induced by eddy mixing of heat at the
ocean's surface. Integration of the circulations of oceans and atmosphere, each with
different temporal response times, is a major focus of current research. The lag in
the effect of processes between them may approach forty years. As a consequence,
Hansen et al. (1984) suggest that even if CO_2 and trace-gases cease to increase at
the present time, we will still have a 1C° warming in the future as a result of past
loadings.
   There are notable gaps in our understanding of northern processes, the model
parameterization of those processes, and the veracity of initialization and verification
data in GCMs. There is lack of conformity between coarse GCM grid-point simulations
and the detailed grid network needed for regional analysis, such as a study of the
effect of enhanced CO_2 concentration on ocean transportation to specific northern
settlements.

REMOTE SENSING

Satellite imagery is proving to be an important data-source for mapping surface variables. The synoptic coverage of an image fills in the spatial detail lacking in traditional observational networks, and for some sensors we have a consistent historical archive of several years from which to draw information. We have gained considerable insight into northern processes as a result of developments in microwave remote sensing (Barry 1981). Because of the cloud-penetration capability of these wave bands, we have almost continuous surveillance of large tracts of previously inaccessible regions.

The total sea-ice concentration and the fraction as multi-year ice has been plotted for the Arctic for the interval from late January to early February in 1979, 1980, and 1981 (Cavalieri and Zwally 1985). These variables were derived from analysis of a scanning multifrequency microwave radiometer (SMMR) – an image scanner on the Nimbus-7 satellite in orbit since 1978 which has provided continuous near-global coverage until very recently.

The significant effect of ocean currents on sea-ice extent can be analysed in this way. The extension of ice down the east coast of Greenland is found to be associated with the cold East Greenland Current. The asymmetric distribution in Davis Strait is attributed to the southward migration of ice with the cold Labrador Current on the west side and the northward flow of the warmer West Greenland Current in the eastern region. Bathymetry also plays an important role, as noted in the relationship between the ice edge and the margin of the continental shelf in the Bering Sea.

Of interest is the notable out-of-phase relationships in ice extent for the various sectors of the Arctic between years. For example, the ice extent is a maximum (for the period of record under discussion) in the Sea of Okhotsk in 1979, yet it was noticeably less in 1981. In the Bering Sea, however, the pattern was reversed, with minimum ice extent in 1979 and greater in 1981. Similar out-of-phase patterns are seen in the multi-year fraction within the Polar Basin: in 1976 the multi-year pack was far away from the North American coast yet close to the coast of the East Siberian Sea, and in 1980 the edge was closer to the North American coast yet had receded in the Chukchi and East Siberian seas.

An explanation for the out-of-phase relationships has been proposed by Cavalieri and Parkinson (1987). They examined areal change in sea ice for the Bering Sea and the Sea of Okhotsk for the winters of 1974–75 and 1975–76 from electrically scanning microwave radiometer (ESMR) imagery. They also decomposed sea-level pressure data for the Northern Hemisphere into the major wave patterns about the pole (see Figure 11.11). On 6 March 1976 the ice area increased in the Sea of Okhotsk and retreated in the Bering Sea, while the reverse was the case on 15 March. Actual sea-level pressure patterns are plotted for these two days in Figure 11.12. The Aleutian Low (harmonic 2) was dominant on 6 March, and the Siberian High (harmonic 1) was dominant on 15 March. On 6 March southerly airflow in the Bering Sea would contribute to recession of the ice edge, and northerly airflow in the Sea of Okhotsk would assist ice advance. On 15 March the surface wind pattern was reversed, with northerly flow over the Bering Sea and southerly flow over the Sea of Okhotsk, which would explain the reversed out-of-phase relationship in the sea-ice extent.

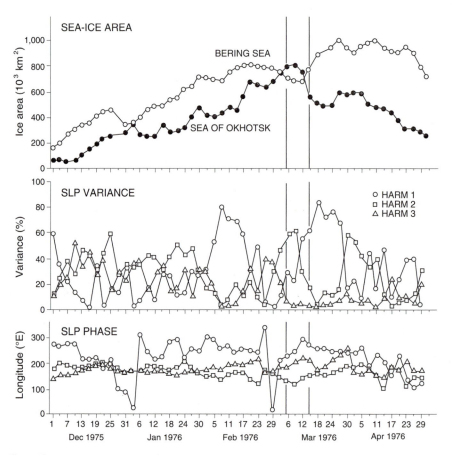

Figure 11.11

Bering Sea and Sea of Okhotsk: (A) time series of ice cover based on interpretation of ESMR imagery; (B) percentage variance; and (C) phase-shift of major harmonics of sea-level pressure. *Source*: Cavalieri and Parkinson (1987).

This analysis highlights the synoptic control on sea ice and the rapid response of ice motion to shifts in the wind field. Demonstration of this link is a step toward understanding the spatial dimensions of atmosphere-cryosphere variability and would not have been possible without extensive coverage of satellite imagery.

UNCERTAINTY AND RISK ANALYSIS

The outside observer's view of climate is that it is inherently unpredictable (Hare 1979). This has evolved from our day-to-day experience in forecasting the travelling depressions in mid-latitudes. Our discussion of sensitivity to change, our limited

6 MARCH 1976

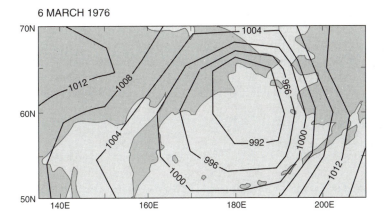

15 MARCH 1976

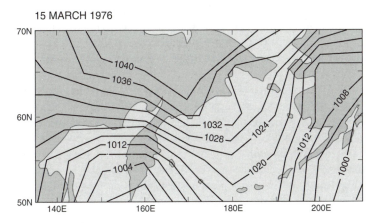

Figure 11.12
Surface pressure charts for 6 and 15 March 1976 in region covered by Figure 11.11. *Source*: Cavalieri and Parkinson (1987).

understanding of atmosphere-cryosphere interactions, and the uncertain data sets from which we work underscore such a view.

If policy is informed selection from several competing strategies aimed at achieving specific results, climate is one of many constraints affecting that selection. Climate change or variability may be considered as change in the risk of impact. This is a concept consistent with policy analysts' emphasis on the frequency of extremes or the return period of hazard, based on the assumption that people react more strongly to real or perceived shifts in the number of extreme events than to continuous but subtle changes in the mean (Parry 1985). Policy may then be adjusted so as to avoid new levels of climatic risk identified through statistical analysis of the climatic record or through simulations of possible future climatic states.

The science and art of risk assessment have evolved as an aid in decision making to improve quality of decisions and provide a framework to probe their acceptability (Macgill and Snowball 1983). The central elements are risk identification, which involves scientific understanding and measurement; risk estimation, which is prediction of consequences in space and time; risk evaluation, which makes social judgments regarding acceptability of identified risks; and risk implementation, which includes setting standards, monitoring, enforcing compliance, and re-evaluation (O'Riordan 1979).

Clearly, this sequential process is critically dependent on successful risk identification. The lesson from our study of northern environments appears to be that we are not sure about the nature and variety of climatic risks. Much less can we assign even nominal scale values to the magnitude of risk. This situation is akin more to the analysis of risk associated with the liquefied natural gas (LNG) hazard, where mere possibilities are considered on the basis of abstract scenarios, than to that of flood hazard, where probabilities may be calculated from the data. Furthermore, as we consider climate change with new boundary conditions, we enter a realm of greater uncertainty. Again the case for greater understanding of climate processes and for more confident data analysis is stressed. We need to progress from possibilities to probabilities.

Experience in the traditional applications of risk assessment has indicated that the overall process has not been as successful as desired. Cases where consequences have uncertainty of an order of magnitude and probabilities have error ranges of one to two orders of magnitude have been cited. Fischhoff, Slovic, and Lichtenstein (1981) list several common problems in risk assessment attributable to the contribution by the "experts." These include overconfidence in current scientific knowledge; failure to appreciate how technological systems function as a whole; slowness in detecting chronic, cumulative effects; and failure to consider ways in which human errors can affect technological systems. It is instructive to consider these observations and review our understanding of cold environments. The authors conclude: "There is no way to get the right answer to many risk problems; all that we can hope to do is to avoid the mistakes to which each of us is attuned; the more perspectives involved, the more local wisdom is brought to bear on a problem" (Fischhoff, Slovic, and Lichtenstein 1981: 197).

A closely allied approach is that of "scenario analysis," in which alternative situations are evaluated. Whereas risk assessment may be applied to changes that are within our experience, such as climate variability, "scenario analysis" may be used to consider new future states that may result from climate change. The purpose is not prediction but rather understanding of the implications of a range of policies. The foci are "what if?" questions. We are not constrained by modelled "facts," but we can consider extended sets of data not now available, the implication of the extreme yet rare event, and socio-economic links. Crucial questions may be asked, such as "What actions, taken when, would help to maximize opportunities and minimize the dangers implicit in the CO_2 (or other) issue?" (Clark et al. 1982).

McKay and Baker (1986) found climatic analogues in the present climate record for a two-times CO_2 warming of many northern stations as predicted by a GCM. With such warming, Whitehorse, Yukon, would have a climate similar to that of present-day Calgary, for example. Using transfer functions for length of frost-free season, biomass production, northern limit of wheat cultivation, mean date of ice clearance, and heating requirements, McKay and Baker speculated on the implications for transportation, agriculture, settlement, wildlife, and use of natural resources.

The greatest difficulty in these risk and scenario analyses is the nonconformity in scale among physical, economic, and social processes, both in time and in space (chapter 13) – see Figure 11.13, developed by Clark (1985). The range of length and time scales for climatic, social, and ecological processes are plotted individually and then combined as a single time-space graph to highlight areas of overlap and disjunction. The e-folding time (time for the system to change by a factor equivalent to the base of natural logarithms) of the proposed greenhouse warming is much longer than the characteristic time scale for local farm activities (crop growth-cycles) and regional agricultural development (crop substitution). Clark concludes that some ecological and social processes may follow climate change rather than be disrupted by it. Figure 11.13 also indicates that social and economic responses to even pro-longed droughts are well below the time scales of major warmings and that the lessons learned from drought scenarios are therefore inappropriate to the CO_2 enrichment question. The disconcerting aspect of Figure 11.13 is the unrealistic isolation of the major climatic hazards from the continuous process cascade of atmospheric events, from tornadoes through atmospheric long waves (on the left of the time-space dia-gram). We have to evaluate the role of process triggers, amplification, and sensitivity in bridging this gap and to understand how modification at a scale that may be affected by policy decisions may nevertheless contribute to hazard events at a different scale.

SUMMARY

In summary, therefore, one must emphasize the totality of cold environments and not just their climates. Our knowledge has progressed enormously since considering the feedbacks between climate and the ice and snow cover. We must now look at the oceans and the biosphere, including the human impact and response, with similar attention, but we are uncertain of the system's response to natural or human-induced perturbations, whether advertent or inadvertent. Our knowledge will come not only from computer model studies but also from detailed field studies of uniquely northern phenomena, such as the North Water polynya of Smith Sound. Finally, we need to integrate remotely sensed data into our analysis. The traditional network of in-situ measurement stations in northern regions was located because of accessibility and

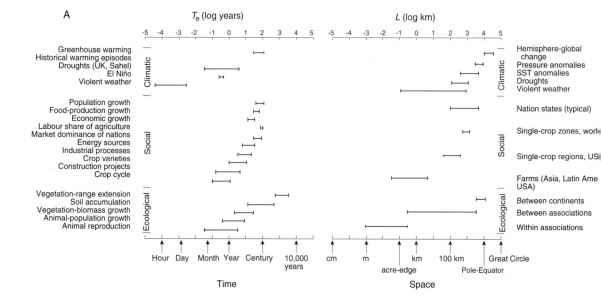

A

T_e (log years)

-5 -4 -3 -2 -1 0 1 2 3 4 5

Greenhouse warming
Historical warming episodes
Droughts (UK, Sahel)
El Niño
Violent weather

Climatic

Population growth
Food-production growth
Economic growth
Labour share of agriculture
Market dominance of nations
Energy sources
Industrial processes
Crop varieties
Construction projects
Crop cycle

Social

Vegetation-range extension
Soil accumulation
Vegetation-biomass growth
Animal-population growth
Animal reproduction

Ecological

Hour Day Month Year Century 10,000 years

Time

L (log km)

-5 -4 -3 -2 -1 0 1 2 3 4 5

Hemisphere-global change
Pressure anomalies
SST anomalies
Droughts
Violent weather

Climatic

Nation states (typical)

Single-crop zones, world

Single-crop regions, US

Farms (Asia, Latin Ame
USA)

Social

Between continents
Between associations
Within associations

Ecological

cm m km 100 km Great Circle
acre-edge Pole-Equator

Space

B Hour Day Month Year Century 10,000 years

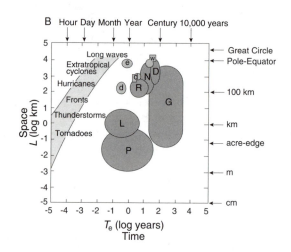

Space L (log km)

T_e (log years)
Time

Great Circle
Pole-Equator
100 km
km
acre-edge
m
cm

Figure 11.13

(a) Characteristic time (*e*-folding time)-and-space scales and (B) space-time diagram for selected climatic, social, and ecological phenomena. Key for (B): D: global political/demographic patterns, G: geographical ecology, L: local farm activities, N: national industrial modernization, P: population ecology, R: regional agricultural development, d: droughts, e: El Niño, w: historical and predicted CO_2 warming. *Source*: Clark (1985).

not for reasons of spatial representativeness. Remotely sensed imagery now provides data about previously inaccessible regions; for example, we have a view of temporal and spatial relationships within the Polar Basin that one could only speculate on in previous decades.

REFERENCES

Barnola, J.M., Raynaud, D., Korotkevich, Y.S., and Lorius, C. 1987. "Vostok Ice Core Provides 160,000-Year Record of Atmospheric CO_2." *Nature* 329: 408–13.

Barry, R.G. 1981. "Trends in Snow and Ice Research." *Transactions, American Geophysical Union* 62 no. 46: 1,139–44.

− 1983. "Arctic Ocean Ice and Climate: Perspectives on a Century of Polar Research." *Annals, Association of American Geographers* 73 no. 4: 485–501.

Barry, R.G., Henderson-Sellers, A., and Shine, K.P. 1984. "Climate Sensitivity and the Marginal Cryosphere." In J.E. Hansen and T. Takahashi, eds., *Climate Processes and Climate Sensitivity*, Geophysical Monograph 29, American Geophysical Union, Washington, DC, 221–37.

Bergthorsson, P. 1969. "An Estimate of Drift Ice and Temperature in Iceland in 1000 Years." *Jokull* 19: 94–101.

Cavalieri, D., and Parkinson, C.L. 1987. "On the Relationship between Atmospheric Circulation and the Fluctuations in the Sea Ice Extents of the Bering and Okhotsk Seas." *Journal of Geophysical Research* 92 C7: 7,141–62.

Cavalieri, D., and Zwally, H.J. 1985. "Satellite Observations of Sea Ice." *Advances in Space Research* 596: 247–55.

Clark, W.C. 1985. "Scales of Climate Impact." *Climatic Change* 7 no. 1: 5–27.

Clark, W.C., Cook, K.H., Marland, G., Weinberg, A.M., Rotty, R.M., Bell, P.R., Allison, L.J., and Cooper, C.L. 1982. "The Carbon Dioxide Question: Perspectives for 1982." In W.C. Clark, ed., *Carbon Dioxide Review, 1982*, Oxford: Clarendon Press, 3–44.

Fischhoff, B., Slovic, P., and Lichtenstein, S. 1981. "Lay Foibles and Expert Fables in Judgements about Risks." In T. O'Riordan and R.K. Turner, eds., *Progress in Resource Management and Environmental Planning 3*, Chichester: John Wiley and Sons, 161–202.

Goody, R. 1980. "Polar Processes and World Climate (A Brief Overview)." *Monthly Weather Review* 108 no. 12: 1,935–42.

Hansen, J., Lacis, A., Rind, D., Russell, G., Stone, P., Fung, I., Ruedy, R., and Lerner, J. 1984. "Climate Sensitivity: An Analysis of Feedback Mechanisms." In J.E. Hansen and T. Takahashi, eds., *Climate Processes and Climate Sensitivity*, Geophysical Monograph 29, American Geophysical Union, Washington, DC, 130–63.

Hansen, J., and Lebedeff, S. 1988. "Global Surface Air Temperatures: Update through 1987." *Geophysical Research Letters* 15 no. 4: 323–6.

Hare, F. Kenneth. 1979. "Climatic Variability, Environmental Change and Human Response." In *Special Environmental Report No. 13*, Meteorology and the Human Environment, WMO No. 517, Geneva: Secretariat of the World Meteorological Organization, 11–30.

Hartman, D.L. 1984. "On the Role of Global-scale Waves in Ice-Albedo and Vegetation-Albedo Feedbacks." In J.E. Hansen and T. Takahashi, eds., *Climate Processes and Climate Sensitivity*, Geophysical Monograph 29, American Geophysical Union, Washington, DC, 18–28.

Jones, P.D. 1988. "Hemispheric Surface Air Temperature Variations: Recent Trends and an Update to 1987." *Journal of Climate* 1: 654–60.

Jones, P.D., and Kelly, P.M. 1983. "The Spatial and Temporal Characteristics of Northern Hemisphere Surface Air Temperature Variations." *Journal of Climatology* 3 no. 3: 243–52.

Kellogg, W.W. 1983. "Feedback Mechanisms in the Climate System Affecting Future Levels of Carbon Dioxide." *Journal of Geophysical Research* 88: 1,263–9.

Kelly, P.M., Jones, P.D., Sear, C.B., Cherry, B.S.G., and Tavakol, R.K. 1982. "Variations in Surface Air Temperatures: Part 2, Arctic Regions, 1881–1980." *Monthly Weather Review* 110 no. 2: 71–83.

Macgill, S.M., and Snowball, D.J. 1983. "What Use Risk Assessment?" *Applied Geography* 3: 171–92.

McKay, G.A., and Baker, W.M. 1986. "Socio-economic Implications of Climate Change in the Canadian Arctic." In H.M. French, ed., *Climate Change Impacts for the Canadian Arctic*, Atmospheric Environment Service, Environment Canada, 116–36.

Manabe, S.R., and Wetherald, R.T. 1975. "The Effects of Doubling the CO_2 Concentration on the Climate of a General Circulation Model." *Journals of Atmospheric Science* 32 no. 1: 3–15.

O'Riordan, T. 1979. "The Scope of Environmental Risk Management." *Ambio* 8 no. 6: 260–4.

Parkinson, C.L., and Kellogg, W.W. 1979. "Arctic Sea Ice Decay Simulated for a CO_2–Induced Temperature Rise." *Climatic Change* 2 no. 2: 149–62.

Parry, M.L. 1985. "Estimating the Sensitivity of Natural Ecosystems and Agriculture to Climatic Change." *Climatic Change* 7 no. 1: 1–3.

Schlesinger, M.E. 1984. "Climatic Model Simulation of CO_2–Induced Climatic Change." In B. Salzman ed., *Advances in Geophysics*, 26, New York: Academic Press, 141–235.

– 1986. "General Circulation Model Simulations of CO_2–Induced Equilibrium Climate Change." In H.M. French, ed., *Climate Change Impacts for the Canadian Arctic*, Atmospheric Environment Service, Environment Canada, Downsview, 15–51.

Schneider, S.H. 1977. "Climate Change and the World Predicament: A Case Study for Interdisciplinary Research." *Climatic Change* 191: 21–43.

– 1984. "On the Empirical Verification of Model-Predicted CO_2–Induced Climatic Effects." In J.E. Hansen and T. Takahashi, eds., *Climate Processes and Climate Sensitivity*, Geophysical Monograph 29, American Geophysical Union, Washington, DC, 187–201.

Walsh, J.E. 1983. "The Role of Sea Ice in Climatic Variability: Theories and Evidence." *Atmosphere-Ocean* 21 no. 3: 229–42.

Washington, W.M., and Meehl, G.A. 1983. "General Circulation Model Experiments on the Climatic Effects Due to a Doubling and Quadrupling of Carbon Dioxide Concentration." *Journal of Geophysical Research* 88 C11: 6,600–10.

CHAPTER TWELVE

Climatic Change and Permafrost

MICHAEL W. SMITH

During the past few thousand years, Earth's climate has been subject to fairly small changes and world temperatures have fluctuated only within a couple of degrees. However, higher levels of carbon dioxide and other "greenhouse" gases in the atmosphere may progressively increase global temperature by as much as $2 + 0 4C°$ over the next several decades.

Current projections suggest that climatic warming in the north polar latitudes would be two or three times greater than this global average, although the precise locations and degree of warming may vary substantially between climatic models (e.g. see Etkin 1990). According to the estimates of Hansen et al. (1984), for example, mean annual air temperature in the western Canadian Arctic could be as much as $7C°$ higher than at present by the mid-twenty-first century, given present rates of atmospheric change. In all models, winter temperatures are predicted to increase much more than summer ones, with some changes exceeding $10C°$ (Table 12.1). Even with increases of only 2 to $3C°$ in summer, however, mean annual temperatures could still rise by as much as $.05 + 0 1C°$ per decade beyond the middle of next century.

In addition to temperature changes, the patterns of precipitation would undoubtedly change. According to Etkin (1990), annual totals would generally increase over the arctic mainland, although current regional projections are again quite variable between models. From the results of Hansen et al. (1984), increases of 10 to 50 per cent in summer and as much as 60 per cent in winter may be anticipated for parts of the western Canadian Arctic (Table 12.1).

Such large and rapid climatic changes would have serious and far-reaching environmental and socio-economic effects in permafrost regions and for the arctic environment as a whole (McBeath 1984; French 1986; Maxwell 1987). Some might look on the transition to a warmer Arctic with happy anticipation: in the long term, it would undoubtedly result in greatly reduced costs of living and operating there. New resources could become available, and mining and agriculture, for example, might well expand. However, the terrestrial environment of the north, in which permafrost plays a major role, would be profoundly disrupted during the transition.

Table 12.1
Predicted climatic changes for the western Canadian Arctic in a $2 \times CO_2$ world

	Temperature		Precipitation	
	Range (C°)	Region of maximum change	Range (%)	Region of maximum change
January	+5.5 to 9.5	125–135°W/70°N	+15 to 55	130°W/66°N
February	+5.0 to 8.0	135–140°W/70°N	+10 to 50	125°W/66°N
March	+4.0 to 6.0	135–140°W/70°N	+10 to 50	125°W/66°N
April	+3.5 to 4.5	140°W/66°N	+8 to 42	125°W/66°N
May	+3.0 to 4.9	120°W/66°N	+20 to 43	120°W/66°N
June	+2.0 to 3.2	120–140°W/66°N	+20 to 50	140°W/66°N
July	+1.7 to 3.0	140°W/62°N	+15 to 45	140°W/66°N
August	+2.5 to 3.0	140°W/62°N	+10 to 35	140°W/66°N
September	+3.0 to 5.5	130–140°W/70°N	+12 to 33	125°W/70°N
October	+4.0 to 9.2	130°W/70°N	+15 to 55	125°W/66°N
November	+5.0 to 11	130°W/70°N	+15 to 60	130°W/66°N
December	+5.0 to 11	130°W/70°N	+15 to 60	125°W/66°N

Source: Hansel et al. (1984).

CLIMATE AND PERMAFROST

The mechanisms of energy exchange at Earth's surface in cold regions are the same as those elsewhere on the planet. Together they determine the surface temperature regime and whether or not frozen ground will exist (see Williams and Smith 1989).

Let us imagine some change in climatic conditions which causes the mean annual surface temperature to fall below 0°C, so that the depth of winter freezing will exceed the depth of summer thaw. A layer of permafrost would grow downward from the base of the seasonal frost, thickening progressively with each succeeding winter. Were it not for the effect of the heat escaping from Earth's interior (the geothermal heat flux), the permafrost would grow to great depths in response to surface temperatures only slightly below 0°C. However, this outward heat flow results in a temperature increase with depth of about 30K/km, the figure varying with regional geologic conditions. Thus the base of permafrost approaches an equilibrium depth where the temperature increase caused by this geothermal gradient just offsets the amount by which the surface temperature is below 0°C (Figure 12.1).

Whereas the base of permafrost is determined by the mean surface temperature and geothermal heat flow, the upper layers of permafrost are influenced more by seasonal and interannual fluctuations of temperature. The major variation in surface temperature has a period of one year, corresponding to the annual cycle of solar radiation (there is also a diurnal variation corresponding to the daily cycle of radiation). Temperature variations experienced with the passage of the seasons at the surface extend in a progressively dampened manner to a depth of some ten or twenty

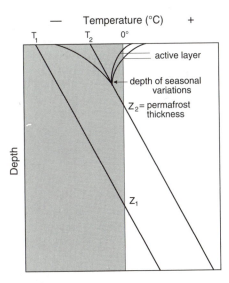

Figure 12.1
Ground temperature profile in a permafrost region. T_1 and T_2 represent different mean annual surface temperature conditions.

metres. Within the layer of annual variation, maximum and minimum figures form an envelope about the mean, and the top of permafrost is that depth where the maximum annual temperature is 0°C. Superimposed on normal periodic variations are other fluctuations with durations from seconds to years; causes may include sporadic cloudiness, variations in weather, and changes in climate.

Let us now imagine some change in climatic conditions which causes mean annual surface temperature to rise. The result would be deepening of the active layer, as both the mean annual temperature and the envelope of maximum (summer) temperatures shift to higher values (Figure 12.1). If climatic warming were sustained, the permafrost table would recede further year by year and the base of the permafrost would begin to rise as surface warming propagated to greater depths. If the progressive warming were great enough, then permafrost could eventually disappear altogether.

Stability of Permafrost

Since permafrost is a thermal condition, it is potentially sensitive to changes in climate. However, changes in the thermal regime of the ground that lead to degradation (or formation) of permafrost can result from environmental changes other than fluctuations in climate. For example, removal, damage, or compaction of surface vegetation, peat, and soil alters the balance of surface energy transfers, generally raising mean summer surface temperature and thawing the upper layer of permafrost (Figure 12.2). Mackay (1970) described the example of an experimental farm site

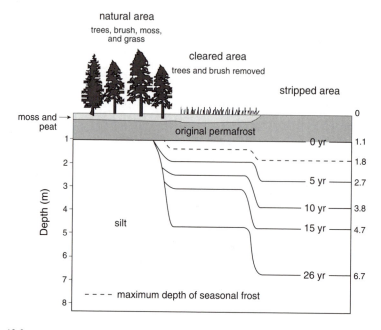

Figure 12.2
Permafrost degradation under surface disturbances in central Alaska, over a twenty-six-year period. *Source*: Linnell (1973).

at Inuvik that was cleared of spruce and birch in 1956. The depth of thaw prior to clearing was about 36 cm, but by 1962 the active layer had deepened to 183 cm.

In winter, increases in snow cover accumulation, as can result from barriers, structures, and depressions or changes in wind patterns, can lead to significant warming of the ground. Figure 12.3 shows how erection of snow fences at Schefferville, northern Quebec, immediately raised ground temperatures. Decreases in snow cover, in contrast, lead to cooling of the ground, other things being equal.

While the effects of surface environmental changes are usually restricted in areal extent, climatic change can affect extensive areas of permafrost. Even modest climatic warming could have drastic effects for terrain conditions and northern engineering, since thousands of square kilometres of warm permafrost would be directly affected. While many centuries would be required for complete degradation of the affected permafrost, thawing from the surface would begin immediately, with many potentially serious results.

Evidence is already available about historic changes in permafrost distribution. For example, Thie (1974) studied permafrost degradation in northern Manitoba, using aerial photographs taken between 1927 and 1967. It can be demonstrated that the rate of collapse of permafrost in peat plateaus in northern Manitoba has been between 0.5 and 1 m/yr. Moreover, degradation was in progress prior to the 1920s and had proceeded most rapidly since about 1850, probably related to the amelioration in

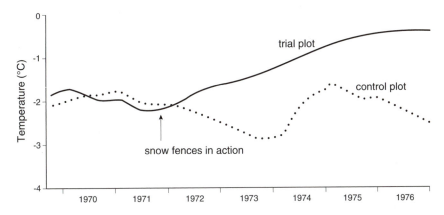

Figure 12.3
Twelve-month running mean ground (10 m) temperatures at adjacent sites with different snow covers, Schefferville, Que. *Source*: Nicholson (1978).

climate following the end of the Little Ice Age, in the mid-nineteenth century. Thie concluded that as much as 60 per cent of the region studied had been underlain by permafrost during the Little Ice Age, but that this had declined to only 15 per cent by 1967.

In a similar vein, Suslov (1961) wrote that the permafrost extent at Mezen, Russia (northeast of Archangel), had retreated northward at an average rate of 400 m per year since 1837; according to Bird (1967), Sumgin, writing in 1934, stated that the southern limit of permafrost in the then USSR was probably receding as a result of climatic amelioration. Mackay (1975) has speculated that permafrost in the Mackenzie valley has become less widespread since 1850.

If degrading permafrost is supersaturated (ice-rich), and if excess water is able to drain away, permanent subsidence of the ground surface results (Figure 12.4). This thaw settlement causes thermokarst topography (see chapter 6). Resulting depressions may trap snow in winter, promoting further warming of the ground. Where depressions collect water, additional degradational effects take over, and thawing on slopes can cause land failures and mass movements of various kinds.

Nature and Significance of Permafrost Distribution

A map of permafrost in Canada (see Figure 1.8) shows broad latitudinal zonation of continuous, widespread, and scattered permafrost. Unfortunately, this concept of permafrost zones has tended to suggest distinct kinds of terrain, although in reality there is gradual transition from the seasonally frozen ground of temperate regions to the extensive, perennially frozen ground of the far north. The zones convey the relative areal dominance of permafrost and permafrost-free conditions.

Although permafrost is temperature dependent, the relation with climate is not straightforward, since the surface temperature regime does not depend solely on

A

B

Figure 12.4
Thermokarst terrain: (A) near Mayo, Yukon, showing landslides caused by thawing of permafrost (foreground) together with many water-filled depressions, *photo*: C.R. Burn; (B) along Beaufort Sea coast, Mackenzie delta, NWT, showing ground ice slumps resulting from thaw of ice-rich sediments, *photo*: H.M. French.

geographic location. Local surface conditions, such as type of vegetation, depth of snow cover, soil type, and moisture content, profoundly affect the surface energy regime, being interposed between the atmosphere and the ground. For example, Price (1971) found that ground temperatures beneath a north-facing slope in south-central Yukon were warmer than an adjacent southeast-facing slope; mosses and thick plant cover on the latter accounted for the difference. Kudriavtsev (1965: 17) presents another example from Siberia, in which the effect of slope aspect is moderated by uneven distribution of snow cover. Prevailing winds remove snow from south- and west-facing slopes and deposit in on north- and east-facing slopes, which, as a result, remain warmer in winter. In summer, the south- and west-facing slopes receive more solar radiation and are warmer. The net result is that the two factors are compensatory, and mean annual soil temperatures are the same for all aspects.

Myriad local variations of vegetation, topography, and soil conditions can cause differences in mean ground temperatures of several degrees over quite small areas. Wherever average temperature is within a few degrees of 0°C, such variation means that permafrost occurs in patches, or discontinuously. These circumstances, together with the scattered nature of direct observations, make precise mapping of permafrost difficult.

Practical Implications of Climatic Warming

While cold is usually seen as the singular feature of high latitudes, problems resulting from thaw are generally of greater practical concern.

Where permafrost contains ground ice, or is ice-rich, considerable thaw settlement can occur. Such action has been responsible for significant damage to buildings, roads, runways, and so on (Figure 12.5), and increased action would undoubtedly cause additional and severe maintenance and repair problems. Some roads and structures might have to be rebuilt or abandoned, and greater depths of gravel padding would be needed to preserve permafrost under roads and structures. Special concern might be directed to existing water-retaining structures, such as reservoirs and hydro-dams, expecially in areas of thaw-sensitive permafrost.

Erosion of lake, river, and reservoir shorelines may increase because of permafrost thawing and a longer open-water season. Greater sediment transport in rivers could shorten the operating life of hydro-electric projects, for example. The expected rise in sea level accompanying global warming could accelerate coastal retreat in permafrost regions and, combined with thaw settlement as permafrost melts, could produce inundation of low-lying areas. Osterkamp (1984) points out that most petroleum-producing areas in arctic coastal Alaska are currently at very low elevations.

The bearing capacity and creep resistance of permafrost itself will decline as its temperature gradually rises. The mechanical behaviour of frozen soils is subject to temperature-dependent effects on the amount and strength of ice bonding in the soil. First, significant amounts of liquid water can coexist with ice in soils that are considerably below 0°C. The quantity of unfrozen water in a soil changes with temper-

Figure 12.5
Human-induced thermokarst: (A) adjacent to Sachs Harbour airstrip, southern Banks Island, NWT; (B) abandoned and unstable house in Dawson City, Yukon, caused by thaw of ice-rich permafrost beneath central part of structure. *Photos*: H.M. French.

ature, and, as a result, ice content and degree of cementation bonding in the soil also vary. This situation leads to a change in the strength properties of the soil. Second, frozen soils are subject to creep deformation, which is strongly affected by temperature change (see chapter 6). Creep rates could rise markedly as the ground warms toward 0°C. Moreover, since fine-grained soils close to 0°C contain sizeable amounts of unfrozen water, the increase in creep deformations could be greater than would be predicted for pure ice.

As a result, the strength of frozen soil will decrease as it warms, further exacerbating creep deformation. Problems could result in the bearing capacity of piles, which are widely used in northern construction, as adhesion forces decline. In addition, with progressive deepening of the active layer the effective length of piles in the permafrost will decrease, while frost heave forces could become stronger (Esch and Osterkamp 1990), leading to further problems. General reductions in soil strength would also affect other forms of foundations, since most existing structures would not have included any allowance for climatic warming. Finally, creep rates of (ice-rich) permafrost slopes would get higher and slope stability would lessen.

THERMAL CONDITIONS IN PERMAFROST

Ground temperatures are strongly influenced by conductive heat transfer, although localized circulation of groundwater can occur, particularly in areas of discontinuous permafrost. Under steady conditions, the mean annual ground temperature profile is linear with depth (assuming constant thermal conductivity), and temperature at any depth, T_z, is given by:

$$T_z = T_s + G_g \cdot z,$$

where T_s is surface temperature and G_g is geothermal gradient (increase of temperature with depth within the ground). In reality, heat conduction in the ground is more complex: steady states are rarely achieved, since surface temperature is continually changing, and natural variations in soil conditions leads to differences in thermal properties. In addition, thermal properties of frozen soils vary with temperature.

A conspicuous feature of temperature profiles observed in many northern boreholes is a distinct inversion in the upper 100 metres or so, with near-surface temperatures being noticeably warmer than those obtained by simple upward extrapolation (Figure 12.6). Such widespread "deviations" can be explained by climatic change (e.g. see Lachenbruch, Cladouhos, and Saltus 1988), although similar effects could be produced locally by changes in surface conditions, such as a forest fire or land disturbance. Other processes that should be considered include relict permafrost (Bonnlander and Major-Marothy 1964), comtemporary snow and ice cover (Ives 1979), and nineteenth-century snow- and ice-cover effects such as those documented by Ives (1962).

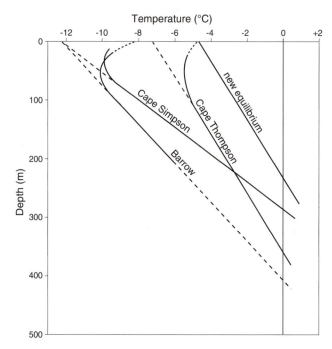

Figure 12.6
Temperature profiles from arctic coast of Alaska. *Source*: Lachenbruch
et al. (1982).

Suppose that a change in climate produces a shift of mean annual surface temperature
from some initial value, T_s, to a new value $(T_s + \Delta T_s)$. During transition toward a
new steady-state condition, the effect of the surface temperature change propagates
into the ground. The transient temperature profile will be given by:

$$T(z, t) = T_s + G_g \cdot z + \Delta T(z, t), \tag{2}$$

where the last term on the right decreases with depth (for any time, t), creating the
characteristic curvature evident in Figure 12.6. The rate at which the temperature at
any depth changes with time depends on the ground's thermal diffusivity (see below).

Lachenbruch and Marshall (1986) analysed borehole temperature data from Alaska
and concluded that the ground's surface temperature has increased by about 2 to 4C°
during the last century or so. This coincides with the amelioration in global temperatures since about 1850, following the Little Ice Age. If maintained, such a change
would eventually lead to thinner permafrost (see Figure 12.6), as the temperature
profile is shifted to warmer values. After thermal equilibrium is re-established in the
ground, the new temperature profile would be:

$$T(z) = T_s + G_g \cdot z + \Delta T_s,$$

and the permafrost would have thinned from the bottom by an amount equal to $(\Delta T_s/G_g)$ (see Figure 12.6). For example, if $G_g = 20\text{K/km}$, and $\Delta T_s = 4\text{C}°$, this would amount to 200 m. Where the permafrost was initially thinner than this, it would eventually disappear altogether.

However, the time required for a new equilibrium to be reached may be very long – perhaps many thousands of years, depending on the thickness of permafrost and its thermal diffusivity. Fluctuations in surface temperature of short period (a century or less) will not significantly affect the base of cold (thick) permafrost. Indeed, Lachenbruch et al. (1982) calculated that climatic warming in the last 100 years or so has melted a total of only 0.8 m at the bottom of permafrost.

Since the main destabilizing effects of climatic change will result from the melting of shallow ground ice as the annual surface temperature regime becomes warmer, we might well consider what non-climatic factors affect evolution of the surface temperature regime.

Significance of Microclimate

The thermal regime in the upper layers of the ground is controlled by exchanges of heat and moisture between the atmosphere and Earth's surface. The processes involved in the energy balance comprise net exchange of radiation (represented by Q^*) between surface and atmosphere, transfer of sensible (Q_H) and latent heat (Q_{LE}) by the turbulent motion of the air, and conduction of heat into the ground (Q_G (see chapter 3). Partitioning of the radiative surplus (or deficit) among Q_H, Q_{LE}, and Q_G is governed by the nature of the surface and the relative abilities of the ground and the atmosphere to transport heat energy. Each term affects surface temperature, and thus the way in which the energy balance is achieved establishes the surface temperature regime (e.g. see Outcalt 1972).

The influence of local conditions on surface energy exchange produces the microclimate of a locale. This microclimate ultimately determines ground thermal conditions, since any differences in microclimate affect the near-surface temperature regime, which in turn influences the depth of the active layer and permafrost thickness itself. Local factors commonly override the influence of larger-scale macroclimatic factors on the thermal conditions of the ground. Luthin and Guymon (1974) visualized boundary-layer interactions in terms of a complex buffer layer interposed between atmosphere and ground (Figure 12.7). Atmospheric mass and energy flows, together with geothermal heat flux, constitute the boundary conditions; the vegetation canopy, snow cover (when present), and surface organic layer act as buffers between atmosphere and ground.

The vegetation canopy reduces solar radiation reaching the ground surface and affects accumulation and persistence of snow cover (e.g. Luthin and Guymon 1974; Rouse 1984). In addition, the canopy's interception of precipitation and its transpiration influence the ground thermal regime through Q_{LE} and the water balance. Comparative observational studies assess the role of the vegetation canopy. For

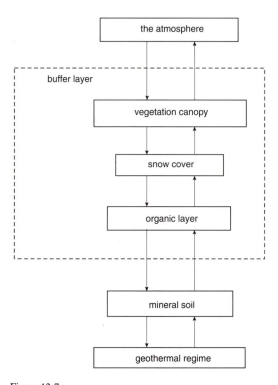

Figure 12.7
Model of climate-permafrost relationship. *Source*: Luthin and Guymon (1974).

example, Rouse (1984) found that summer soil temperatures beneath an open spruce forest were lower than adjacent tundra (Figure 12.8) as a result of interception of radiation by the canopy, higher evaporation from the wetter surface, and the greater roughness of the forest which increases turbulent exchange with the atmosphere.

Smith (1975) concluded that vegetation's direct effect was far less important than its role in snow accumulation – a conclusion supported by Rouse (1984). Figure 12.8 also reveals that near-surface soils in the forest, though cooler in summer, are considerably warmer in winter than those of the nearby tundra: the forest acts as a snow fence and traps a deep blanket of snow. As a result, forest soil temperatures are more than 3C° warmer on an annual basis.

Snow profoundly affects the ground thermal regime, since it presents a barrier to heat loss from the ground to the air. In the Mackenzie delta, where mean daily air temperature is below −20°C for almost six months in winter, the 1-m ground temperature beneath 120 cm of snow did not fall below −0.2°C (Smith 1975). Calculations showed that outflow of ground heat during winter was only one-fifth to one-tenth of that at a site with only 25 cm of snow and that permafrost was actively degrading because of increased snow accumulation. In the Schefferville area of

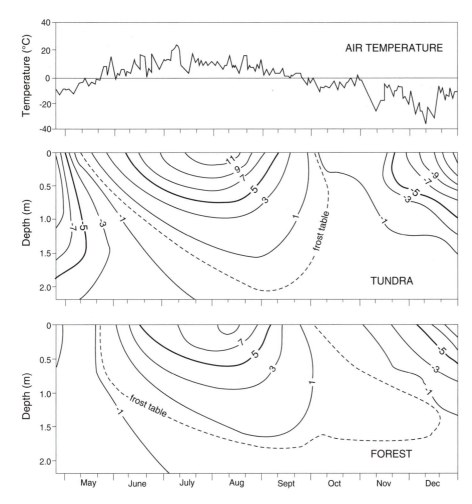

Figure 12.8
Ground thermal regime at tundra and forest sites near Churchill, Man. *Source*: Rouse (1984).

Quebec, Nicholson and Granberg (1973) found variation in mean annual ground temperatures to be determined primarily by snow depth, with variation in summer conditions being less important.

In marginal areas of permafrost distribution, snow cover alone may be the critical local factor determining the presence of permafrost. In the colder regions of more widespread permafrost, it influences the depth of the active layer. Also, in regions of heavy snowfall, lake and river ice will not be so thick, so that even bodies with shallow water may not freeze through – as in the Mackenzie delta, where snow cover shapes local distribution of permafrost (Smith 1976).

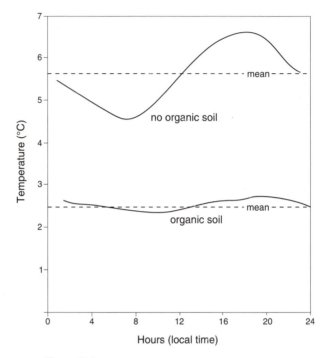

Figure 12.9
Mean diurnal 10-cm ground temperature regimes at adjacent sites, Mackenzie delta. *Source*: Smith (1975).

The influence of the organic layer on ground thermal conditions has also been well documented in the literature, and the presence of permafrost in marginal areas is frequently associated with peat (e.g. Lindsay and Odynsky 1965; Zoltai and Tarnocai 1975; Fitzgibbon 1981). Brown (1973) concluded that variations in the vegetation canopy, compared to the surface organic layer, little influenced the ground thermal regime. He reported that in the Yellowknife area the greatest local extent of permafrost is in peatlands and that mean annual temperature at a depth of 15 m ranges from about 2°C in granite to − 1.0°C in spruce peatland.

The lower mean annual temperatures in association with peat are generally attributed to its insulation of the ground from the heat of summer. In summer, when the surface tends to be dry (because of evaporation), peat's low thermal conductivity inhibits warming of the ground. Figure 12.9 shows temperatures measured at a depth of 10 cm over the summer at two sites only 1 m apart, in a spruce forest in the Mackenzie delta. One site had a 10-cm layer of organic material at the surface; the other, simply bare mineral soil. The mean daily temperature is 3C° warmer at the bare site and the diurnal range five time greater.

In autumn, peat can become quite moist, because of slower evaporation. Further, when it freezes, its conductivity becomes even higher and the ground can cool rapidly.

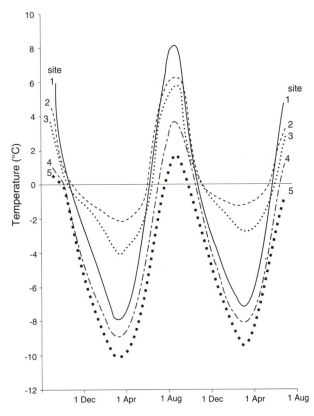

Figure 12.10
Annual temperature regime at depth of 50 cm for five sites, Mackenzie
delta. *Source*: Smith (1975).

The net effect – in contrast to snow – is mean annual ground temperatures much
lower than under adjacent areas without peat.

Buffer-layer effects can result in widely varying ground thermal conditions within
small areas of uniform climate, although in areas with little vegetation or snow cover
– as can be found in the Canadian arctic islands, for example – the link between air
and ground temperatures is more direct. Figure 12.10 shows substantial differences
in annual ground temperature regimes between neighbouring sites in the Mackenzie
delta, where mean annual air temperature is close to − 10°C. Soil conditions are
fairly uniform among sites, and differences in ground temperature are caused principally by local variations in snow accumulation and organic material, with shading
by vegetation of secondary importance. The observed range in ground thermal conditions is equivalent climatically to several degrees of latitude. Such results are neither
special nor unique but may be considered quite typical.

The variable linking of ground thermal conditions to the present climate guarantees
differential buffering from the effects of changes in climate. The response to climatic

warming may be slow to develop and permafrost will persist, particularly where organic material is present, especially when combined with a forest canopy. Vegetation itself will alter in response to climatic change, with resultant effects on shading, evaporation, and snow accumulation.

Lithologic Conditions

Heat conduction theory expresses the fact that ground temperatures respond to surface changes according to the thermal properties of the ground, specifically thermal diffusivity – the ratio of thermal conductivity to volumetric heat capacity. The higher the diffusivity, the faster the response. For example, in Figure 12.6 we can see that the effects of the post-1850s climatic warming have penetrated most deeply at Cape Thompson (highest diffusivity) and least at Cape Simpson (lowest).

Diffusivity (and hence thermal response) vary with soil type, but most significantly with the moisture content of the ground. In frozen soils, temperature changes are dominated by the phase-change relations of the water (latent heat), especially within a few degrees below 0°C. When ice changes to water it absorbs heat equivalent to that required to raise the temperature of an equal volume of rock by about 150° (Gold and Lachenbruch 1973). Melting of permafrost therefore requires the exchange of far more energy than the warming of cold permafrost (below -5°C) or unfrozen ground. Furthermore, if excess ice is present in the permafrost, latent heat effects dominate even more and ground temperatures become anchored near the melting point for a long period. (This is analogous to the so-called zero curtain effect observed seasonally in the active layer; see chapter 6). Total moisture content will therefore be decisive in the thermal response of permafrost.

Figure 12.11 illustrates how lithologic conditions can moderate the effect of surface climatic changes on ground thermal conditions. Changes in ground temperature were calculated for an increase in the surface temperature of 0.5C° per decade over a sixty-year period, starting from an initial value of -2°C. This model might be considered representative for areas of discontinuous permafrost. Mean annual temperature profiles at the end of the period are shown for three values of thermal diffusivity. An initial profile is included for reference.

The lowest diffusivity value (1×10^{-7} m^2/s) represents thawing of ice-rich permafrost, while the value of 1×10^{-6} applies to rock or unsaturated frozen (sandy) soil. There are striking differences in the calculated rates of permafrost degradation, ranging from only 6 or 7 m in sixty years in ice-rich permafrost to complete thawing for the highest diffusivity. In the latter case, permafrost actively degrades from below as well as from above. Even after 500 years, the permafrost table in the ice-rich case would degrade to only about 40 m, and the base of permafrost would rise by only 4 or 5 m. Complete degradation of permafrost could take many centuries, at least, because of the thermal inertia of ice-rich materials and the considerable thickness of permafrost in some areas.

The above calculations are supported in a general way by the field results of Burn and Smith (1988), who reported thawing to a depth of about 15 m in ice-rich perma-

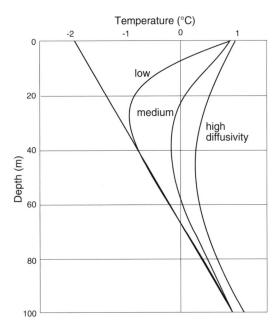

Figure 12.11
Estimated ground temperature profiles in materials of differing thermal diffusivity following climatic warming.

frost following initiation of a thermokarst lake some 100 years ago. They estimate associated change in mean surface temperature of about 3 or 4C°, although they believe the actual cause in this case was forest fire.

Changes in Precipitation

Various ground surface processes in the permafrost environment would be affected by possible increases in rainfall. Changes in snow cover would also alter ground thermal conditions.

Goodrich (1982) showed that doubling of snow cover from 25 to 50 cm increased minimum ground surface temperature by about 7C° and mean annual surface temperature by 3.5C°. If the 50 cm of snow accumulates within thirty days in autumn, mean temperature would rise above 0°C and permafrost would degrade. Nicholson and Granberg (1973) found that erection of snow fences leads to immediate warming of the ground, which is then maintained for years even against natural cooling trends (see Figure 12.3). Mackay (1984) found that snow depth of 60 cm was sufficient to prevent ice-wedge cracking in an area of active ice wedges.

Precipitation increases of as much as 60 per cent in autumn and early winter projected in some climate models would therefore help accelerate permafrost degradation, particularly in marginal areas.

EFFECTS ON COLD CLIMATE PROCESSES

Climatic changes would also affect various ground surface processes characteristic of permafrost, such as mass movements and creep, ice-wedge cracking, and frost heave.

More slope failure would probably result from changes in climate and degradation of permafrost. Deepening of the active layer (with consequent melting of ice-rich soil) could lead to instability and failure, as also could greater summer rainfall. In addition to there being mass displacements of thawed ground, frozen ground subject to warming might experience increased deformation, as mentioned previously.

In ice-wedge cracking, the relation of frost cracking to (mean) air or ground temperatures is only approximately known. Early studies (e.g. Péwé 1966) suggested that ice-wedges form only where mean annual temperature is −6°C or colder. However, Mackay and Mathews (1973) have reported active ice-wedges near Fort Good Hope, and Burn (1990) has documented thermal-contraction cracking at Mayo; both localities are in the discontinuous zone where mean annual air temperature is warmer than −6°C. In any event, the authors seem to agree that rapid drop in the ground temperature is necessary to cause cracking. This would certainly be impeded by the insulating properties of deep snow cover, as Mackay (1984) has demonstrated at Inuvik, NWT.

Ultimately there will be long-term changes in the distribution of permafrost, as tens of thousands of square kilometres of permafrost eventually disappear. In some areas complete degradation could take hundreds of years, because of the thermal inertia of ice-rich materials and the considerable thicknesses of permafrost affected. In the short term there would be various rapid-onset effects associated with progressive deepening of the active layer. Melting of shallow ground ice might lead to widespread thermokarst and decreased slope stability, creating in some cases severe maintenance and repair problems for all manner of structures. Differential settlement would cause large internal stresses in structures, producing distortions and possibly even failure. Higher ground temperatures would cause operational difficulties associated with the reduced strength of warmer permafrost.

While many practical problems would undoubtedly arise from widespread permafrost degradation, rate of change from present to future conditions will often determine their severity. The relevant practical questions to ask concern what makes natural and human systems vulnerable to change and what actions will minimize problems and develop new or expanded opportunities.

REFERENCES

Bird, J.B. 1967. *The Physiography of Arctic Canada*. Baltimore: Johns Hopkins Press.
Bonnlander, B., and Major-Marothy, G.M. 1964. "Permafrost and Ground Temperature Observations." McGill Subarctic Research Paper, Montreal, 16, 33–50.

Burn, C.R. 1990. "Implications for Palaeoenvironmental Reconstruction of Recent Ice-wedge Development at Mayo, Yukon Territory." *Permafrost and Periglacial Processes* 1: 3–14.

Burn, C.R., and Smith, M.W. 1988. "Thermokarst Lakes at Mayo, Yukon Territory, Canada." *Fifth International Conference on Permafrost*, Trondheim, Norway, vol. 1, Trondheim: Tapir Publishers, 700–5.

Esch, D.C., and Osterkamp, T.E. 1990. "Arctic and Cold Regions Engineering: Climatic Warming Concerns." *Journal of Cold Regions Engineering* 4: 6–14.

Etkin, D.A. 1990. "Greenhouse Warming: Consequences for Arctic Climate." *Journal of Cold Regions Engineering* 4: 54–66.

Fitzgibbon, J.E. 1981. "Thawing of Seasonally Frozen Ground in Organic Terrain in Central Saskatchewan." *Canadian Journal of Earth Sciences* 18: 1,492–6.

French, H.M., ed. 1986. *Climate Change Impacts in the Canadian Arctic: Proceedings of a Canadian Climate Program Workshop*, 3–5 March, 1986, Geneva Park, Ont.

Gold, L.W., and Lachenbruch, A.H. 1973. "Thermal Conditions in Permafrost: A Review." *Proceedings of the Second International Conference on Permafrost*, Yakutsk, USSR, North American Contribution, Washington, DC: National Academy of Sciences, 3–25.

Goodrich, L.E. 1982. "The Influence of Snow Cover on the Ground Thermal Regime." *Canadian Geotechnical Journal* 19: 421–32.

Hansen, J.E., Lacis, A., Rind, D., Russell, G., Stone, P., Fung, I., Ruedy, R., and Lerner, J. 1984. "Climate Sensitivity: Analysis of Feedback Mechanisms." In J. Hansen and T. Takahashi, eds., *Climate Processes and Climate Sensitivity*, Maurice Ewing no. 5, American Geophysical Union, Washington, DC, 130–63.

Ives, J.D. 1962. "Indications of Recent Extensive Glacierisation in North-Central Baffin Island." *Journal of Glaciology* 4: 197–205.

– 1979. "A Proposed History of Permafrost Development in Labrador-Ungava." *Géographie physique et quaternaire* 33: 233–4.

Kudriavtsev, V.A. 1965. *Temperature, Thickness and Discontinuity of Permafrost*. National Research Council of Canada, Technical Translation 1187.

Lachenbruch, A.H., Cladouhos, T.T., and Saltus, R.W. 1988. "Permafrost Temperatures and the Changing Climate." *Fifth International Conference on Permafrost*, Trondheim, Norway, vol. 3, Trondheim: Tapir Publishers, 9–17.

Lachenbruch, A.H., and Marshall, B.V. 1986. "Changing Climate: Geothermal Evidence from Permafrost in the Alaskan Arctic." *Science* 234: 689–96.

Lachenbruch, A.H., Sass, J.H., Marshall, B.V., and Moses, T.H., Jr. 1982. "Permafrost, Heat Flow and Geothermal Regime at Prudhoe Bay, Alaska." *Journal of Geophysical Research* 87 B11: 9301–16.

Lindsay, J.D., and Odynsky, W. 1965. "Permafrost in Organic Soils of Northern Alberta." *Canadian Journal of Soil Science* 45: 265–9.

Linnell, K.A. 1973. "Long Term Effects of Vegetation Cover on Permafrost Stability on an Area of Discontinuous Permafrost." *Proceedings of the Second International Conference on Permafrost*, Yakutsk, USSR, North American Contribution, Washington, DC: National Academy of Sciences, 688–93.

Luthin, J.N., and Guymon, G.L. 1974. "Soil Moisture–Vegetation-Temperature Relationships in Central Alaska." *Journal of Hydrology* 23: 233–46.

McBeath, J.H., ed. 1984. *The Potential Effects of Carbon Dioxide–Induced Climatic Changes in Alaska*. School of Agriculture and Land Resources Management, University of Alaska–Fairbanks, Miscellaneous Publication 83–1.

Mackay, J.R. 1970. "Disturbances to the Tundra and Forest Tundra Environment of the Western Arctic." *Canadian Geotechnical Journal* 7: 420–32.

– 1975. "The Stability of Permafrost and Recent Climatic Change in the Mackenzie Valley." In Current Research, Part A, Geological Survey of Canada, Paper 75–1A; 173–6.

– 1984. "The Direction of Ice-Wedge Cracking in Permafrost: Downward or Upward." *Canadian Journal of Earth Sciences* 21: 516–24.

Mackay, J.R., and Mathews, W.H. 1973. "Geomorphology and Quaternary History of the Mackenzie River Valley near Fort Good Hope, N.W.T., Canada." *Canadian Journal of Earth Sciences* 10: 26–41.

Maxwell, J.B. 1987. "Atmospheric and Climatic Change in the Canadian Arctic: Causes, Effects and Impacts." *Northern Perspectives* 15 no. 5: 2–6.

Nicholson, F.H. 1978. "Permafrost Modification by Changing the Natural Energy Budget." *Proceedings of the Third International Conference on Permafrost*, Edmonton, Canada, vol. 1, Ottawa: National Research Council of Canada, 61–7.

Nicholson, F.H., and Granberg, H.B. 1973. "Permafrost and Snow Cover Relationships near Schefferville." *Proceedings of the Second International Conference on Permafrost*, Yakutsk, USSR, North American Contribution, Washington, DC: National Academy of Sciences, 151–8.

Osterkamp, T.E. 1984. "Potential Impact of a Warmer Climate on Permafrost in Alaska." In J.E. McBeath, ed., *The Potential Effects of Carbon Dioxide–Induced Climatic Changes in Alaska*, School of Agriculture and Land Resources Management, University of Alaska–Fairbanks, Miscellaneous Publication 83–1, 106–13.

Outcalt, S.I. 1972. "The Development and Application of a Simple Digital Surface Climate Simulator." *Journal of Applied Meteorology* 11: 629–36.

Péwé, T.L. 1966. "Palaeoclimatic Significance of Fossil Ice Wedges." *Biuletin Peryglacjalny* 15: 65–72.

Price, L.W. 1971. "Vegetation, Microtopography and Depth of Active Layer on Different Exposures in Subarctic Alpine Tundra." *Ecology* 52: 638–47.

Rouse, W.R. 1984. "Microclimate of Arctic Tree Line. 2. Soil Microclimate of Tundra and Forest." *Water Resources Research* 20: 67–73.

Smith, M.W. 1975. "Microclimatic Influences on Ground Temperatures and Permafrost Distribution, Mackenzie Delta, Northwest Territories." *Canadian Journal of Earth Sciences* 12: 1,421–38.

– 1976. *Permafrost in the Mackenzie Delta, Northwest Territories*. Geological Survey of Canada, Paper 75–28.

Suslov, S.P. 1961. *Physical Geography of Asiatic Russia*. Translated by N.D. Gershevsky. San Francisco: W.H. Freeman.

Thie, J. 1974. "Distribution and Thawing of Permafrost in the Southern Part of the Discontinuous Permafrost Zone in Manitoba." *Arctic* 27: 189–200.

Williams, P.J., and Smith, M.W. 1989. *The Frozen Earth: Fundamentals of Geocryology.* Cambridge: Cambridge University Press.

Zoltai, S.C., and Tarnocai, C. 1975. "Perennially Frozen Peatlands in the Western Arctic and Subarctic of Canada." *Canadian Journal of Earth Sciences* 12: 28–43.

CHAPTER THIRTEEN

Cold Environments
and Global Change

OLAV SLAYMAKER AND
HUGH M. FRENCH

The causal links between climate change and global environmental change are not well established. Chapter 11 has explored some of the practical difficulties of establishing the fact of climate change and the need for risk and "scenario" analysis in the face of uncertainty. If the empirical evidence for contemporary climate change is difficult to find, then that for global change is even more elusive because of the lagged response of environmental systems to climate change (Rosswall, Woodmansee, and Risser 1988). This final chapter identifies some technical problems that restrict the ability to predict global change quantitatively. It then examines possible effects of any change on Canada's cold environments and the ambiguity of some of the evidence for present change.

Climate change caused by the greenhouse effect would alter Canada's cold environments more dramatically than the country's other environments. Global climate models (GCMs) predict consistently higher shifts in temperature for polar latitudes than elsewhere. Moreover, in Canada's cold mountain environments, horizontal and vertical environmental gradients are greater than elsewhere in the country. It should therefore be possible to detect the effects of small changes very rapidly in mountain environments.

The context of our discussion is one of cold environmental systems. Unique changes in behaviour of environmental systems occur as the ambient temperature moves through the freezing point. The latter threshold determines presence or absence of snow, ice, and permafrost; infiltrability and permeability of soils; hydrologic response of rivers; geomorphic erosional response; and the ability of vegetation to transpire and of water to evaporate. Other significant thresholds concern amount, intensity, and sequencing of precipitation; amount and intensity of drought; and temperature, precipitation, wind, and snow necessary to maintain the timberline. Table 13.1 suggests the wide range of variables that must be explored in order to assess global environmental changes resulting from climatic change.

Table 13.1
Environmental effects of climatic change: data requirements

Variable	Statistics	Impact
Temperature	Mean	Threshold value for all variables
Wind and wind-chill	Variance	Climatic
Precipitation	Persistence	Hydrologic
Surface runoff	Sequencing	Glaciologic
Snow cover		Vegetational
Permafrost	*Temporal scale*	Soils
Glacier ice	Long term (greater than annual)	Geomorphic systems
Available soil moisture	Annual	Human use systems
Groundwater	Seasonal	
Evapotranspiration	Short term	
Storm events		
Sediment movement	*Spatial scale*	
Timberline	Regional (10^5–10^6 km^2)	
Soils	Drainage basin or watershed	
Landforms	(10^2–10^5 km^2)	

TECHNICAL PROBLEMS

Time Scales

Most environmental research has traditionally explored short-term (one week or less) or long-term (thousands of years or more) effects of climate on environmental systems. The intermediate scale of months to centuries is comparatively poorly understood (e.g. see Figure 4.11). In ecology, for example, traditional ecological experiments have provided short-term results, and paleoecology has examined Holocene and Pleistocene changes over tens of thousands of years (see chapters 4 and 9), but little is known about the actual rate of migration of, for example, vegetation assemblages on intermediate time scales. In contemporary cold environments one can only guess how rapidly timberline migrates, permafrost degrades, or glaciers retreat. In a few areas, techniques exist to ameliorate this problem. For example, tree ring chronologies from the mountains of western Canada (Luckman 1990) can assist in linking climate with data on vegetation response (Figure 13.1A). For available lines of evidence, and their relations to the time scales involved, see Figure 13.1.

Transient Responses

All simulations of future climates responding to greenhouse gases carried out on global climate models are equilibrium solutions – i.e. they represent environmental conditions beyond which no adjustments in system state can be ascribed to the increase in greenhouse gases. Unfortunately the time taken to achieve this equilibrium response

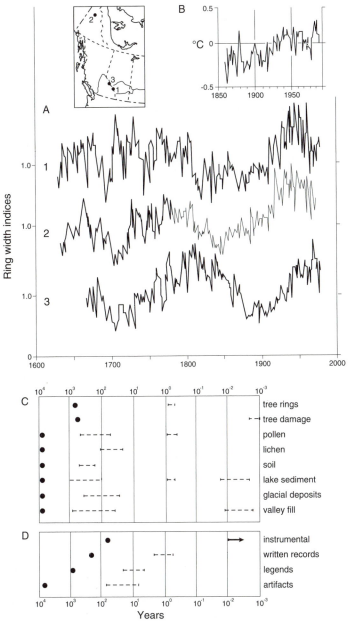

Figure 13.1

(A) Tree-ring chronologies for spruce trees at treeline localities in western Canada: (1) Larch Valley, Alta, 1630–1986, *Picea engelmannii* (Luckman 1990); (2) Twisted Tree–Heartrot Hill, Yukon, 1630–1975, *Picea glauca* (Jacoby and Cook 1981); and (3) Athabasca Glacier, Alta, 1660–1980, *Picea engelmannii* (Luckman 1990). (B) Standardized mean annual Northern Hemisphere temperatures 1861–1984 (from Jones, Wigley, and Wright 1986 Figure 5a). (C) Biological and geographical attributes and time scales used for environmental reconstruction. (D) Historical and archaeological sources and time scales used for environmental reconstruction.

is completely unknown. We do know, for example, that while the transient response to CO_2 doubling is occurring, further changes in the input conditions are also taking place. An equilibrium solution is an entirely theoretical construct (see chapter 11).

Erosional systems take even longer to respond to climate change. Climate provides radiant energy and moisture for erosional systems, but resistance and buffers substantially delay initiation of erosion processes and are highly variable over space. For example, local available relief, thermal diffusivity, and hydraulic conductivity of regolith help control depth of penetration into the ground of a temperature wave or of a slug of moisture in a short-term, single rainfall event. Storage by snow and lakes can delay any climate effect over seasonal time scales, while glaciers, aquifers, vegetation, and soils may introduce lags of decades to centuries. Church and Slaymaker (1989), for example, claim that sediment carried by British Columbia's larger river systems is a transient response to hydrology and climatic controls. Attempts to establish equilibrium geomorphic responses to climatic change of the magnitude of Ice Age cooling are misguided (Figure 13.2; see chapter 7). Likewise, few of Canada's cold northern environments possess landscapes in equilibrium with prevailing cryogenic processes, and much permafrost is relict, because the influence of Quaternary events persists even to the present.

Spatial Scales

Much environmental research focuses on small systems. Statistical models of widely differing levels of sophistication favour site-scale predictions of response to climate change. GCMs, in contrast, provide global-scale predictions. However, in order to discuss the response of Canada's cold environments to climate change we need regional-scale predictions. Undoubtedly the method discussed by LeDrew in chapter 11 will provide some answers, but the regional predictions provided by GCMs differ greatly. Additional work is needed to link global climate models with more detailed regional models (Klemes 1985; Rosswall, Woodmansee and Risser 1988).

LESSONS FROM THE HOLOCENE

Global environmental changes during the Holocene should provide some clues to future patterns of change. Past changes were not complicated, however, by the effects of modern industrialized society. But Holocene changes provide the most reliable source of information on the way in which the time factor has modified or mediated the relationship between climate and global change in the past.

Early Holocene periods when conditions were warmer and/or drier than present – in particular, the Hypsithermal – provide only limited analogues: timberlines reached their highest known Holocene levels, and glaciers were less extensive than now. There are also crude analogies from Pleistocene interglacial periods when temperatures were as much as 4–12C° higher, sea ice was absent, rainfall in northern regions was much greater, and forests extended to the present Arctic Ocean in North America

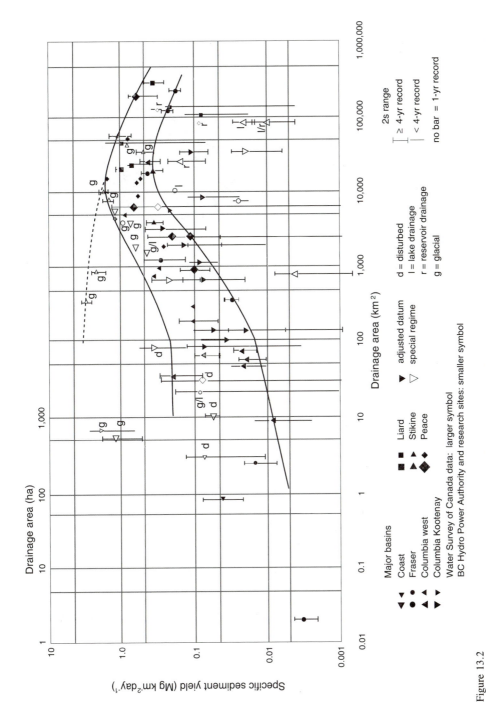

Figure 13.2
Specific sediment yield as function of drainage area for fluvial suspended sediment transport record in BC rivers. *Source*: Church and Slaymaker (1989).

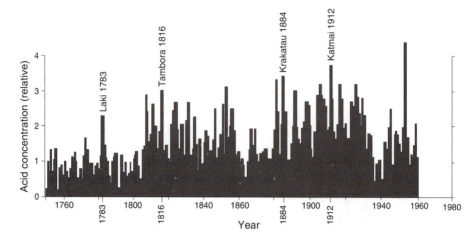

Figure 13.3

Record of solid electrical conductivity from Agassiz Ice Cap core, Ellesmere Island. Each bar represents average electrical conductivity for one year. *Source*: Koerner, Dubey, and Parnandi (1989).

Table 13.2

Change in dimensions of glaciers in Premier Range between Little Ice Age maximum position and August 1970

	Icefields	Glaciers (km³)				Median altitude of glaciers (m)
		<1	1–10	>10	Total	
Little Ice Age maximum	11.9	56.3	128.0	80.5	276.9	2,301
1970	9.8	34.5	101.7	70.8	216.8	2,432
Change	2.1	21.8	26.3	9.7	−59.9	+131

Source: Luckman (1990).

and Siberia. We contend, however, that the late Holocene record is the geological time scale most relevant for inferring global warming effects in the next century.

The glacier record provides one of the clearest signals of climatic trends over the last century or so. Most Rocky Mountain glaciers reached maximum Little Ice Age positions in the early eighteenth or mid-nineteenth century (Luckman 1986) and have since receded (Gardner 1972). The overall pattern is one of rapid frontal recession between the 1920s and 1960s followed by sharp reduction in recession rates which mirrors, though it lags behind, available temperature records.

Luckman (1990) reports a glacier inventory of the Premier Range of the BC Cariboo Mountains which covered 1,120 km² (19.5 per cent glacierized) using 1970 aerial photography. An equivalent inventory was compiled for the Little Ice Age maximum

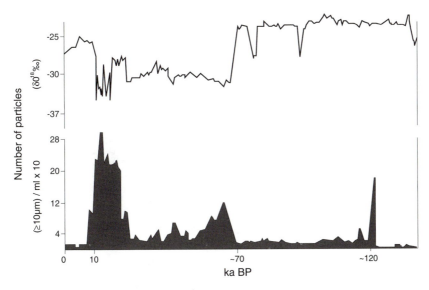

Figure 13.4
Microparticle concentrations in ice from Devon Ice Cap, from more than 120,000 years ago (<120 ka BP) to present. *Source*: Koerner, Dubey, and Parnandi (1989).

extent of over 200 glaciers based on the mapping of lateral and terminal moraines. This inventory work (Table 13.2) indicates loss of about 22 per cent of glacier cover between Little Ice Age maximum positions and 1970, with the greatest proportional losses in the smallest glaciers.

The large permanent ice sheets and glaciers in high latitudes also provide a mechanism for monitoring long-term climate change and pollution because they act as storage banks of atmospheric data. Drilling on the Agassiz Ice Cap of Ellesmere Island, for example, permitted collection of ice cores from which atmospheric pollutants were extracted. (See Figure 13.3). Most impurities are probably H_2SO_4. Before the Industrial Revolution, this type of pollutant was almost always associated with volcanic eruptions.

Some ice cores, at depth in these ice sheets, data back to the last interglacial period, over 100,000 years ago. These cores demonstrate that the atmosphere was, at certain times during the Pleistocene, even dirtier than it is today. For example, Figure 13.4 plots microparticle concentration in ice from the Devon Ice Cap over the last 100,000 years together with δO^{18} values as an indirect indicator of temperature change.

Changes in timberline in alpine regions can also indicate changes in past climate. Distribution of fossil tree remnants in the southern Rocky Mountains indicates that timberline was at least 110 m higher at 8000–8800 yr BP than today, and at Maligne Pass conditions were about 1.2–1.6C° warmer (Luckman and Kearney 1986). The present vertical zonation of ecological zones and paleoenvironmental records from mountain areas indicate that ecosystem displacement in response to environmental changes is predominantly vertical (upslope) and involves relatively small horizontal distances. For example, increase in temperature of 1C° may produce a 150–200-m rise in timberline but a latitudinal displacement of at least 100 km northward for the western Canadian treeline (Davis 1988; Roberts 1989).

EVIDENCE FROM THE PRESENT DAY

One can argue that global warming patterns might be detected in present-day events and trends. Others would argue that these so-called trends are merely short-term fluctuations around a mean. It is appropriate therefore to consider this evidence, albeit briefly.

Figure 13.5 shows the annual average global-surface air temperature since 1900 plotted against the 1950–79 average. These data indicate that temperatures have increased during the present century. If we consider a smaller area, however, such as the southern Canadian Rockies, data are more ambiguous (Figure 13.6). Over a common period of record (1916–88), mean annual temperature records from five valley-floor locations show synchronous variation, although the amplitude and long-term trends differ among stations. Only Banff, with the longest record, shows an overall positive trend (Luckman 1990).

Records at Norman Wells, NWT (Figure 13.7A), give winter mean temperatures since 1944 that suggest a warming trend since 1966. Moreover, the data also demonstrate that the winters of 1985–86 and 1988–89 were the two warmest on record. It is extremely difficult, however, to draw generalizations from such data, since there is considerable regional variability. For example, summer records for the Mackenzie valley during 1988 for Yellowknife and Inuvik (Figure 13.7B) suggest that while Inuvik experienced a warm summer, Yellowknife had a wet one. Collectively, these climate data illustrate both the regional variability and the temporal-scale problems highlighted earlier in this chapter.

CLIMATE CHANGE AND CANADA'S
NORTHERN ENVIRONMENTS

Over the next half-century, many scientists anticipate, human activities will substantially alter global climate through increase in atmospheric CO_2 and other gases from the burning of fossil fuels, air pollution, deforestation, and agricultural practices. In the Arctic, the greenhouse effect is compounded by a number of factors. The concentration of carbon dioxide is rising slightly faster in the Arctic than in southern locations. Moreover, thinning of the ozone layer over arctic regions permits more solar radiation to enter the atmosphere. The north is also believed to be a major source of natural greenhouse gases because it contains one of the world's largest areas of wetlands, which generate and release methane gas. Finally, reduction in sea-ice cover and snowfall consequent on a warming trend may result in substantial feedback (chapter 11).

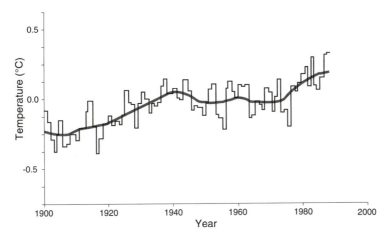

Figure 13.5
Global surface air temperatures (1900–88) plotted against 1950–79 average.

By the year 2050, average global surface temperatures may have increased by 1.5 to 4.5C° (Ferguson 1988). During the Pleistocene, when continental ice sheets and glaciers periodically expanded to cover most of Canada, average global temperatures were only 4–5C° cooler than today. During the last 1,000 years, a period that encompasses the Viking colonization of Greenland and the Little Ice Age in Europe, average temperatures have not varied by more than 1C°. In the Arctic, warming is anticipated to be less than the global average during summer – perhaps as little as 0.5C°. However, during winter a dramatic increase in temperature is expected – as much as 8 to 10C° – at least twice the global average. In addition, rainfall that currently falls in southern Canada is expected to shift northward, increasing arctic precipitation by 20 to 30 per cent. We can summarize under six headings the anticipated changes resulting from this degree of climate alteration in Canada's northern regions.

Sea Ice

There is a strong possibility that the Arctic Ocean's ice cover, currently about 1.5 metres thick on average and with a mean annual area of about four million square kilometres, will disappear. The change in net albedo, the increased source of atmospheric moisture from an ice-free ocean, and the consequent increase in cloudiness and precipitation will alter the environment of the high arctic islands, many of which currently receive little precipitation and technically are deserts.

Since about 1840, when reliable observations on sea ice began, its maximum extent has fluctuated by about 50 per cent (see chapter 2). General experience suggests that rapid climatic warming in northern Canada would produce an almost instantaneous

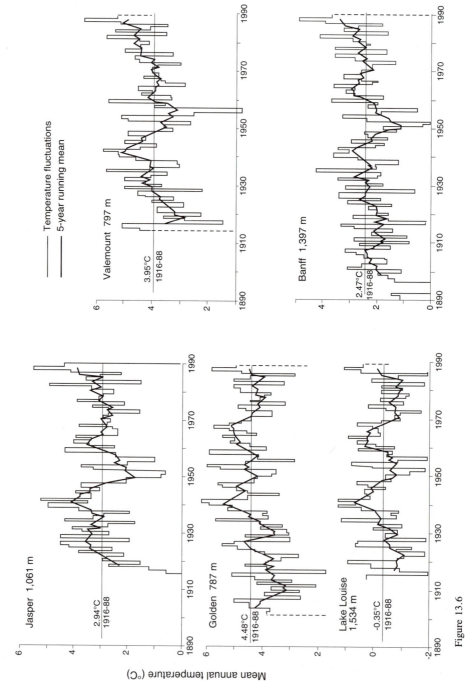

Figure 13.6
Annual temperature data for five valley-floor sites in southern Rocky Mountains. *Source*: Luckman (1990).

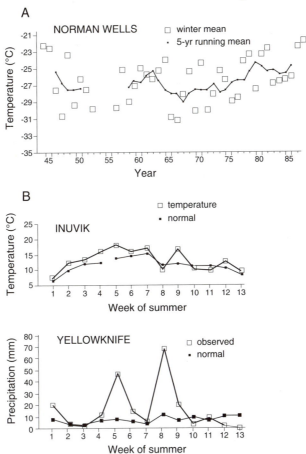

Figure 13.7
(A) Winter mean temperature, Norman Wells, NWT (1944–87), and five-year running means; (B) time series of temperature for Inuvik, NWT, and of precipitation for Yellowknife, Yukon, over 1988 summer (31 May–28 Aug.).

response of lighter sea-ice conditions in the marginal areas of the Arctic Ocean. Depending on storm patterns, cloudiness, and precipitation, sea ice in the central Arctic Ocean might take a decade or more to respond; but warming of the amounts postulated would lead to instability and eventual disappearance of the ice.

These changes will have mixed economic implications. Reduction in thickness, extent, and duration of sea ice in the southern Beaufort Sea would assist oil and gas exploration. Sea transportation routes westward and northward toward the Sverdrup Basin would also be open much longer. Iceberg generation in the eastern Arctic, however, would be significantly expanded, interfering with marine transportation and east-coast oil production. Also, melting of sea ice between the high arctic islands

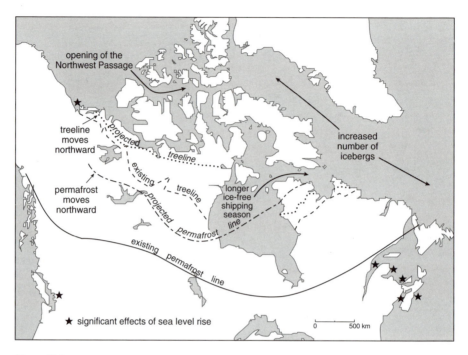

Figure 13.8
Projected changes in northern Canada following climate warming.

would open the Northwest Passage during summer as the quickest sea route between Europe and Japan, raising questions of Canada's sovereignty over the Arctic.

Permafrost

As explained in chapter 12, global warming will result in northward migration of the southern limits of permafrost (Figure 13.8). This will be accompanied by erosion, subsidence, and extreme slope instability. By contrast, in high latitudes the much colder permafrost will not degrade but instead the active layer will thicken, leading to thawing of the ice-rich zone which typifies the upper few metres of permafrost (see chapter 6). A more general concern is that the melting of permafrost releases methane, one of the gases that contributes to the greenhouse effect.

The anticipated degradation or thawing of permafrost would damage roads, buildings, and pipelines now constructed on permafrost, especially those in discontinuous permafrost zones. Everywhere, existing structures would need to be reinforced and new engineering design and construction techniques introduced. The costs cannot be overemphasized.

Snow Cover and Glaciers

Higher average winter temperatures will probably result in greater winter snowfall throughout most of the cold regions of Canada. At the same time the snow season will be reduced and there will be more winter cloudiness and increased runoff from seasonal snowmelt. While glaciers will disappear in subpolar regions, they may well become larger in the coastal Cordillera and in the more polar regions. Iceberg production from Ellesmere Island and Greenland might increase with greater glacier activity, accentuating the hazard to shipping and offshore oil platforms in the eastern Arctic posed by changes in sea-ice cover described earlier.

Sea Level

Climate change in high latitudes will result in major changes in Greenland and Antarctica's ice caps, with world-wide increase in sea level of as much as 1.0 m predicted. However, in northern Canada relative sea level is already very mobile because of isostatic rebound after recent deglaciation and coastal changes associated with thermokarst and melting of ground ice. Furthermore, in contrast to certain parts of the world, relatively few people reside in coastal communities near sea level in northern Canada so that the effects of coastal flooding will not be large.

Of special concern, however, will be the coastal lowlands of the Beaufort Sea, where melting of unconsolidated and ice-rich sediments is already widespread. A climate-induced rise in sea level will increase coastal erosion and flooding and influence inshore marine sedimentation. In the Mackenzie delta region, Tuktoyaktuk and other coastal communities would have recurring floods. Elsewhere, industrial structures built at shoreline or offshore (e.g. drilling platforms and artificial islands for offshore oil or gas pipelines) will have to deal with changes in water depths, shoreline configurations, and sea-ice conditions.

Boreal Forest Migration

Higher temperatures in winter and changed precipitation will significantly affect vegetation zonation and agricultural practices. The boreal forest and the northern treeline will move northward. For example, Figure 13.9 shows the possible shift in the position of the thermal boundaries of the forest for the Mackenzie valley lowlands, using the 600– and 1,300–growing-degree–day isolines as approximations of the northern and southern boundaries of the boreal forest. While the shift in the northern boundary ranges from about 100 to 700 km, the shift in the southern boundary is much greater, about 250 to 900 km (Table 13.3). Such predictions are based on climate-change models of two-time CO_2 using data from the Goddard Institute for Space Studies (GISS) and data from the Geophysical Fluid Dynamics Laboratory (GFDL). One can question whether the rate of change of the forest will be a direct

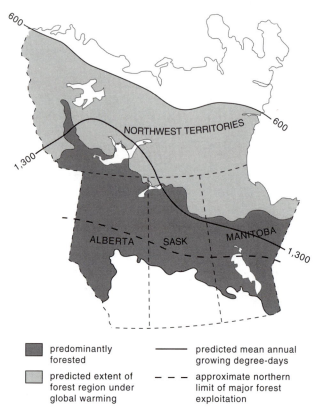

Figure 13.9
Mean annual growing degree–days (above 5°C) for GFDL-based models of climatic change in western Canada.

response to a change in thermal conditions and whether the growing-degree–day method is suitable for examining forest zone shifts, but such predictions highlight the major changes in vegetation, land use, and agricultural practice that will follow warming in northern Canada. In the subarctic in particular, the potential for agriculture would increase in a warmer climate, with the growing season lengthening by 30–40 per cent. Growing conditions in Whitehorse and Yellowknife, for example, would become similar to those today in Edmonton and Calgary, approximately 1,000 km to the south. As a consequence, agriculture might become viable in certain areas of the Mackenzie valley and interior Yukon, where soil conditions permit.

Changes to the Tundra and Polar-Desert Ecosystems

Limiting factors for viable plant communities in the High Arctic appear to be not severity of temperature but availability of nutrients and water. Climatic warming will

Table 13.3
Predicted changes in vegetation zonation for western and northwestern Canada as implied by GISS- and GFDL-based climatic-change models

A   Northward shift (km) in northern and southern boundaries of boreal forest
    (Delimited by 600– and 1,300–growing-degree–day isolines)

| | Climatic-change model | |
Forest boundary	GISS	GFDL
Northern	80–720	100–730
Southern	470–920	250–900

B   Changes in vegetation zonation according to GFDL-based model for three locations

Location	Present zonation	GFDL zonation
64°4′N, 120°W (southern NWT between Great Bear and Great Slave Lake)	Subarctic	Aspen parkland
59°9′N, 120°W (northeastern BC/northwestern Alta)	Boreal	Aspen parkland
55°5′N, 105°W (northern Sask.)	Boreal	Boreal temperate

thus influence arctic vegetation according to its effect on soil moisture and nutrients, especially nitrates. These vegetation changes are difficult to quantify at present, but we might expect a subtle but significant change in the distribution of tundra species, and their assemblages.

Closely linked to changes in vegetation zonation will be shifts in wildlife migratory patterns and aquatic and marine habitats. Settlements chosen for proximity to hunting or fishing grounds may no longer be well located for such activities. Warmer, wetter winters, for example, may decimate populations of caribou and muskoxen, since heavier snowfall will bury the tundra mosses and lichens on which these animals depend. In contrast, most marine life and migratory birds are expected to flourish while warmer temperatures would increase fish populations in rivers and lakes and the ocean. Native settlements – such as Sachs Harbour, which depends on fox, caribou, and muskoxen; Aklavik, which relies on fur trapping in the Mackenzie delta; and coastal communities such as Tuktoyaktuk and Paulatuk, which exploit marine resources of the Beaufort Sea – may all experience change as wildlife and mammal migratory patterns evolve.

The structure of northern biological communities can indeed change rapidly, often within decades. Two documented examples are the fluctuations in species mix of major fishes in Ungava Bay since the International Polar Year of 1882–83 and the growth, then decline, of the West Greenland cod fishery in the middle of the present century. Both examples appear to be natural and related to minor fluctuations in climate; they indicate the extent and nature of biological changes that might be expected if predicted warming trends for northern Canada are correct.

CLIMATE CHANGE AND CANADA'S MOUNTAIN ENVIRONMENTS

Any rise in the near-surface temperature of Earth would affect Canada's mountain environments. Only some of these effects can be explored here, especially as there is disagreement on the extent of change that will result within each of the highly variable mountain regions. For example, for the western mountains between 50° and 60°N, the mean prediction from five GCMs for AD 2050 is an increase of 4.2C° in winter and 2.4C° in summer, with approximately 1.8C° uncertainty around the means.

In mountain environments, where present climates are so spatially variable, it is crucial to consider not only mean estimates of change but also extreme boundary conditions. For example, the warmest and driest outcome predicted for the western mountains involves temperature increases of 6C° for winter and 4.2C° for summer and no change in precipitation; the coolest and wettest suggests increases of 2.4C° in winter and 0.6C° in summer and 300 mm more precipitation. In spite of the uncertainty, all models predict warming and most anticipate greater precipitation.

Snowfall, Snow Cover, and Glaciers

Of the five hydrologic regions recognized in the Cordillera (Figure 7.9), four would probably experience more snowfall under all but the driest models. The exception would be the Insular Mountains, where snowfall is a relatively minor part of the precipitation even under present temperatures. The biggest increase would be in the St Elias and Coast Mountains. The extent and duration of snow cover would decrease, however, throughout the region in response to raised temperature, and the climatic snowline would be up to 600 m higher than at present. Under the influence of greater snowfall, many glaciers in the Coast Mountains and Interior Alpine Mountains would expand; at the very least their "activity ratios" and dynamism would grow if snowfall in the accumulation zone were to increase along with amount of melt in the ablation zone.

Mountain Permafrost

Extensive degradation of permafrost would be likely in the Subarctic Dry Interior Mountains of Yukon and Northwest Territories; permafrost, now widely scattered in the southern Cordillera, would disappear from mountain tops. Presumably similar change would occur in the Shickshock Mountains in Quebec's Gaspé Peninsula, where permafrost is found near the summits of Mt Jacques Cartier and Mt Albert.

A

B

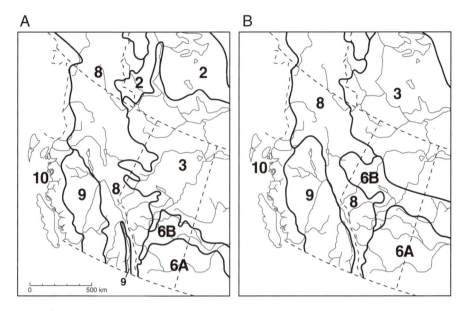

Figure 13.10
Ecoclimatic provinces of Canada's Cordillera: projected changes consequent upon climate warming:
(A) present ecoclimatic provinces; (B) after CO_2 doubling. See Table 13.4 for details and key.

A rise in summer temperature of as much as $4.2C°$ would certainly degrade bodies of permafrost in the subarctic mountains. But strong surface drying would decouple the permafrost thermally from the atmosphere, thus slowing degradation.

Timberline and Ecozones

Timberline would probably advance to higher elevations, and areas of open forest would fill in to the closed crown state. Only minor areal changes are predicted in the Pacific Cordilleran ecoclimatic province, which covers the coastal region of British Columbia. (Figure 13.10), but major eastward and northward expansion of the Interior Cordilleran province is anticipated ($+34$ per cent, according to Table 13.4). The big issue once again is how long it would take for these vegetation adjustments to be established. Even if the areal extent of some vegetation zones were not greatly altered, upslope migration of timberline would substantially affect acreage of commercial timber, undoubtedly increasing pressure to log in higher-elevation zones, which commonly have steeper slopes and are potentially less stable.

Fish Resources

There are two quite separate sets of considerations surrounding mountain climate change and the fish resource. For freshwater fisheries, changes in distribution of

Table 13.4
Predicted change in area of BC ecoclimatic provinces after doubling of atmospheric CO_2

Province*	Present Area (km^2)	Area (km^2) after doubling of CO_2[†]	Amount and nature of change
Pacific Cordilleran (10)	176,934	198,734	Slight increase
Interior Cordilleran (9)	230,242	350,311	About 34-per-cent eastern and northern expansion
Cordilleran (8)	482,492	329,343	Eastern shift in south and about 32-per-cent reduction
Transitional Grassland (6B)	0	19,931	Northwestern shift into British Columbia
Boreal (3)	58,132	49,481	About 15-per-cent decrease

* Approximate values measured from Zoltai (1988: Figures 1 and 2).

[†] Numbers in parentheses indicate number of region as recorded in Figure 13.10.

ecoclimatic provinces are probably most important; for sea-going salmonids, temperature changes in many locations in the eastern Pacific may dramatically influence mountain fish resources (Figure 13.11). With change in climate highly probable, and given variability of climate, fisheries management needs to become more opportunistic, adaptive, flexible, and experimental to avoid loss of commercial and recreational fisheries.

Recreational Skiing Resorts

In a study of the effects of climatic change on downhill skiing in Quebec (Lamothe and Périard 1988), a warming of 4–5C° throughout the winter was considered. Table 13.5 summarizes resulting reduction in number of skiable days. Similar problems might be experienced in western Canada, at Whistler and Blackcomb in the Western Cordillera and at Banff in the Eastern Cordillera. Snow-avalanche hazard might increase for mountain-top skiers, along with slope instabilities on access roads to the resorts (see the next subsection). Possible disappearance of permafrost from the higher summits such as Plateau Mountain would eliminate one set of problems in construction of stable foundations for ski lifts.

Natural Hazards

Warming, more precipitation in the mountains, greater snowfall, and more intense precipitation events will affect the nature and incidence of natural hazards. In larger river basins, such as that of the Fraser River, dominated by a single spring snowmelt flood, a higher spring freshet can be anticipated. This would place new pressures on the elaborate dyking system of the lower Fraser valley (chapter 10). In the small,

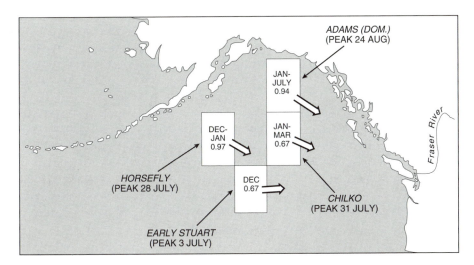

Figure 13.11
Return of the salmon and sea surface temperatures: areas and times of highest correlation coefficient for various stocks of Fraser River sockeye. *Source*: Blackbourn (1987).

steep river basins of the Insular Mountains and the Coast Mountains, rain-on-snow floods, concentrated during the winter six months, would probably be bigger and more frequent. This small-basin flood hazard is often associated with the debris-torrent hazard (chapter 10; Slaymaker 1988). In the Interior Alpine and Dry Interior Mountains, heavy summer and autumn rains could intensify small-basin flooding and debris-torrent hazard. Danger from the jökulhlaup hazard, or glacier dam-burst flooding, would probably increase, because of warming and the greater dynamism of glaciers in the St Elias and Coast Mountains. The most likely effects of accelerated glacier melt are hydrological: more short-term discharge but less long-term storage of water in glaciers. This situation clearly has implications for downstream irrigation or power projects. Accelerated glacier melt could also lead to jökulhlaup activity from ice-marginal lakes and ice avalanching, as described in Gilbert and Desloges (1987). Snow-avalanche hazard, associated with warming and increased snowfall, may well spread from established snow-avalanche hazard sites at Rogers Pass, in the Interior Alpine Mountains, and in the Coast Mountain passes of the Western Cordillera toward the Northern and Eastern Cordillera (see chapter 10).

Finally, natural hazards associated with permafrost degradation are expected to intensify throughout southern Yukon and southwestern Mackenzie District, affecting particularly road networks, pipeline, rights of way, and construction projects.

Accuracy of predictions about incidence of natural hazards caused by climatic change will be confounded by shifting population distribution and intensification of land use over the next half-century. However, trends in climate change and land use in Canada's mountain environments point in the same directions – toward bigger and more frequent natural hazards.

Table 13.5
Mean annual number of skiable days for actual and projected conditions, southern Quebec

Snow conditions	Sainte-Agathe-des-Monts			Sherbrooke			Quebec City		
	Actual	Projected	% reduction	Actual	Projected	% reduction	Actual	Projected	% reduction
Snow cover									
of 10 cm	109	36	67	86	27	69	105	51	51
of 30 cm	84	11	87	54	7	87	84	25	70
Feasibility of snow with minimum snow cover									
of 10 cm	121	60	50	97	51	47	109	63	42
of 30 cm	111	58	48	87	49	44	100	58	42

Source: Lamothe and Périard (1988).

CONCLUSION

In this chapter we have explored the hypothesis that Canada's climate will probably become warmer over the next hundred years. The country's colder regions, both northern and mountain, are expected to be disproportionately affected. Although we cannot predict details, we are convinced of the urgency of defining the general problem, so that people can understand in advance the likely range of environmental changes.

If "sustainable development" is maintenance of an adequate quantity of land with required qualities to support indefinitely the full range of societal demands that depend on the territorial resource base, then creative thinking is crucial if we are to reduce the likelihood of unacceptable outcomes deriving from climate change. This chapter alerts the reader to the likely range of the results of climate change in Canada's cold environments. It may well be necessary to modify demand for, and manage supply of, environmental resources in Canada's cold regions. Our objective will have been achieved if we contribute to that understanding which is a prerequisite to effective stewardship of two-thirds of Canada's landmass.

REFERENCES

Blackbourn, D.J. 1987. "Sea Surface Temperature and Pre-Season Prediction of Return Timing in Fraser River Sockeye Salmon." *Canadian Special Publications, Fisheries and Aquatic Science* 96: 296–406.

Church, M., and Slaymaker, O. 1989. "Disequilibrium of Holocene Sediment Yield in Glaciated British Columbia." *Nature* 337: 452–4.

Davis, M.B. 1988. "Ecological Systems and Dynamics." In *Towards an Understanding of Global Change*, Washington, DC: National Academy Press, 69–106.

Ferguson, H.M. 1988. "The Changing Atmosphere: Implications for Global Security," Toronto, 27–30 June 1988.

Gardner, J.S. 1972. "Recent Glacier Activity and Some Associated Landforms in the Canadian Rockies." In O. Slaymaker and H.J. McPherson, eds., *Mountain Geomorphology*, Vancouver; Tantalus Press, 55–62.

Gilbert, R., and Desloges, J.R. 1987. "Sediments of Ice Dammed, Self Draining Ape Lake, B.C." *Canadian Journal of Earth Sciences* 24: 1,735–47.

Jacoby, G.C., and Cook, E.R. 1981. "Past Temperature Variations Inferred from a 400 Year Tree Ring Chronology from Yukon Territory." *Arctic and Alpine Research* 13: 409–18.

Klemes, V. 1985. *Sensitivity of Water Resource Systems to Climate Variations*. World Climate Programme Report 18, World Meteorological Organization.

Koerner, R., Dubey, R., and Parnandi, M. 1989. "Scientists Monitor Climate and Pollution from Ice Caps and Glaciers." *Geos* 18 no. 3: 33–8.

Lamothe and Périard. 1988. *Implications of Climate Change for Downhill Skiing in Quebec.* Climate Change Digest CCD 88–03, Environment Canada.

Luckman, B.H. 1986. "Reconstruction of Little Ice Age Events in the Canadian Rocky Mountains." *Géographie physique et quaternaire* 40: 17–28.

– 1990. "Mountain Areas and Global Change: A View from the Canadian Rockies." *Mountain Research and Development* 10: 171–82.

Luckman, B.H., and Kearney, M.S. 1986. "Reconstruction of Holocene Changes in Alpine Vegetation and Climate in the Maligne Range, Jasper National Park, Alberta." *Quaternary Research* 26: 244–61.

Roberts, L. 1989. "How Fast Can Trees Migrate?" *Science* 243: 735–7.

Rosswall, T., Woodmansee, R.G., and Risser, P.G. 1988. *Scales and Global Change: Spatial and Temporal Variability in Biospheric and Geospheric Processes.* SCOPE 35.

Slaymaker, O. 1988. "The Distinctive Attributes of Debris Torrents." *Hydrologic Sciences Journal* 33: 567–75.

– 1990. "Climate Change and Erosion Processes in Mountain Regions of Western Canada." *Mountain Research and Development* 10: 183–95.

Wheaton, E.E., and Singh, T. 1988. *Exploring the Implications of Climatic Change for the Boreal Forest and Forestry Economics of Western Canada.* Climate Change Digest CCD 89–02, Environment Canada.

Zoltai, S.C. 1988. "Ecological Provinces of Canada and Man-Induced Climatic Change." Canada Committee on Ecological Land Classification, *Newsletter* 17: 12–15.

Index